Unit Operation in Food Processing Fundamentals and Application

a division of
NIPA GENX ELECTRONIC RESOURCES & SOLUTIONS P. LTD.
New Delhi-110 034

About the Author

Prabhat K. Nema is a Professor in the Department of Food Engineering and Dean Research at NIFTEM, India. He obtained B. Tech. in Agricultural Engg from JNKVV, Jabalpur (1995) and M. Tech. in Dairy & Food Engineering from IIT, Kharagpur in India (1997). He has experience of working in Food Industry for about four years. He completed his PhD from IIT Kharagpur in Dairy and Food Engg (2008). He has more than twenty years of experience at various universities including NERIST (Itanagar, Arunachal Pradesh), JNKVV (Jabalpur, MP), RVSKVV (Gwalior, MP), AAU (Anand, Gujarat) and serving NIFTEM (Kundli, Haryana) since 2011. He was National Vice President of AFSTI in 2019. He has published scientific research articles in the journals of national and international repute. He is Editor of Journal of Food Science and Technology (Springer). He has been instrumental in developing labs, conducting funded research projects, guiding research projects for number of food professionals at UG, PG and PhD level. He offers consultancy services in the area of food processing. He is also recipient of Seligman APV Award by Society of Chemical Industry, London (UK) which enabled him to have research experience in the University of Reading (UK) in 2007. As a member of NIFTEM delegation, he has visited various academic and research institutions in Germany and the Netherland in 2013. He was awarded Travel Grant by Department of Science and Technology, Government of India to deliver Keynote Speech and Chair Technical Session in Asia Pacific Drying Conference in Wuxi, China in 2017. He has organized First International Conference (2013), First and Second Drying Workshop (2014 & 2015) and various training programmes at NIFTEM.

Unit Operation in Food Processing Fundamentals and Application

Prabhat K Nema
Dean Research and Professor (Food Engineering)
National Institute of Food Technology Entrepreneurship and Management
(NIFTEM)
(Deemed to be University (De-novo Category)
and an Autonomous Institution under Ministry of Food Processing Industries
Govt. of India
Kundli-131028 Sonepat, Haryana, INDIA

nipa
a division of
NIPA GENX ELECTRONIC RESOURCES & SOLUTIONS P. LTD.
New Delhi-110 034

a division of

NIPA GENX ELECTRONIC RESOURCES & SOLUTIONS P. LTD.

101,103, Vikas Surya Plaza, CU Block
L.S.C.Market, Pitam Pura, New Delhi-110 034
Ph : +91 11 27341616, 27341717, 27341718
E-mail:newindiapublishingagency@gmail.com
www: www.nipabooks.com

For customer assistance, please contact
Phone: + 91-11-27 34 17 17 Fax: + 91-11- 27 34 16 16
E-Mail: feedbacks@nipabooks.com

ISBN: 978-93-87973-13-8

INR 3995.00

Composed and Designed by NIPA.

Preface

Fundamental knowledge of various unit operations of food processing which adds value and enhances the shelf life of the food products is very important for students and teachers. The main objective of this book is to present in a progressive and systematic form, the knowledge of fundamental food processing unit operations majorly required specially for undergraduate students. The applications of the unit operations in food processing and their explanations with numerical examples will help the students to understand the topic clearly. It will help the students to prepare for various competitive examinations also.

This book has 24 chapters which present comprehensive yet concise coverage of basic principles and major unit operations. In this book, fundamental principles of material and energy balance are explained with suitable examples. Transfer processes for fluid and heat are covered which are the basis for many food processes. Chapters on Psychrometry and Refrigeration are useful topics for food process engineering students. Different unit operations like cleaning and grading, size reduction would be more useful for students interested in post harvest processing. Similarly other relevant topics as per the recommendations of government and industrial bodies are also included in this book.

These chapters are prepared considering the convenience of understanding concepts and aspects of the topics especially by undergraduate and post graduate students as well as by faculty members for teaching purposes. We hope this book will be a meaningful contribution to the teaching-learning of food processing/ engineering in the country.

The editor wishes to sincerely acknowledge the authors for their significant contributions to this book. These authors are among the leading scientists, engineers and experts from academia. Each of the invited contributors in this book has taught extensively in the respective area of food processing and hence their strong experience in the form of chapter will benefit the readers. We are pleased to thank them all for their valuable contributions in this book. Constructive inputs for further improving the quality of the book are always welcome. We also thankfully acknowledge the support received from the Vice Chancellor of NIFTEM in the process of completing this book.

Prabhat K Nema
NIFTEM, India

Contents

List of Contributors

Ajit Kumar Singh
Department of Food Technology, School of Chemical Technology, Harcourt Butler Technical University, Kanpur-208002, UP, India
Email: ajitksingh.iitkgp@gmail.com

Akash Pare
Indian Institute of Food Processing Technology (IIFPT), Ministry of Food Processing Industries, Govt. of India, Pudukkottai Road, Thanjavur 613005, TN, India
Email: akashpare@iifpt.edu.in

Akhil K.G.
Indian Institute of Food Processing Technology (IIFPT), Ministry of Food Processing Industries, Govt. of India, Pudukkottai Road, Thanjavur 613005, TN, India
Email: akhilkalathil1996@gmail.com

Anit Kumar
Department of Food Technology, School of Chemical Technology, Harcourt Butler Technical University, Kanpur 208002, Uttar Pradesh, India
Email: anitkumarsingh09@gmail.com

Ashutosh Upadhyay
Department of Food Science and Technology, National Institute of Food Technology Entrepreneurship and Management, Kundli, 131028 Sonepat, Haryana, India
Email: ashutosh.niftem@gmail.com

Babar Onkar Avinash
Department of Food Engineering, National Institute of Food Technology Entrepreneurship and Management, Kundli, 131028 Sonepat, Haryana, India.
Email:omkarbabar@gmail.com

Brijesh Srivastava
Department of Food Engineering & Technology, School of Engineering, Tezpur University Tezpur 784028, Assam, India
Email: brijesh@tezu.ernet.in

Debabandya Mohapatra
Agro-Produce Processing Division, ICAR-Central Institute of Agricultural Engineering Bhopal 462038, Madhya Pradesh, India
Email : debabandya@gmail.com

D.C. Joshi
Kota Agriculture University, Borkhera, Baran Road, P B No 20, Kota 324001, Rajasthan
Email: dcjoshi258@gmail.com

Dipsikha Kalita
Department of Agricultural and Food Engineering, IIT Kharagpur-721302, West Bengal
Email: dkalita4@gmail.com

G.P. Sharma
Department of Processing & Food Engineering, College of Technology & Engineering
MPUAT, Udaipur 313001, Rajastahan, India
Email: gpsharma1961@gmail.com

H.N. Mishra
Agricultural and Food Engineering Department, IIT Kharagpur-721302, West Bengal
Email: hnm@agfe.iitkgp.ac.in

Harsh Prakash Sharma
Department of Food Processing Technology, College of Food Processing Technology and Bio-energy, Anand Agricultural University, Anand 388 110, Gujarat, India
E-mail: harsh187@gmail.com

Hilal A Makroo
Department of Food Technology, Islamic University of Science and Technology
Awantipora, Jammu & Kashmir, 192122 India
Email: hilal.makroo@islamicuniversity.edu.in

Hridyesh Pandey
Department of Food Processing Technology, College of Food Processing Technology and Bio-energy, Anand Agricultural University, Anand 388 110 Gujarat, India
Email : hridyeshpandey@gmail.com

Khursheed A. Khan
Department of Agricultural Engineering, College of Horticulture,
R.V.S. Agriculture University, Mandsaur, 458001Madhya Pradesh, India
Email: khan_undp@yahoo.ca

Kirtiraj K. Gaikwad
Department of Paper Technology, Indian Institute of Technology, Roorkee -247667
Uttarakhand, India.

Kshitiz Kumar
Department of Food Processing Technology, ADIT, Vallabh Vidyanagar, Anand, Gujarat, India
Email: kshitizyp@gmail.com, kshitizns@gmail.com

Kumar Sandeep
Food Safety Officer, Department of Health and Family Welfare, Government of West Bengal
Moulpur Rural Hospital, Malda-732142, West Bengal, India
Email: sandeepfoodeng@gmail.com

Kumaresh Halder
Department of Dairy Engineering, Faculty of Dairy Technology, West Bengal University of Animal and Fishery Sciences, Mohanpur Campus, P.O.-B.C.K.V., Mohanpur
Nadia, West Bengal -741252, India
Email: kumar.halder@gmail.com

Madaraboina Venkateswara Rao
Indian Institute of Food Processing Technology (IIFPT), Ministry of Food Processing Industries, Govt. of India, Pudukkottai Road, Thanjavur, TN-613005, India
Email: venkatmadanaboina123@gmail.com

Parag Prakash Sutar
Department of Food Process Engineering, National Institute of Technology
Rourkela, Odisha 769008, India
Email: sutarp@nitrkl.ac.in, paragsutar@gmail.com

Pramod K Prabhakar
Department of Food Science and Technology, National Institute of Food Technology Entrepreneurship and Management, Sonipat 131028, Haryana, India
Email : pramodkp@niftem.ac.in

Prodyut Kumar Paul
Department of Pomology and Postharvest Technology, Uttar Banga Krishi Viswavidyalaya
Pundibari, Cooch Behar, West Bengal-736165, India
E-mail: prodyut24@yahoo.com

Radha Charan Verma
Department of Processing and Food Engineering, College of Technology and Engineering
Maharana Pratap University of Agriculture and Technology, Udaipur 313 001 Rajasthan, India
Email: verma.drrc@gmail.com, rcvermashp@yahoo.com

Rachna Sehrawat
National Institute of Technology, Rourkela, Odisha-769008, India
Email: sehrawatrachna2017@gmail.com

Ramanpreet Kaur
Dr. S S Bhatnagar University Institute of Chemical Engineering and Technology
Panjab University, Chandigarh-160014. India

R.K. Jain
A D Patel Institute of Technology, New V V Nagar, Anand 388121, Gujarat, India
Email: rajeevpjain12@gmail.com

Rupesh Chavan
Operations Quality Assurance, Abbott Healthcare Pvt Ltd., Jhagadia, India
Email: rupesh.chavan@abbott.com

S.S. Shirkole
Department of Food Process Engineering, National Institute of Technology, Rourkela, 769008
Odisha India
Email:shivanandshirkole@gmail.com

Sajad A Sofi
Department of Food Technology, Islamic University of Science and Technology
Awantipora, 192122 Jammu & Kashmir, India

Santanu Basu
SS Bhatnagar University Institute of Chemical Engineering and Technology, Panjab University, Chandigarh-160014. India
Email: santbasu@gmail.com

Shivmurti Srivastav
Department of Food Processing Technology, A D Patel Institute of Technology
New V V Nagar, Anand 388121, Gujarat
Email: shivmurtis@gmail.com, shivamurtis@yahoo.com

Smriti Pillai
Indian Institute of Food Processing Technology (IIFPT), Ministry of Food Processing Industries, Govt. of India, Pudukkottai Road, Thanjavur 613005, TN, India
Email: smriti.rajiv0891@gmail.com

Somya Tiwari
Product Manager, Britannia Industries Limited, Prestige Shantiniketan, Tower C, The Business Precinct, 16th & 17th Floor, Whitefield Main Road, Mahadevapura, Bengaluru
Karnataka 560048
Email: somyatewari4@gmail.com

Subir Kumar Chakraborty
Agro Produce Processing Division, ICAR - Central Institute of Agricultural Engineering, Nabibagh, Berasia Road, Bhopal 462038, M.P., India
E-mail: subir8275@gmail.com

Sunil C.K.
Indian Institute of Food Processing Technology (IIFPT), Ministry of Food Processing Industries, Govt. of India, Pudukkottai Road, Thanjavur 613005, Tamil Nadu, India
Email: sunil.ck@iifpt.edu.in

U.S. Shivhare
Dr. S S Bhatnagar University Institute of Chemical Engineering and Technology,
Panjab University, Chandigarh 160014, India
Email: shivhareus@pu.ac.in

Vijay Singh Sharanagat
Department of Food Engineering, National Institute of Food Technology Entrepreneurship and Management, Sonipat 131028, Haryana, India
Email: Vijaysinghs42@gmail.com

Vinkel Kumar Arora
Department of Food Engineering, National Institute of Food Technology Entrepreneurship and Management, Sonipat 131028, Haryana, India
Email : vinkelarora17@gmail.com

1

Material Balance

Shivmurti Srivastav and R. K. Jain

1. Introduction

Law of conservation of mass says that material cannot be created nor destroyed, which is the basic fundamentals of physical science. Many engineering analyses involve material balances. These balances are simply accounting processes that account for all material in processes of interest. The material balance is one of the most useful tools fundamental in the "engineering approach" to design new process. The law of conservation of mass tells that the total mass of all material entering any process must equal the total mass of materials leaving plus the mass of any material accumulated in the process.

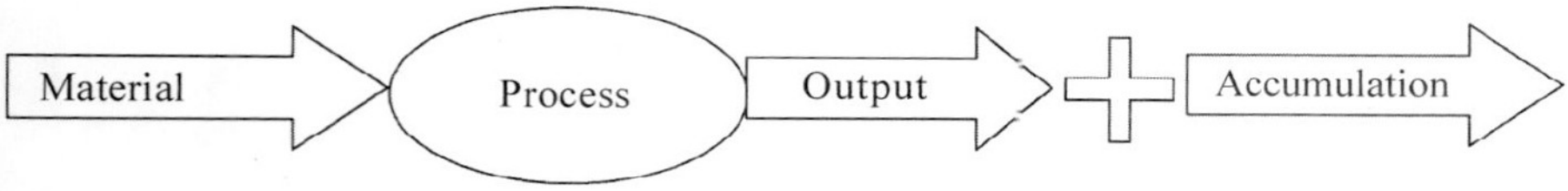

In the steady state condition or continuous process in material balance means no accumulation will be there of the material in the process. In this condition sum of the input is equal to sum of the output. What material goes in must comes out. If the accumulation term is not zero, then the quantity and concentration of components in the system could change with time and the process is an unsteady state i.e. scaling in the pipes. Material balances are essential in processes involving material flow like drying, evaporation, distillation and liquid handling systems.

2. Overall Material Balance

In a process there can be more than one input and output streams. If there are two different input streams I_1 and I_2 and one output O_1, then the sum of the mass flow rates of all streams entering the process must equal the sum of the flow rates of all streams leaving the process i.e. $I_1+I_2=O_1$

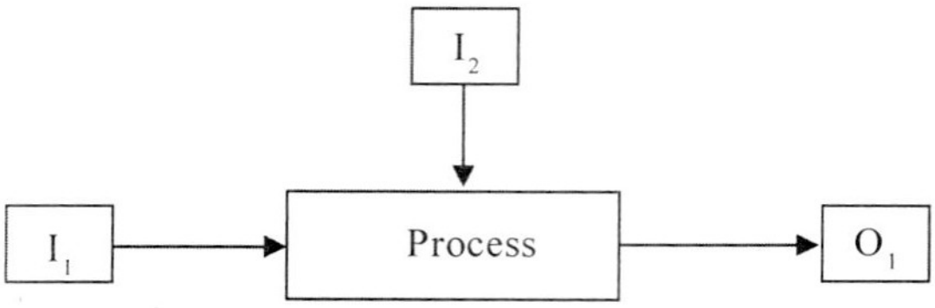

Steps of material balance

In order to solve the material balance problems of different processing systems, take the following steps;

a) Prepare a simple flow diagram of the process by showing entering of each stream with an arrow and also leaving of each stream pointing out by each arrow. Each input and output must be included by composition, amount and all other data pertaining to the process flow.

b) Express the chemical equation, if required.

c) If there is no chemical reaction then start making the material balance. The material balance can be done based on total mass and mass of each component present in the material.

All we know that each food processing operations depends upon the product development like mixing, evaporation, drying, recycling, freezing, separation etc. Each processing operations may be grouped into different categories for the purpose of material balance calculations.

Example-1 A solution contains 2.5 kg of ethanol in 3.75 kg of water. Express the concentration of ethanol as both a mass fraction and a mole fraction.

Solution

Molecular weight of ethanol and water is 46 and 18

Molar masses of ethanol and water are 3.75/46 and 2.5/18 kmol

The total mass of the solution = 2.5+3.75 = 6.25 kg

Mass fraction of ethanol (x) is

$$x = \frac{2.5}{6.5} = 0.40$$

Mole fraction of ethanol is (y)

$$y = \frac{\left(2.5/_{46}\right)}{\left(\frac{2.5}{46}\right) + \left(\frac{3.75}{18}\right)} = 0.207$$

Therefore, the concentration of ethanol is 40% by mass and 20.7% by molar.

Example-2 Determine the amount of a juice concentrate containing 65% solids and single strength juice containing 15% solids that must be mixed to produce 100 kg of a concentrate containing 45% solids.

Solution

Total mass balance is

X + Y = 100- (1)

Total component balance is

X (0.65) + Y (0.15) = 100 (0.45) (2)

Solving equations, (1) and (2) we get X = 60 kg and Y = 40 kg

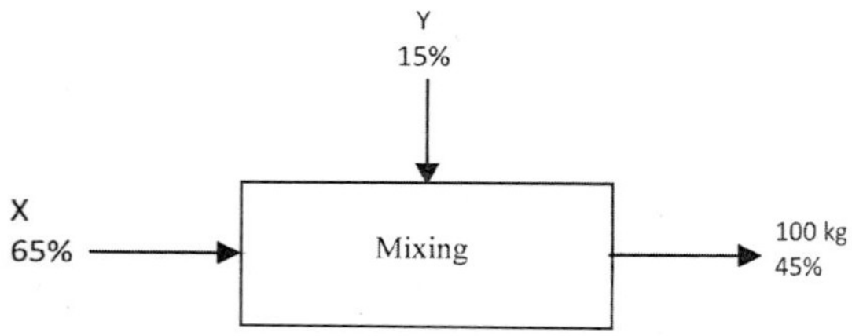

Example-3 How much dry sugar must be added in 100 kg of aqueous sugar solution in order to increase its concentration from 20% to 50%?

Solution

Note: Dry sugar contains 100% solid.

Total mass balance

100+Y=Z (1)

Component mass balance

100 (0.20) + Y (1) = Z (0.50) (2)

Solving above two equations, we get Y= 60kg and Z=160kg

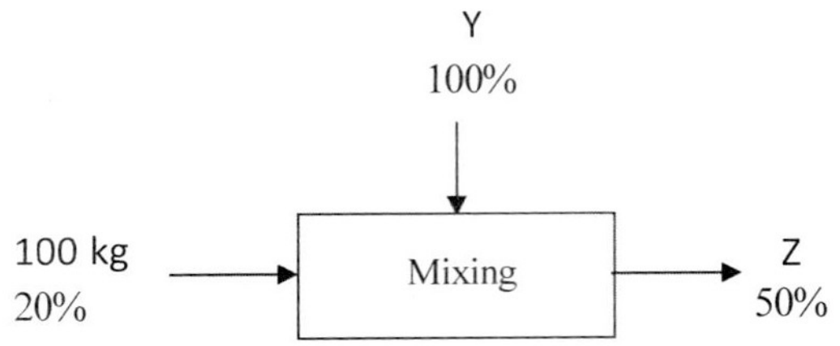

Example-4 Tomato juice containing 7% solids by mass is fed to an evaporator and water is removed at a rate of 500 kg/h. If the concentrate is to contain 35% solids, determine the necessary feed rate.

Solution

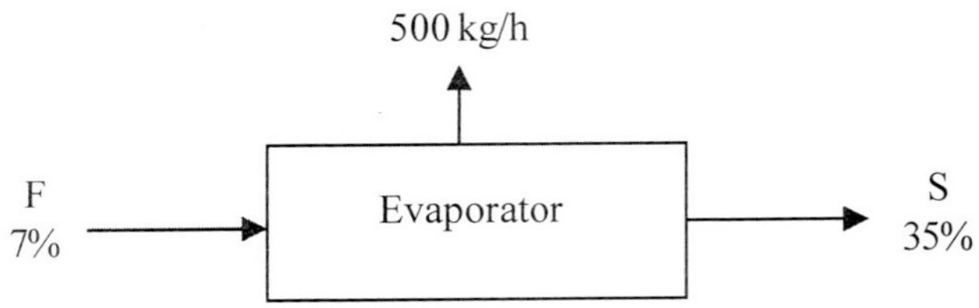

The overall material balance is

F= 500+S (1)

Component balance is

F (0.07) = 500 (0) + S (0.35) (2)

Solving the above equations we get, F = 625 kg/h, S = 125 kg/h

Example-5 Fresh orange juice containing 8% solid is concentrated in an evaporation to 55% solids of the juice is entering at 500kg/h calculate the amount of water of removed and concentrated juice produced.

Solution

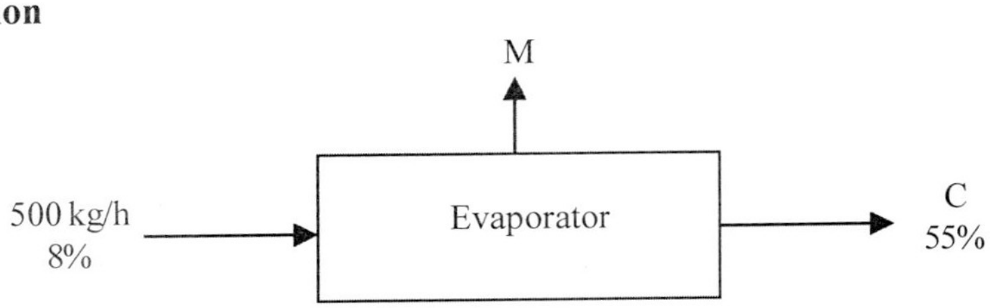

The overall material balance is

500 = M + C (1)

Component balance is

500 (0.08) = M (0) + C (0.55) (2)

Solving the above equations we get, M = 427.27 kg/h, C = 72.73 kg/h

Example-6 How much water would be evaporated from 100kg of milk of 15% solid to from 40% concentrated milk.

Solution

The overall material balance is

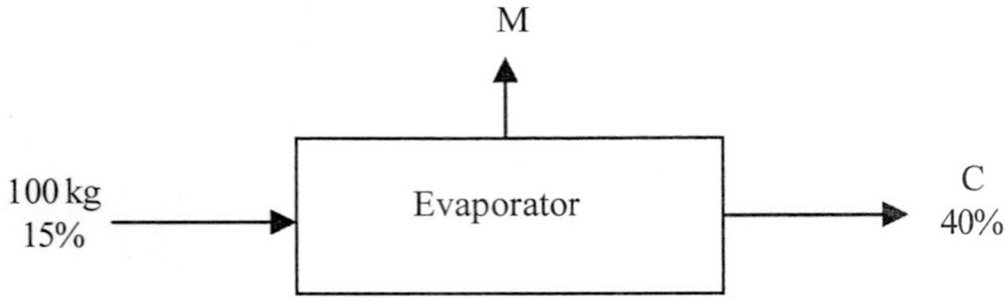

$$100 = M + C \qquad (1)$$

Component balance is

$$100\,(0.15) = M\,(0) + C\,(0.40) \qquad (2)$$

Solving the above equations, we get, M = 62.50 kg/h, C = 37.50 kg/h

Example-7 1000kg of fruit juice with 10% solid is freeze concentrated to 40% solid. The dilute juice is fed to a freezer where ice crystals are formed and them slush is separated in a centrifugal separator into ice crystals and concentrated juice. An amount of 500kg/h of liquid is recycled from the separator to freezer. Calculate the amount of ice that in removed in the reparation and amount of concentrated juice produced.

Solution

The overall material balance is

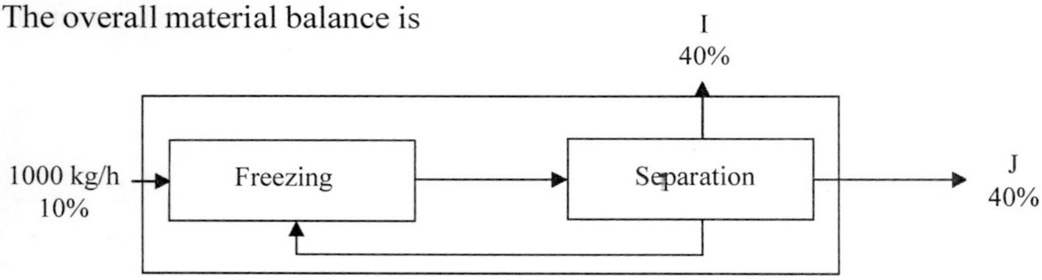

$$1000 = I + J \qquad (1)$$

Component material balance (solid)

$$1000\,(0.1) = I\,(0) + J\,(0.40) \qquad (2)$$

Solving the above equations we get, J = 250 kg/h and I = 750 kg/h

Example-8 How many kg/h of sugar syrup with 10% sugar must be fed to an evaporator to producer 10000kg/h of sugar syrup with 65% sugar?

Solution

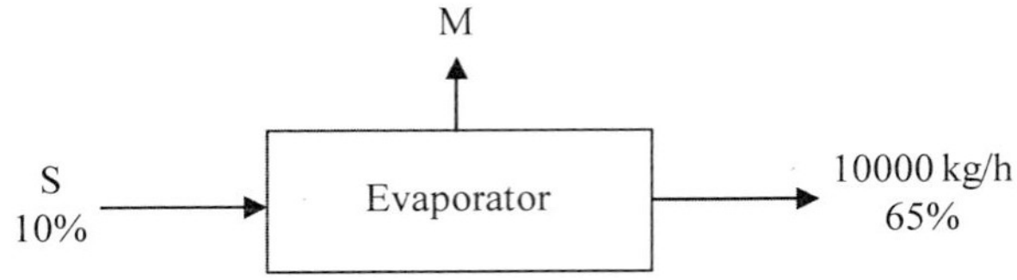

The overall material balance is

S = M+10000 (1)

Component balance is

S (0.10) = M (0) + 10000 (0.65) (2)

Solving the above equations we get, S = 65000 kg/h, C = 55000 kg/h

Example-9 How much water must be added to 200 kg of concentrated juice with 65%solid to produced orange juice with 12% solids?

Solution

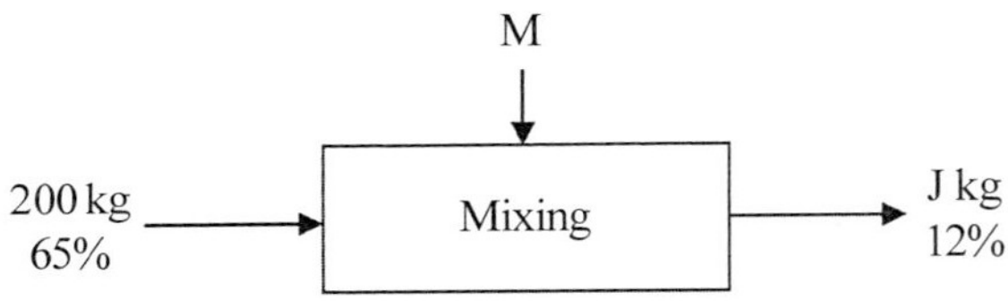

The overall material balance is

200 = M + J (1)

Component balance is (solid)

200 (0.65) = M (0) + J (0.12) (2)

Solving the above equations we get, J = 1083 kg, M = 883 kg

Example-10 Milk with 3.8% fat and 8.1% Fat Free Solids (FFS) is used for the production of canned concentrated milk. The process includes separation of cream in a centrifuge and concentration of the partially defatted milk in an evaporator. If the cream that is produced is the centrifuge contain 55% water, 40% fat and 5% Fat Free Solids, calculate how much milk is necessary in order to produce a can of concentrated milk that contain 410g milk with 7.8% fat and 18.1% FFS. How much cream and how much water must be removed in the centrifuge and the evaporation respectively.

Solution

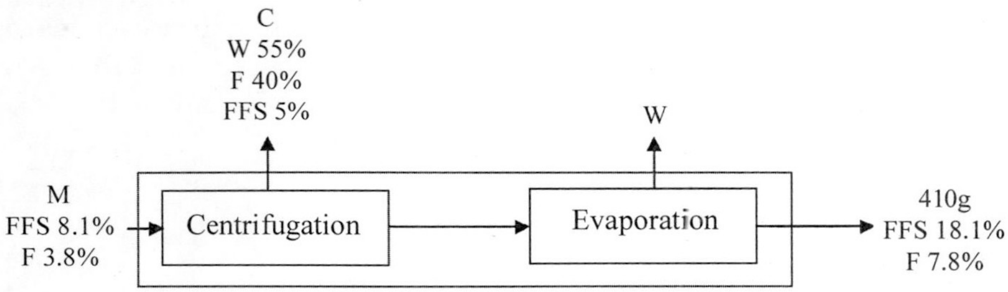

The overall material balance is

M = C + W + 410 (1)

Component balance for fat free solid is

M (0.081) = C (0.05) + 410 (0.181) + W(0) (2)

Component balance for fat is

M (0.038) = C (0.40) + 410 (0.078) + W(0) (3)

Solving the above equations we get, M = 920.83g, C = 7.56g and W = 503.26g

Example-11 Calculate the total mass balance and component mass balance for mixing ingredients to make 25 kg of beef sausages having a fat content of 30%. Typically beef meat contain protein 18%, fat 12%, water 68% and beef fat contain 78% fat, 12% water and 5% protein.

Solution

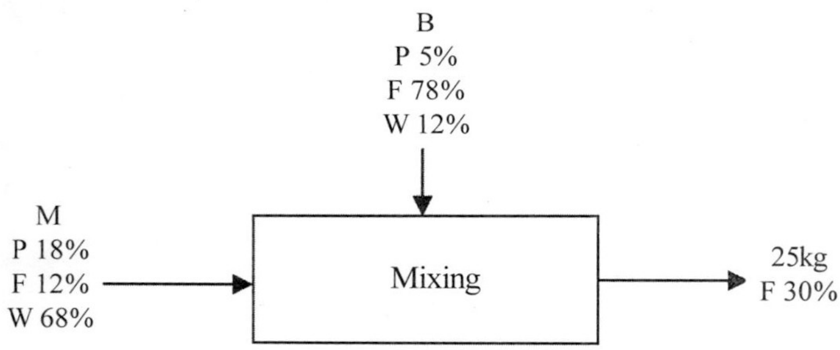

The overall material balance is

M + B = 25 (1)

Component balance is (Fat)

M (0.12) + B (0.78) = 25 (0.30) (2)

Solving the above equations we get, M = 18.18 kg and B = 6.82 kg

Example-12 Milk containing 3.7% fat and 12.8% total solids to be evaporated to produce a product containing 7.9% fat. What is the yield of product from 100 kg of milk and what is the total solid concentration in the final product.

Solution

The overall material balance is

$100 = M + Y$ (1)

Component balance is

(Total Solids) $100\,(0.128) = M\,(0) + Y\,(0.079)$ (2)

(Fat) $100\,(0.037) = M\,(0) + Y\,(Z)$ (3)

Solving the above equations we get, Y = 46.8 kg and Z = 27.3 kg

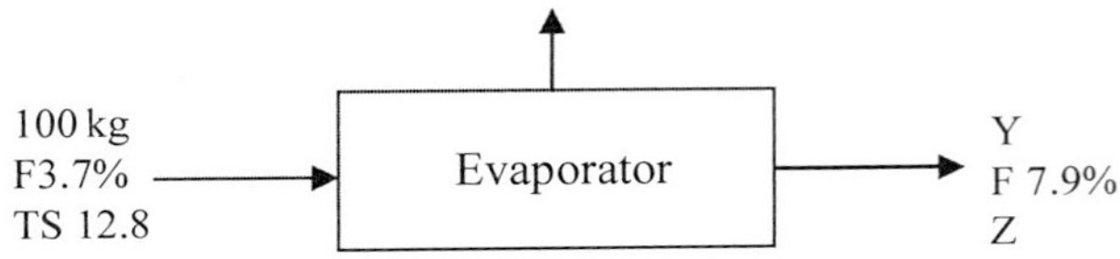

Example-13 Fresh orange juice with 12% soluble solids content is concentrated to 60% in a multiple effect evaporator. To improve the quality of the final product, the concentrated juice is mixed with an amount of fresh juice so that the concentration of the mixture is 42%. Calculate how much water per hour must be evaporated in the evaporation how much fresh juice per hour must be added back and how much final product will be produced it the inlet feed flow late is 10000 kg/h fresh juice.

Solution

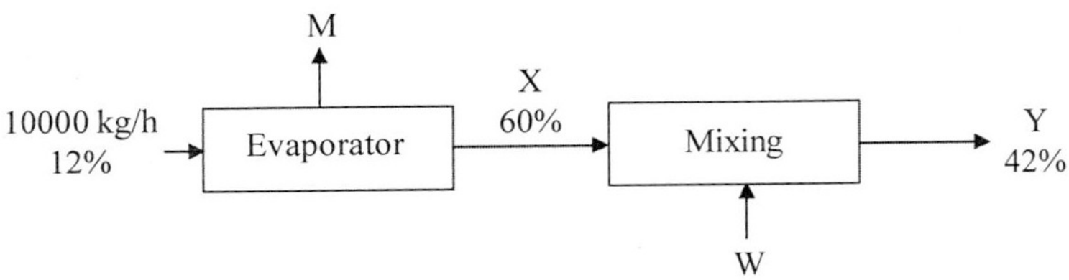

The overall material balance of stage one is

$10000 = M + X$ (1)

The overall material balance of stage two is

$X + W = Y$ (2)

Component balance of stage one is

10000 (0.12) = M (0) + X (0.06) (3)

Component balance of stage two is

X (0.6) + F (0.12) = Y (0.42) (4)

Solving the above equations we get, X = 2000 kg/h, M = 8000 kg/h, W = 1200 kg/h and Y = 3200 kg/h

Example-14 A membrane separation system is used to concentrated total solids (TS) in a liquid food from 10% to 30%. The concentration is accomplished in two stages with first stage resulting in release of low total solid liquid stream. The second stage separates the final contraction product from a low total solid steam, which is returned to the first stage. Determine the magnitude of the recycle stream when the recycle contains 2% TS, the waste stream contains 0.5% TS and the stream between stages one and two contains 25% TS. The process should produce 100 kg/min of 30% TS.

Solution

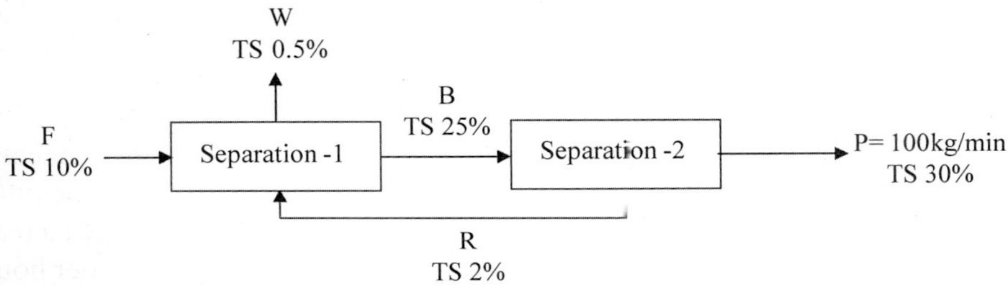

The overall material balance of whole system is

F = 100 + W (1)

Component balance of whole system is (Total Solids)

F (0.1) = 100 (0.3) + W (0.005) (2)

The overall material balance of system one is

F + R = W + B (3)

Component balance of stage one is (Total Solids)

F (0.1) + R (0.02) = W (0.005) + B (0.25) (4)

The overall material balance of system two is

B = 100 + R (5)

Component balance of stage two is (Total Solids)

B (0.25) = 100 (0.3) + R (0.02) (6)

Solving the above equations we get, F = 210 kg/min, W = 310 kg/min, B = 6400 kg/min and R = 6300 kg/min

Example-15 1000 kg of mixture of benzene (B) and toluene (T), containing 40% by mass of benzene to be separated into two streams in a distillation column. The top output stream contains 375 kg of benzene and the bottom output stream contains 515 kg of toluene.

A) Perform the mass balance for B and T

B) Determine the composition of the top and bottom stream

Solution

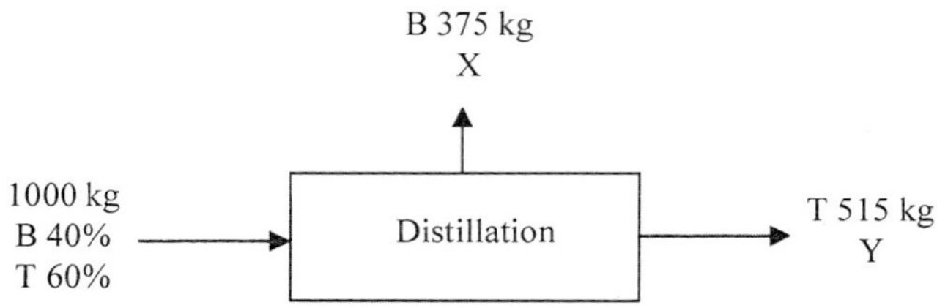

In input stream benzene is 400kg and toluene is 600kg

Overall material balance is

$400 = 375 + Y$ (1)

$600 = 515 + X$ (2)

We get X = 85kg and Y = 25kg

Example-16 Two methanol and water mixture are contained in the separate tank. The first mixture contains 40%wt methanol and the second contains 30%wt water. Mix 200gm of the first mixture with 150gm of second one. What are the total mass and composition of the product?

Solution

The overall material balance is

$200 + 150 = X$ (1)

Component balance is

$200\,(0.40) + 150\,(0.70) = X\,(M)$ (2)

$200\,(0.60) + 150\,(0.30) = X\,(W)$ (3)

Solving the above equations we get, X = 350g, M = 52.9% and W = 47.1%

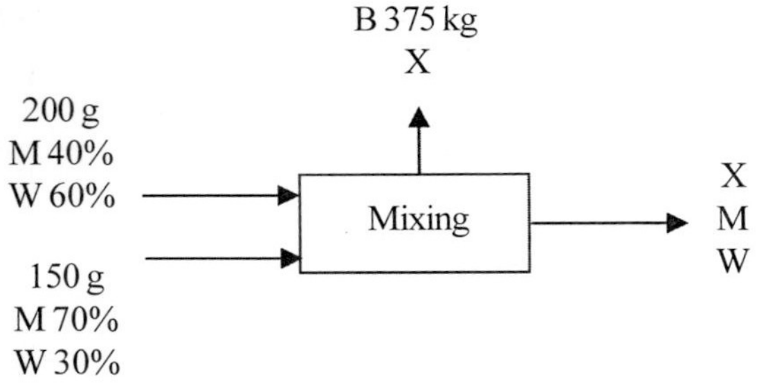

Example-17 Production of low fat cream (18% fat) from double cream (48%fat) and milk (3.5% fat). How much double cream and milk are required to produce 100kg low fat cream.

Solution

The overall material balance is

$$M + X = 100 \quad (1)$$

Component balance is

$$M\,(0.035) + X\,(0.48) = 100\,(0.18) \quad (2)$$

Solving the above equations we get, X = 32.55kg, M = 67.42kg

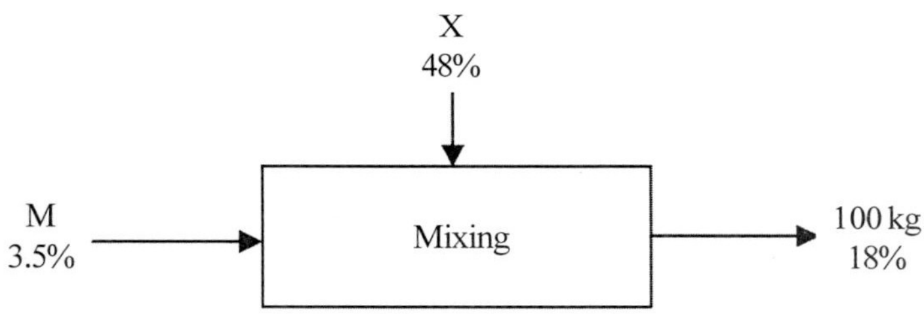

Example-18 Falling film evaporator capacity 10kg water per hour. Feed enter at 5.5% solids and 25% solids are desired. To do that, system of recycles was used at which recycling pump moves 20 kg/hr cf fluid into evaporator. Calculate the followings

1. The feed rate
2. Concentrate production rate
3. Amount of concentrate recycled
4. Concentration of mixture of feed and recycled concentrate

Solution

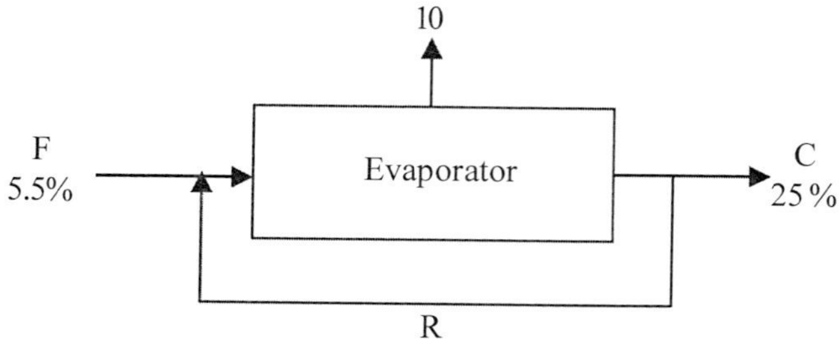

The overall material balance is

F = 10 + C (1)

Component balance is (Total Solids)

F (0.055) = 10 (0) + C (0.25) (2)

Solving the above equations we get, F = 12.82kg/h and C = 2.82 kg/h

When there is recycling process then

The overall material balance is

F + R = 10 + C i.e. = 20 (3)

Component balance is

F (0.055) = 10 (0) + C (0.25) (4)

Solving the above equations we get, F = 12.82kg/h and C = 2.82 kg/h

Example-19 Set up a total mass and component balance equation for a crystallizer where 100 kg of a concentrated sugar solution containing 85% sucrose and 1% inert, water-soluble impurities (balance, water) enters. Upon cooling, the sugar crystallizes from solution. A centrifuge then separates the crystals from a liquid fraction, called the mother liquor. The crystal slurry fraction has, for 20% of its weight, a liquid having the same composition as the mother liquor. The mother liquor contains 60% sucrose by weight.

Solution

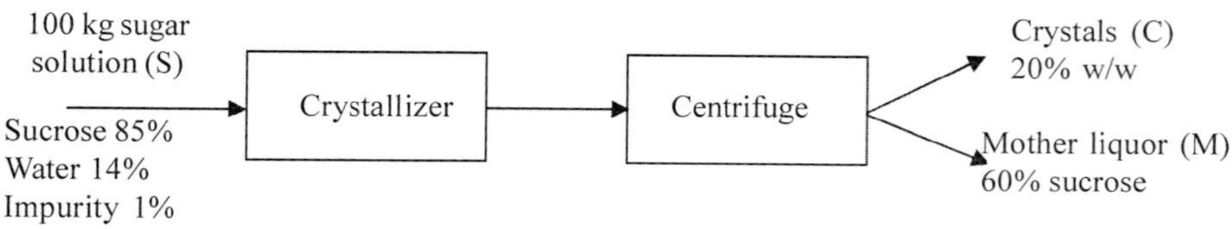

Total mass balance is

$$S = C + M \quad (1)$$

Component balance of sucrose is

$$S(0.85) = M(0.6) + C(0.2)(0.6) + C(0.8) \quad (2)$$

Component balance on water is

Let x = mass fraction of impurity in the mother liquor

$$S(0.14) = M(0.4 - x) + C(0.2)(0.4 - x) \quad (3)$$

Component balance on impurity is

$$S(0.01) = M(x) + C(0.2)(x) \quad (4)$$

Solving above equations we get; C=78.12kg, M=21.8kg

Example-20 In a process of producing KNO_3 salt, 1000 kg/h of the feed solution containing 20 wt % KNO_3 is fed to an evaporator, which evaporates some water at 422 K to produce a 50 wt % KNO_3 solution. This is then fed to a crystalizer at 311 K, where crystals containing 96 wt % KNO_3 are removed. The saturated solution containing 37.5 wt % KNO_3 is recycled to the evaporator. Calculate the amount of recycle stream R in kg/h and the product stream of crystals P in kg/h.

Solution

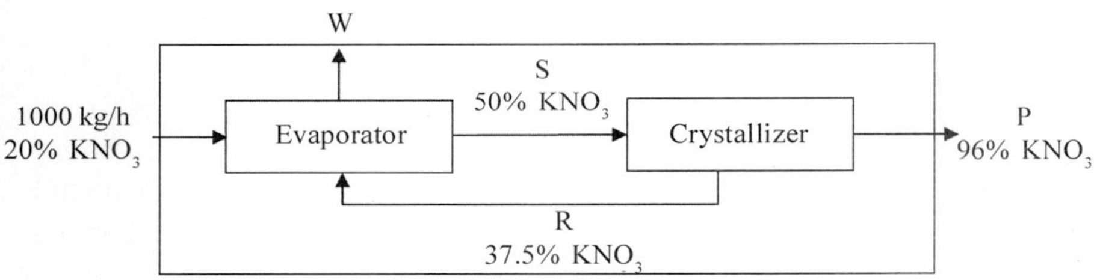

Total mass balance is

$$1000 = W + P \quad (1)$$

Component balance of KNO_3 is

$$1000\,(0.20) = W\,(0) + P\,(0.96) \quad (2)$$

Solving the above equation we get P = 208.3 kg/h

Total mass balance in the recycle stream can be done in evaporator or crystallizer i.e.

$S = 208.3 + R$ (3)

Component balance of KNO_3 in crystallizers is

$S\,(0.50) = 208.3\,(0.96) + R\,(0.375)$ (4)

Solving above equations we get S = 974.9 kg/h and R = 766.6 kg/h

Example-21 In the concentration of orange juice a fresh extracted and stained juice containing 7.08% solid is fed to an evaporator. In the evaporator, water is removed and the solids content increased to 58% solids. For 1000 kg/h entering, calculate the amounts of the outlet streams of concentrated juice and water.

Solution

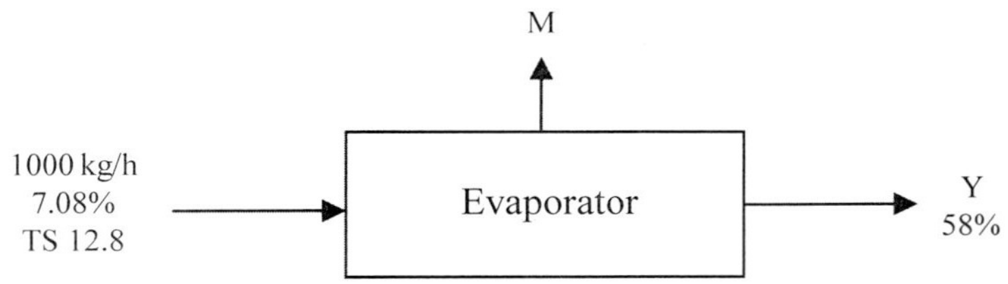

Total mass balance is

$1000 = M + Y$ (1)

Component mass balance is

$1000\,(0.0708) = M\,(0) + Y\,(0.58)$ (2)

Solving the above equations we get M = 877.9 kg/h and Y = 122.1 kg/h

Example-22 Determine the amounts of lean beef, pork fat, and water that must be used to make 100 kg of a frankfurter formulation. The compositions of the raw materials and the formulations are;

Lean beef: 14% fat, 67% water, 19% protein.

Pork fat: 89% fat, 8% water, 3% protein.

Frankfurter: 20% fat, 15% protein, 65% water.

Solution

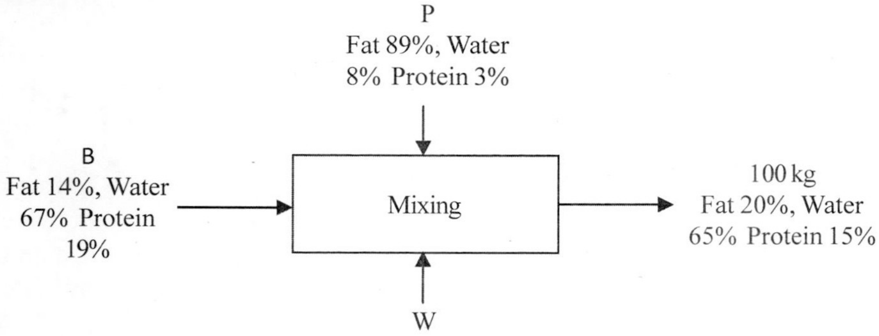

Total mass balance is

B + P + W = 100 (1)

Component balance for fat

B (0.14) + P (0.89) + W (0) = 100 (0.20) (2)

Component balance for Protein

B (0.19) + P (0.03) + W (0) = 100 (0.15) (3)

We have three equations with three unknowns, namely, B, P and W. We will solve these equations using a spreadsheet procedure to solve simultaneous equations. The method described in the following was executed on spread sheet.

1. The equations are rewritten in such a way that all unknown variables are collected on the left-hand side.

2. The equations are rewritten so that the coefficients can easily be arranged in a matrix. The spreadsheet method will use a matrix inversion process to solve the simultaneous equations.

3. As shown in **Fig. 1**, enter the coefficients of the left-hand side of the preceding equations in array C4:E6; enter the right-hand side coefficients in a column vector G4:G6.

4. Select another array C9:E11. Type=MINVERSE(C4:E6) in cell C9 and press the CTRL, SHIFT, and ENTER keys simultaneously. This procedure will invert the matrix C4:E6 and give the coefficients of the inverted matrix in array C9:E11.

5. Highlight cells G9:G11 by dragging the cursor starting from cell G9. Type= MMULT(C9:E11,G4:G6) into cell G9; press the CTRL, SHIFT, and ENTER keys simultaneously. The answers are displayed in the column vector G9:H11.

By solving the simultaneous equations with above steps, we get B = 77.32kg, P = 10.31kg and W = 12.37kg

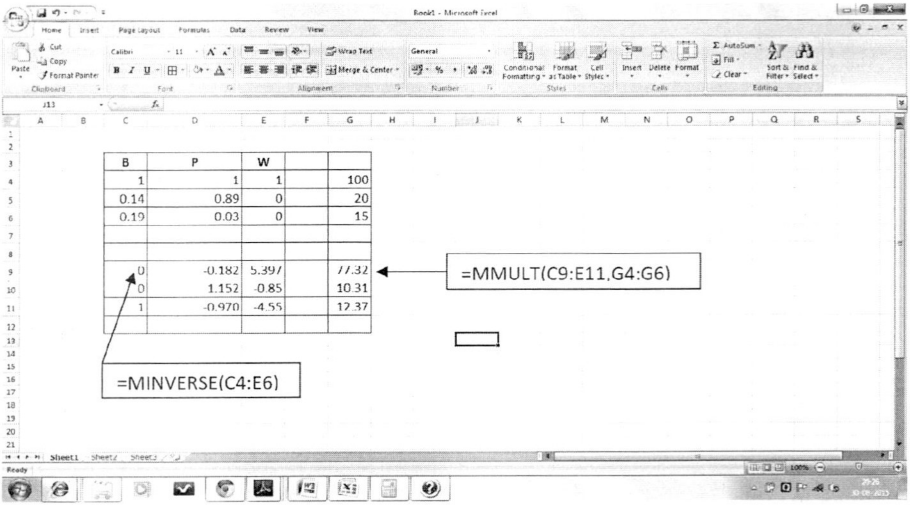

Fig. 1: Spread Sheet

Example-24 A blend of sorghum syrup is to be produced by mixing pure syrup with corn syrup and semi-syrup to produce 100 kg of blended product. Properties of the three syrups are shown in the following table. All percentages are by mass. Calculate amount of Sorghum syrup, Corn syrup and Semi-syrup.

Component	Water, %	Sugar, %	Flavour, %
Sorghum syrup (*S*)	23.0	75.5	1.5
Corn syrup (*C*)	20.0	80.0	0.0
Semi-syrup (*P*)	40.0	59.0	1.0
Desired blend (D)	24.0	75.0	1.0

Solution

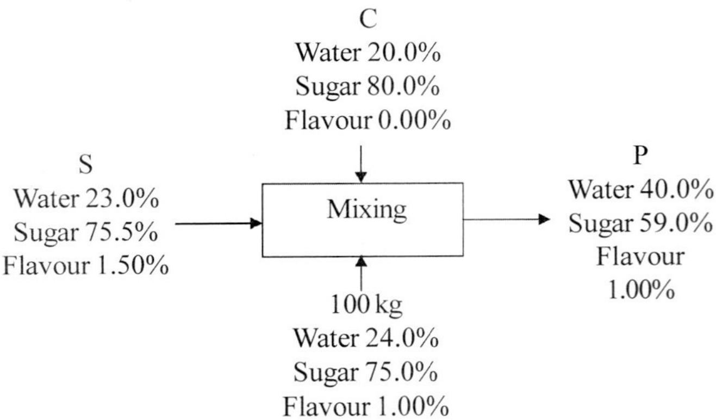

Total mass balance is

$$S + C + P = 100 \quad (1)$$

Component balance for water is

$$S\,(0.23) + C\,(0.20) + P\,(0.40) = 100\,(0.24) \quad (2)$$

Component balance for sugar

$$S\,(0.755) + C\,(0.80) + P\,(0.59) = 100\,(0.75) \quad (3)$$

Component balance for flavour is

$$S\,(0.015) + C\,(0.00) + P\,(0.10) = 100\,(0.10) \quad (4)$$

Solving the above equations we get $S = 59.26$kg, C = 29.63kg and P =11.11kg

Example-25 A continuously operated rotary drier is used to dry 12 kg/min of a starch-based food containing 25% moisture (wet weight basis) to give a product containing 10% moisture. However, the drier cannot handle feed material with moisture content greater than 15% and therefore a proportion of dry product must be recycled and mixed with the fresh feed. Calculate the evaporation rate and the recycle ratio.

Solution

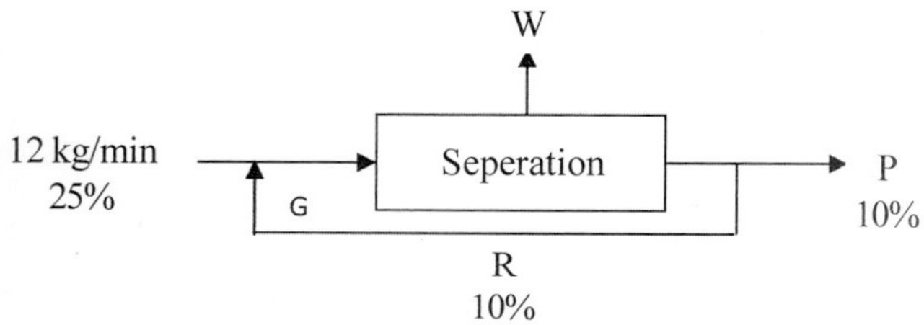

Total mass balance before recycling (water/moisture)

$$12 = W + P \quad (1)$$

Component mass balance is

$$12\,(0.25) = W + P\,(0.10) \quad (2)$$

Solving above equations we get W = 2kg/s

Total mass balance after recycling

$$12 + R = G \quad (3)$$

Component mass balance is

$$12\,(0.25) + R\,(0.10) = G\,(0.15) \quad (4)$$

Solving above equations we get R = 24kg/s, Recycle ratio is = 24/12 = 2

Example-26 A food mix is to be made that would balance the amount of methionine (MET), a limiting amino acid in terms of food protein nutritional value, by blending several types of plant proteins. Corn, which contains 15% protein, has 1.2 g MET/100 g protein; soy flour with 55% protein has 1.7 g MET/100 g protein; and non fat dry milk with 36% protein has 3.2 g MET/100 g protein. How much of these ingredients must be used to produce 100 kg of formula that contains 30% protein and 2.2 g MET/100 g protein.

Solution

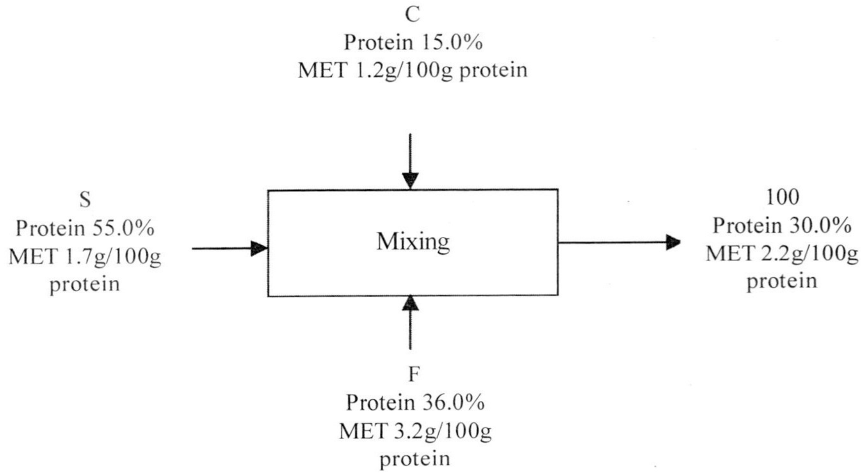

Overall mass balance is

$$S + C + F = 100 \qquad (1)$$

Component balance for protein is

$$S\,(0.55) + C\,(0.15) + F\,(0.36) = 100\,(0.30) \qquad (2)$$

Component balance for MET is

$$S\,(0.55 \text{x} 1.7)/100 + C\,(0.15 \text{x} 1.2)/100 + F\,(0.36 \text{x} 3.2)/100 = 100\,(0.30 \text{x} 2.2)/100$$

$$S\,(.00935) + C\,(0.0018) + F\,(0.01152) = 0.66 \qquad (3)$$

Solve the above equation with the procedure given in **Fig. 1**, we get S = 19.54kg, C = 46.25 kg and F = 34.20kg

Unsolved Problems

1. Two tones of paddy is dried from 24% moisture content (wet basis) to 12% in three stage. Moisture content after each stage is 18%, 15% and 12 % respectively. Calculate the final weight of paddy after each stage.

Ans.

Wet basis moisture content (%)	24	18	15	12
Dry basis moisture content (%)	31.58	21.95	17.65	13.64
Total weight (Kg)	2000	1853.64	1788.28	1727.33

2. Groundnut (44% oil) is expelled in the 8-bolt expeller. The residual oil in the cake is 13%. If total 200 kg oil is collected. Calculate the total amount of groundnut is crushed and amount of oil remain in the deoiled expeller cake.

Ans. Ground Nut crushed : 516.8 kg; Oil in cake 47.19 Kg

3. 25 kg mustard (34% oil) crushed in ghanis Output of such 8 ghanies fed in the expeller. If the average oil recovered from each ghani is 6 kg and oil in cake produced from expeller is 7.5%. Calculate the total oil recovered and oil wasted in the expeller cake.

Ans. Recovery 84.259%; Remain as waste in cake: 15.741%

4. Seasmum 200kg, crushed in ghani having 48% oil. Residual ghani cake (18%) oil is pressed in expeller. The cake comes out of the expeller contain 8% oil which is put in solvent extraction plant which leaves 1% oil in cake. Calculate the total oil recovered.

Ans. Total Oil recovery 93.92 Kg

5. Fruit Juice concentrate 25% solids is available with the company. It is to be diluted to 10 % solids. Calculate how much water is to be added to make 1000 liter of juice.

Ans. 400 Kg Concentrate and 600 Kg water

6. Fruit pulp containing 70% moisture and 30 % solids is used to make fruit drink. To adjust the acidity 0.25% citric acid powder is adder. Final fruit-drink contains 12% solids which have been brought by adding sugar 35% sugar. Calculate amount of drink is prepared from 1 tonne pulp.

Ans. 400 Kg Concentrate and 600 Kg water

7. Fresh orange juice (8% solids) is concentrated in an evaporator to 40 and 56% solids. If juice inflow is 500 kg/h, calculate the out flow of juice for every solid contents of the concentrated juice.

Ans. (i) For 40% concentration: 100kg/h; (ii) 56% concentration: 71.43 Kg/h

8. Potato flakes 75% moisture content (wet basis) is dried in dryer. If ¾ th of available water is removed then calculate the mass of water removed per tonne of the potato flakes and composition of dried flakes.

Ans. 250 Kg dry matter; 187.5 Kg moisture

9. A feed containing 1t/h sugar solution (25% Solids) is fed in evaporator producing concentrate of 60% solids and this concentrate is fed in Crystallizer , where solids of 97% solid content is taken as product and saturated solution of 42 % is recycled to evaporator. Calculate the amount of recycle stream and out of solid from the crystallizer.

Ans. Recycle stream 280.359 kg/h solids 136.311 Kg/h

10. A process of producing jam from crushed fruit containing 14% solids is mixed with 1.2kg sugar per kg of fruit and 0.0025kg pectine per kg of fruit in a mixer. Resultant mixer is heated to produce Jam of 67% solids in steam kettle. Calculate the amount of Jam prepared per tonne of fruits.

Ans. 2000 kg Jam

2

Energy Balance

Kshitiz Kumar and Vijay Singh Sharanagat

1. Introduction

Energy balance means keeping an account of how the flow of energy is taking place in and out of any system or process. It helps the engineers to estimate the total energy requirement and identify the area for conservation of energy in any process. In food processing the concept of energy balance is applicable in unit operation such as blanching, drying, evaporation, freezing, refrigeration, distillation, crystallization. The scopes of energy balance in food industry are:

i) Streamlining and reduction of energy consumption in the existing process.

ii) Selection and sizing of any equipment and process.

iii) Economic analysis of alternate process/technology.

iv) Designing new economically viable process/technology

v) Waste reduction and utilization

vi) Mechanization and digitalization of any process.

2. Basic Energy Terms

2.1 Energy: It is a scalar quantity that must be transferred to an object in order to perform work on, or to heat, the object. The common forms of energy are kinetic energy, potential energy, electrical energy, magnetic energy, chemical energy and internal energy. SI unit of energy is joule (J). Energy is transferred as either heat or work.

2.2 Heat: Heat is energy in transfer to or from a thermodynamic system because of a temperature difference between the system and surroundings. SI unit of heat is joule (J). The conventional symbol for heat transferred in a thermodynamic process is Q. If heat is transferred from a system into its surroundings, then Q is by convention a negative quantity ($Q < 0$); when a system absorbs heat from its surroundings, Q is positive ($Q > 0$).

2.3 Sensible heat and latent heat

Sensible heat is heat exchanged by a body in which the exchange of heat changes the temperature of the body. Sensible heat can be calculated as:

$$Q = mc_p\Delta T \quad (1)$$

Where m is mass (kg), c_p is specific heat capacity of the material over the specified temperature range (kJ kg^{-1} K^{-1}) and ΔT is change in temperature (K)

Latent heat is heat exchanged by a body during a constant-temperature process usually a first-order phase transition. It can be understood as hidden form of energy which is exchanged to change the state of a substance without changing its temperature. Examples are latent heat of fusion and latent heat of vaporization which are required during melting of ice and vaporization of water respectively. Specific latent heat (L) indicates the amount of energy in the form of heat (Q) required to completely affect the phase change of a unit of mass (m). Latent heat is calculated as:

$$Q = mL \quad (2)$$

Where Q is the amount of energy released or absorbed during the change of phase of the substance (kJ), m is the mass of the substance (kg), and L is the specific latent heat of the substance (kJ kg^{-1}), either L_f for fusion, or L_v for vaporization. The latent heat of fusion for water at 0°C is 333.2 kJ/kg. The latent heat of vaporization of water is function of temperature and pressure. At 100 °C, the latent heat of vaporization of water is 2257.06 kJ/kg.

2.4 Work: Works is energy transferred between the system and its surrounding as a result of any driving force other than temperature difference. SI unit of work is joules (J). The conventional symbol for work is W. The sign convention for work is opposite to that of heat transfer. If work is done by a system into its surroundings, then W is by convention a positive quantity ($W > 0$); when a work is done on system by its surroundings, W is negative ($W < 0$).

2.5 Intensive and Extensive Properties

Intensive quantity is one whose magnitude is independent of the size of the system whereas an extensive quantity is one whose magnitude is additive for subsystems. Temperature, density, specific heat capacity, specific volume, thermal conductivity are examples of intensive property. Mass, volume, energy, enthalpy, internal energy are examples of extensive property.

Extensive variables can be converted to specific quantities by dividing by the mass; for example, specific heat capacity is heat capacity divided by mass. Because specific properties are independent of the mass of the system, they are also intensive variables. Extensive properties denoted by an upper-case symbol while the specific property is given in lower-case notation. For example if C_p is heat capacity with unit J/K, c_p denotes specific heat capacity with unit $JK^{-1}kg^{-1}$.

2.6 Steady and Transient State Process/System

A process/system is said to be steady state if variables such as temperature, flow rate, or pressure are independent of time. If these variables changes with time the process/system is called transient state.

2.7 System and Boundary

A system is defined as a region in space or quantity of matter. The interface between the system and its surroundings is known as the boundary. A system may comprise of a complete processing line, one unit operation in the line, or a portion of that unit operation. Boundary which is drawn around the system and defines it may be real or imaginary

2.8 Open or Closed System

An open system allows exchange of mass and energy with the surroundings. A closed system allows only transfer of energy across the system boundary, the system boundary being impervious to mass transfer.

2.9 Specific Heat: Specific heat is the amount of heat required to increase the temperature of a unit mass of the substance by unit degree. SI unit of specific heat is J kg^{-1} K^{-1}. Specific heat of ffod materials depends on their composition. Siebal (1892) proposed following equation for the estimation of specific heat for non-fat fruit and vegetable product above their freezing point:

$$C_p = 0.837 + 3.349X_w \tag{3}$$

Where c_p is specific heat in kJ kg^{-1} K^{-1}, X_w is the mass fraction of moisture within the sample.

The equation for estimation of specific heat above freezing when fat is present is:

$$C_p = 1.674\, X_F + 0.837\, X_{SNF} + 4.186\, X_W \tag{4}$$

Where X_F is mass fraction of fat, X_{SNF} is mass fraction of solid non-fat

Heldman (1975) suggested a method to estimate the specific heat of food materials based on mass fraction of the constituent (water, protein, fat, carbohydrate, and ash):

$$Cp = 4.180X_W + 1.547X_p + 1.672X_F + 1.42X_{CHO} + 0.836X_{ash} \quad (5)$$

Where X_p is mass fraction of protein, X_{CHO} is mass fraction of carbohydrate, X_{ash} is mass fraction of ash, cp is given in kJ/kg K.

Choi and Okos (1986) proposed a more accurate estimation of specific heat of food material containing n components as:

$$c_p = \sum_{i=1}^{n} X_i c_{pi} \quad (6)$$

Where X_i is mass fraction of component i and c_{pi} is specific heat of component i (J kg^{-1} K^{-1}). They also suggested the equation for the temperature dependence of specific heat of major food components in J kg^{-1} K^{-1} as:

$$C_{p'}\ water = 4081.7 - 5.3062\ T + 0.99516\ T^2\ (\text{for - 40 to } 0^{o}C) \quad (7)$$

$$C_{p'}\ water = 4176.2 - 0.0909\ T + 5.4731 \times 10^{-3}\ T^2\ (\text{for 0 to 150 } ^{o}C) \quad (8)$$

$$C_{p'CHO} = 1548.8 + 1.9625\ T - 5.9399 \times 10^{-3}\ T^2\ (\text{for -40 to 150 } ^{o}C) \quad (9)$$

$$C_{p,\ protein} = 2008.2 + 1.2089\ T - 1.3129 \times 10^{-3}\ T^2\ \text{for -40 to 150 } ^{o}C) \quad (10)$$

$$C_{p,fat} = 1984.2 + 1.4373\ T - 4.8008 \times 10^{-3}\ T^2\ (\text{for -40 to 150 } ^{o}C) \quad (11)$$

$$C_{p,ash} = 1092 + 1.8896\ T - 3.6817 \times 10^{-3}\ T^2\ (\text{for -40 to 150 } ^{o}C) \quad (12)$$

$$C_{p,ice} = 2062.3 + 6.0769\ T$$

where temperature (T) is in °C in these equations.

2.10 Specific Heat of Gases and Vapours

Specific heat of gases are designated as cp or cv depending on whether the process in carried out at constant pressure or constant volume respectively. If a gas is heated from a reference temperature T_o to final temperature T_2 the heat required is equal to change in enthalpy for this temperature range. We can write:

$$\Delta H = mc_{pm} (T_2 - T_o) \quad (13)$$

Where c_{pm} is the mean specific heat over the temperature range T_o to T_2.

If instead of reference temperature T_0 the gas is heated from any temperature T_1 to final temperature T_2 the change in enthalpy can be calculated as:

$$\Delta H = mc_{pm}(T_2 - T_0) - mc'_{pm}(T_1 - T_o) \quad (14)$$

Where c'_{pm} is mean specific heat from the reference temperature T_o to T_1. Mean specific value of gases based on reference temperature 25 °C are available. If there is very little change in c_{pm} for any two temperature range, equation (13) can be used instead of equation (14) for sake of simplification.

Example 1: Calculate the heat required to raise the temperature of 20 m^3/s air at 1 atm from 50 °C to 150 °C. Given: $c_{pm}^{'}$ at 50 °C and 150 °C is 1008 J/(kg K) and 1014 J/(kg K) respectively based on reference temperature 25 °C.

Solution: Here reference temperature T_o is 25 °C, initial temperature of air T_1 is 50 °C and final temperature T_2 is 150 °C.

$c_{pm}^{'}$ at T_1 = 50 °C = 1008 J/(kg K) and $c_{pm}^{'}$ at T_2 = 150 °C = 1014 J/(kg K)

As $c_{pm}^{'}$ from To to 50 °C and from To to 150 °C are different we have to use equation (14) to determine ΔH.

Mass of the gas is determined using ideal gas equation.

$$m = \frac{PVM}{RT} = \frac{1(atm)\times 20\left(m^3 s^{-1}\right)\times 29\left(kg/kgmole\right)}{0.08206\left(\dfrac{m^3 atm}{kg\,mole\,K}\right)\times\left(50+273\right)\left(K\right)} = 21.88\,kg/s$$

Using equation (14)

$$\Delta H = mc_{pm}^{'}\left(T_2 - T_o\right) - mc_{pm}^{'}\left(T_1 - T_o\right)$$

$$\Delta H = (21.88)(1014)(150-25) - (21.88)(1008)(50-25) = 2221.91\,kJ/s$$

2.11 Enthalpy: Enthalpy is expressed as summation of internal energy and a work term, which is given by product of pressure and volume.

$$H = U + PV \tag{15}$$

where H is enthalpy (kJ), U is internal energy (kJ), P is pressure (kPa), and V is volume (m^3).

Enthalpy may also be expressed per unit mass as follows:

$$h = u + Pv \tag{16}$$

where h is **specific enthalpy** (kJ.kg^{-1}), u is specific internal energy (kJ.kg^{-1}), P is pressure (kPa), and v is specific volume (m^3.kg^{-1}).

Since internal energy cannot be measured in absolute term, it is not possible to measure the absolute value of enthalpy. Enthalpy value is always expressed in form of change in enthalpy (ΔH) relative to a reference state, where the value of enthalpy of all the components that enter or leave the system is arbitrarily selected as zero. The reference temperature for determining the enthalpy of water in the steam tables is 0.01 °C. It can be shown mathematically that for heating process at constant pressure, which is most commonly encountered in food processing, change in enthalpy (ΔH) is equivalent to the heat content (Q), i.e.

$$\Delta H = Q \tag{17}$$

The change in enthalpy of a substance at constant pressure due to a change in temperature ΔT is:

$$\Delta H = Q = mc_p \, \Delta T = mc_p \, (T_2 - T_1) \tag{18}$$

The change in enthalpy due to phase change is calculated from the latent heat as:

$$\Delta H = Q = mL \tag{19}$$

Example 2. Calculate the total change in enthalpy for heating 2 kg ice at -15 °C to superheated steam at 110 °C. Specific heat of ice is 2.05 kJ/(kg K). Specific heat of water is 4.182 kJ/(kg K), latent heat of fusion is 333.2 kJ/kg, and latent heat of vaporization at 100 °C is 2257.06 kJ/kg, specific heat of steam is 1.996 kJ/(kg K)

Solution: Given: $c_{p'ice} = 2.05$ kJ kg^{-1} K^{-1}; $c_{p'water} = 4.0182$ kJ kg^{-1} K^{-1}; $c_{p'steam} =$ 1.996 kJ kg^{-1} K^{-1}; $L_{f,ice} = 333.2$ kJ kg^{-1}; $L_{v,steam} = 2257.06$ kJ kg^{-1}

Let the temperature of ice at -15 °C be T_1, water/ice at 0 °C be T_2, water/ steam at 100 °C be T_3 and superheated steam at 110 °C be T_4.

Heating of ice from -15 °C to superheated steam at 110 °C can be broken down into five stages as follows:

i. Sensible heating of ice from -15 °C (258.15 K) to ice at 0 °C (273.15 K)

   $\Delta H_1 = mc_{p,ice} \, (T_2 - T_1) = 2(\text{kg}) \times 2.05 \, (\text{kJ kg}^{-1} \text{K}^{-1}) \times (273.15\text{-}258.15)(\text{K})$

   $\Delta H_1 = 61.5$ kJ

ii. Absorption of latent to convert ice at 0 °C to liquid water at 0 °C

   $\Delta H_2 = mL_{f,ice} = 2(\text{kg}) \times 333.2(\text{kJ kg}^{-1}) = 666.4$ kJ

iii. Sensible heating of liquid water from 0 °C (273.15 K) to liquid water at 100 °C (373.15 K).

   $\Delta H_3 = mc_{p,water} \, (T_3 - T_2) = 2(\text{kg}) \times 4.0182 \, (\text{kJ kg}^{-1} \text{K}^{-1} \times (373.15\text{-}273.15)(\text{K})$

   $\Delta H_3 = 803.64$ kJ

iv. Absorption of latent to convert water at 100 °C to saturated steam at 100 °C

$$\Delta H_4 = mL_{v,steam} = 2(kg) \times 2257.06\ (kJ\ kg^{-1}) = 4514.12\ kJ$$

v. Sensible heating of saturated steam from 100 °C (373.15 K) to superheated steam at 110 °C (383.15 K).

$$\Delta H_5 = mc_{p,steam}\ (T_4 - T_3) = 2(kg) \times 1.996\ (kJ\ kg^{-1}\ K^{-1}) \times (383.15\text{-}373.15)(K)$$

$$\Delta H_5 = 39.92\ kJ$$

Hence, total change in enthalpy

$$\Delta H = \Delta H_1 + \Delta H_2 + \Delta H_3 + \Delta H_4 + \Delta H_5$$

$$\Delta H = 61.5 + 666.4 + 803.64 + 4514.12 + 39.92 = 6085.58\ kJ$$

3. Energy Balance

The basis of energy balance is first law of thermodynamics which is a statement of

conservation of energy. The law states that energy can neither be created nor be destroyed but can be transferred from one form to another. This can be expressed in form of general energy balance equation for a system as:

Energy in – Energy out = Energy Accumulation (20)

For steady state energy accumulation is zero. thus

Energy in = Energy out (21)

When equation (21) is used all the energy terms which are changing within the system must be considered. This may be kinetic energy, potential energy, internal energy. The heat content is expressed as enthalpy change (ΔH) from a set reference temperature. If the effects of mechanical, electrical, or electromagnetic on the total heat content are significant they must be accounted also. For a system containing only two components across which energy transfer is taking place the energy balance equation becomes:

Energy lost by component 1 = Energy gained by component 2 (22)

3.1 General Method for Energy Balances Calculation

The energy balance calculation for any process can be done by proceeding in systematic way. These calculation helps us to know the energy associated with any unknown stream as well as total efficiency of the process. The stepwise procedure for energy balance calculation are:

a) Construct a labeled flow diagram of the process and draw the boundary (dashed line) identifying the system and surrounding. The label should indicate all the inlet and outlet stream with their composition, phase (solid, liquid, vapour), temperature and pressure.

b) State the *units* selected and *basis* for calculation clearly.

c) State all the assumptions made to solve the problem. Some common assumption in energy calculation are: the system is homogeneous, heat of mixing is negligible, without vigorous agitation shaft work is negligible, heat loss to surrounding is negligible.

d) Determine the specific enthalpy of each stream with respect to reference state which is generally water at 0°C.

e) Write the appropriate energy balance equation and solve for unknown parameters.

Example 3. 2000 kg/h of milk is being pasteurized in heat exchanger from 35 °C to 72 °C with water as heating medium. Hot water at 95 °C enters the pasteurizer and leaves at 75 °C. The heat loss to the surrounding is 1 kJ/s. Calculate the mass flow rate of water in the pasteurizer. The composition of the milk is: 4% Fat, 4.6% carbohydrate, 3.5% protein, 0.7% ash, 87.2% moisture. Specific heat capacity of water is 4.2 kJ/kg °C.

Solution:

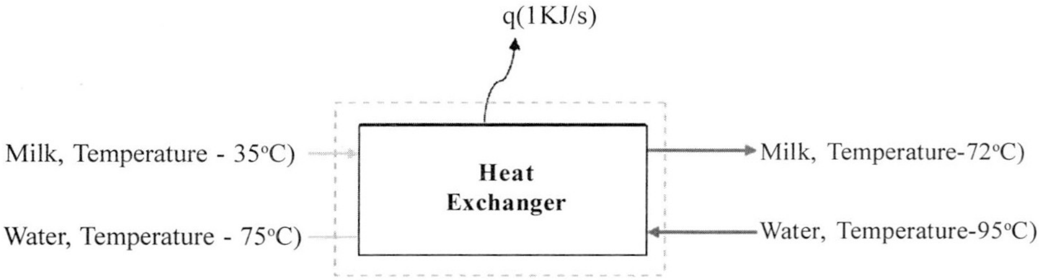

Step 1: Construct the flow diagram:

Step 2: The basis of calculation is flow of milk per hour. All calculations to be done in SI unit.

Step 3: The assumptions are:

- The system is at steady state.
- The specific heat capacity of the liquid streams remains constant over given temperature range.
- The terms of kinetic and potential energy and shaft work in the energy balance equation are negligible.

Step 4: Write the enthalpy of each stream entering and leaving the system with respect to reference state, which is water at 0°C. Let the subscript M, W, I and O in parameters mass (m), temperature (T), enthalpy (H) and specific heat (c_p) denote milk, water, input and output stream respectively.

i. Enthalpy of water stream is:

Input: $\Delta H_{W,I} = m_{W,I}\, c_{p,W}\, (T_{W,I} - T_{ref}) = m_{W,I} \times 4.2 \times (95\text{-}0) = m_{W,I}$ 399 kJ kg^{-1}

Output: $\Delta H_{W,O} = m_{W,O}\, c_{p,W}\, (T_{W,O} - T_{ref}) = m_{W,O} \times 4.2 \times (75\text{-}0) = m_{W,O}$ 315 kJ kg^{-1}

ii. Enthalpy of milk stream is:

Input: $\Delta H_{M,I} = m_{M,I}\, c_{p,M}\, (T_{M,I} - T_{ref})$

The specific heat of milk is found using equation as-

$c_{p,M} = 4.180\, X_w + 1.547 X_P + 1.672 X_F + 1.42 X_{CHO} + 0.836 X_{ash}$

$c_{p,M} =$

(4.180) (kJ kg^{-1} K^{-1}) (0.872) + (1.547) (kJ kg^{-1} K^{-1}) (0.035) + (1.672) (kJ kg^{-1} K^{-1}) (0.04) + 1.42 (0.046) + 0.836 (0.007) = 3.84 (kJ kg^{-1} K^{-1})

$\Delta H_{M,I}$ = 2000 (kg) (3.84) (kJ kg^{-1} K^{-1}) (35-0) (K) = 268800 kJ

Output: $\Delta H_{M,O} = m_{M,O}\, c_{p,M}\, (T_{M,O} - T_{ref})$

$\Delta H_{M,O}$ = 2000 (kg) (3.84) (kJ kg^{-1} K^{-1}) (72-0) (K) = 552960 kJ

Step 5: Write the appropriate energy balance equation and solve for unknown parameters

$$\Delta H_{M,I} + \Delta H_{W,I} = \Delta H_{M,O} + \Delta H_{W,O} + q_{lost}$$

$\Rightarrow$ 268800 (kJ) + $m_{W,I}$ 399 (kJ kg^{-1})

= 552960 (kJ) + $m_{W,O}$ 315 (kJ kg^{-1}) + (1×3600) (kJ)

Using $m_{W,I} = m_{W,O}$ and solving for $m_{W,I}$ yields

$m_{W,I}$ = 3425.71 kg

Therefore, mass floe rate of water required will be 3425.71 kg/h.

4. Properties of Steam

Water and steam are the two most common heat transfer medium used in food processing. A basic knowledge of properties of steam is must for energy balance calculations. The different forms in which water exists near its boiling point are:

4.1 Saturated Liquid: It is the condition when liquid water is at equilibrium with its vapor. Saturated liquid may exist at any temperature and the corresponding pressure when the liquid is at the boiling point.

4.2 Saturated vapor: It is the vapor (or steam) at equilibrium with liquid water. Like saturated liquid, saturated vapor may also exist at any temperature and the corresponding pressure when the liquid is at the boiling point. If the temperature of saturated vapour is lowered at constant pressure, the vapour condenses to liquid.

4.3 Vapor-Liquid Mixtures: It indicates continuous range of states that exists between that of a saturated liquid and that of a saturated vapor. It also exists at temperature and pressure corresponding to boiling point of liquid. When heat is added to vapour-liquid mixture, first the saturated liquid is converted to vapour without any change (increase) in temperature as heat is taken up as latent heat of vaporization. Similarly when heat is removed from vapour-liquid mixture, first of all the vapours are converted to liquid, again without change (lowering) in temperature as heat is being removed in form of latent heat of condensation.

4.4 Steam Quality: It indicates the percentage (or fraction) of vapour in a vapor-liquid mixture. For example, steam quality of 80% or 0.8 means that 80% of mixture is vapour and remaining 20% is liquid.

The enthalpy of steam with steam quality less than 100% is determined using the equation:

$$H = H_c + x_s (H_v - H_c) \qquad (23)$$

Where H is enthalpy of steam (kJ/kg), x_s is steam quality in fraction, H_c is enthalpy of liquid water at specified boiling condition (kJ/kg), H_v is enthalpy of saturated vapour at specified boiling condition (kJ/kg)

Equation (23) can also be written as:

$$H = (1 - x_s) H_c + x_s H_v \qquad (24)$$

The specific volume of steam with quality less than 100% is determined using the equation:

$$V' = (1 - x_s) V_c' + x_s V_v' \qquad (25)$$

Where V' is enthalpy of steam (m^3/kg), is specific volume of liquid water at specified boiling condition (m^3/kg), is specific volume of saturated vapour at specified boiling condition (m^3/kg).

4.5 Superheated steam: It is vapor at a temperature higher than the boiling point. The heat content is greater than saturated vapor. Addition of heat to superheated steam at constant pressure will increase its temperature. *Degrees superheat* indicates number of degrees the temperature exceeds the boiling temperature. Removal of heat from superheated steam will lower the temperature to saturated vapour or boiling point.

Example 4: 2000 kg/h milk is to be heated from 40 to 120 °C using steam at 125 °C as heating medium. The heat exchanger has efficiency of 90% and the steam quality is 95%. Condensate at 120 °C flows from the steam trap. Determine the mass and volume flow rate of the steam required. The supportive data are: specific heat of milk is 3.86 kJ/(kg °C), Specific heat of water at 120 °C is 4.232 kJ/(kg °C), From steam table the properties of steam at 125 °C is: H_c = 524.99 kJ/kg; H_v = 2713.5 kJ/kg; = 0.0010649 m^3/kg; = 0.7706 m^3/kg.

Solution

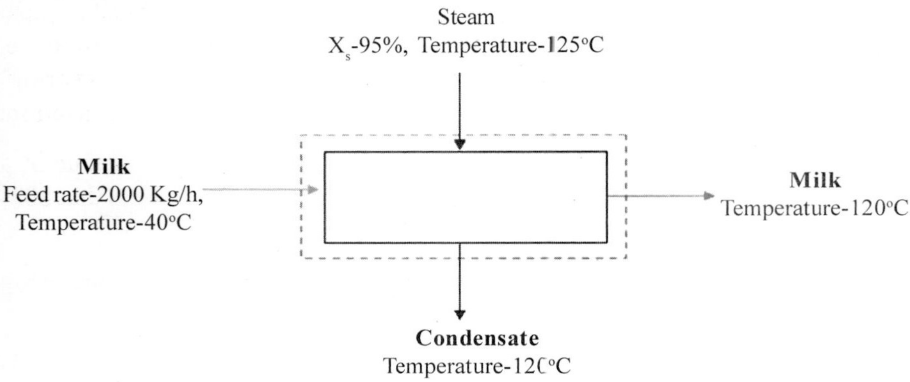

Step 1: Construct the flow diagram:

Step 2: The basis of calculation is flow of milk per hour. All calculations to be done in SI unit.

Step 3: The assumptions are:

- The system is at steady state.
- The specific heat capacity of the liquid streams remains constant over given temperature range.
- The terms of kinetic and potential energy and shaft work in the energy balance equation are negligible.

Step 4: Write the enthalpy of each stream entering and leaving the system with respect to reference state, which is water at 0°C. Let the subscript M, S, I and

O in parameters mass (m), temperature (T), enthalpy (H) and specific heat (c_p) denote milk, steam, input and output stream respectively.

i. Enthalpy of input steam with steam quality of 95% is:

$$H_s = m_s\ [(1 - x_s)\ H_c + x_s\ H_v] = m_s\ [(1\text{-}0.95)\ (524.99) + (0.95)(2713.5)]$$

$$H_s = m_s\ 2604.07\ \text{kJ kg}^{-1}$$

ii. Enthalpy of condensate leaving the heat exchanger is

$$H_c = m_s\ c_{p,W}\ (T - T_{ref}) = m_s\ (4.232)\ (120\text{-}0) = m_s\ 507.84\ \text{kJ kg}^{-1}$$

iii. Enthalpy of input milk

$$H_{M,I} = m_{M,I}\ c_{p,M}\ (T_{M,I} - T_{ref}) = (2000)\ (3.86)\ (40\text{-}0) = 30880\ \text{kJ}$$

iv. Enthalpy of output milk

$$H_{M,O} = m_{M,O}\ c_{p,M}\ (T_{M,O} - T_{ref}) = (2000)\ (3.86)\ (120\text{-}0) = 926400\ \text{kJ}$$

Step 5: Write the appropriate energy balance equation and solve for unknown parameters

$$H_{M,I} + H_S = H_{M,O} + H_C$$

$$30880\ (\text{kJ}) + m_s\ 2604.07\ (\text{kJ kg}^{-1}) = 926400\ (\text{kJ}) + m_s\ 507.84\ (\text{kJ kg}^{-1})$$

Solving for m_s yields

$$m_s = 427.20\ \text{kg}$$

For 90% heat exchanger efficiency

$$m_s = \frac{427.20}{0.85} = 474.66\,kg$$

Therefore mass flow rate is 474.66 kg/h

For 95% steam quality

$$V' = (1 - x_s)V_c' + x_s V_v' = (1 - 0.95)(0.0010649) + (0.95)(0.7706) = 0.7321\,m^3 kg$$

$$Volumetric\ flow\,rate = 474.66\left(kgh^{-1}\right) \times 0.7321\left(m^3 kg\right) = 347.49\,m^3 h^{-1}$$

5. Simultaneous Mass and Energy Balance

Several unit operations in the food processing such as drying, evaporation, flash cooling involves simultaneous mass and energy transfer. In such case linking the mass transfer and energy transfer equations acts as powerful tool to estimate the unknown stream as well as overall energy requirement. The approach for solving simultaneous mass and energy transfer remains the same as discussed in energy balance or mass balance alone, only difference being taking both of them together. Following example illustrate the solution of mass and energy balance simultaneously.

Example 5: A fruit juice is being concentrated in single effect evaporator at the rate of 4000kg/h. At steady condition, the dilute juice is fed at 25 °C with concentration 8% total solid. The juice is to be concentrated to 75% total solid. The evaporator is operated at sufficient vacuum to allow the product to boil at 70 °C. The steam is supplied at 313 kPa and the condensate is released at 70 °C. A small amount of energy 1000 kJ/h is lost to the environment due to lack of efficient insulation. The specific heat of dilute juice and concentrate are 4.0 kJ/kg °C and 3.0 kJ/kg °C. Assume negligible boiling point elevation. Calculate mass flow rate of product, vapour and condensate.

Solution

Given

Mass flow rate of feed m_F = 4000 kg/h

Feed Temperature T_F = 25 °C

Concentration of feed $X_F^S = 0.08$

Specific heat of feed c_{pF} = 4.0 kJ/kg °C

Concentration of product $X_P^S = 0.75$

Specific heat of product c_{pF} = 3.0 kJ/kg °C

Product Temperature T_P= 70 °C (Since product boils at 70 °C in evaporator)

Steam pressure = 313 kPa. From steam table, this corresponds to steam temperature of 135 °C

Temperature of vapour leaving the evaporator T_V= 70 °C

Temperature of condensate leaving the evaporator T_C= 70 °C

Energy lost to the surrounding q_{lost} = 1000 kJ/h

From steam table for steam at 313 kPa or 135 °C: $H_V^{135} = 2727.3\,kJkg^{-1}$,

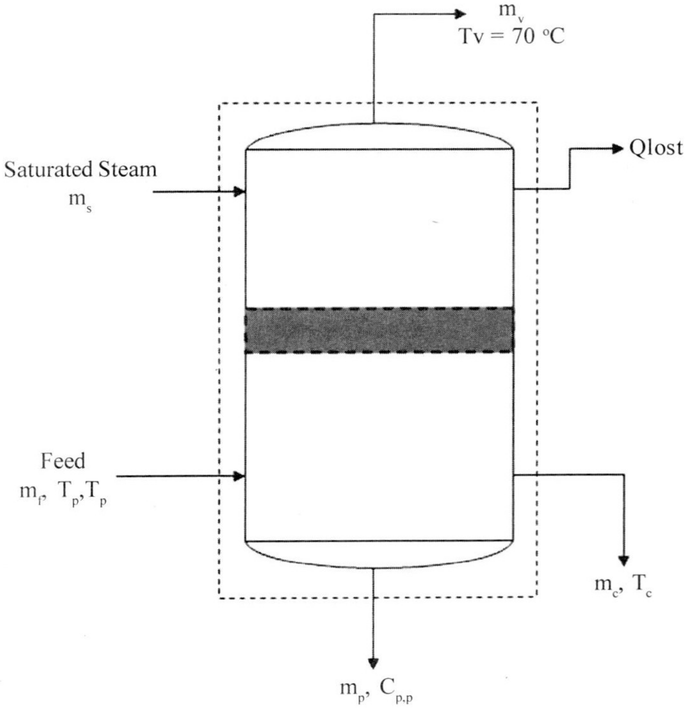

For steam at 70 °C: $H_V^{70} = 2626.8\,kJkg^{-1}$, $H_C^{70} = 292.98\,kJkg^{-1}$

Step 1: Construct a labeled flow diagram:

Step 2: Write the general mass and energy balance equation around the system

Total mass balance equation: Total mass in = Total mass out

$m_F + m_S = m_P + m_V + m_C$

Since, $m_s = m_c$

So, $m_F = m_P + m_V$

$$4000\left(\frac{kg}{h}\right) = m_P + m_V \tag{i}$$

Writing component mass balance for solid content in feed and product $X_F^S m_F = X_P^S m_p$ (Solid content in steam, condensate and vapour stream is zero)

$$(0.08)(4000)\left(\frac{kg}{h}\right) = (0.75)(m_P) \tag{ii}$$

Solving (i) and (ii)

$m_P = 426.66\ (kg\ h^{-1})$ and $m_V = 3573.33\ (kgh^{-1})$

Total energy balance equation: Total energy in = Total energy out

$$H_F + H_S = H_P + H_V + H_C + q_{lost} \qquad \text{(iii)}$$

Step 3: Determine enthalpy of each stream: Reference state is water at 0 °C.

$H_F = m_F\, c_{p,F}\, (T_F - T_{ref}) = (4000)\,(4.0)\,(25\text{-}0) = 400000\ kJh^{-1}$

$H_S = m_S\, H_V\, 135 = m_S\, 2727.3\ kJ\ kg^{-1}$

$H_P = m_P\, c_{p,P}\, (T_P - T_{ref}) = (426.66)\,(3.0)\,(70\text{-}0) = 89598.6\ kJh^{-1}$

$H_V = m_V\, H_V\, 70 = (3573.33)\,(2626.8) = 9386423.24\ kJh^{-1}$

$H_C = m_C\, H_C\, 70 = m_C\, (292.98)\ kJ\ kg^{-1}$

qlost = 1000 kJ h^{-1}

Put all enthalpies value in equation (iii)

$400000\ (kJh^{-1}) + m_S\, 2727.3\ (kJ\ kg^{-1}) = 89598.6\ (kJh^{-1}) + 9386423.24\ (kJh^{-1}) + m_C\, (292.98)\ (kJ\ kg^{-1}) + 1000\ (kJh^{-1})$

Since $m_S = m_C$ solving for m_S yields

$m_S = m_C = 3728.77\ kgh^{-1}$

References

Farkas, B.E. and Farkas D.F. (1997). Material and Energy Balances. In: Rotstein E., Singh R.P., Valentas, K.J. (Eds.), *Handbook of Food Engineering Practices*. CRC Press, Boca Raton, Florida, USA.

Fellows, P. (2000). *Food Processing Technology Principles and Practice Second Edition* (pp. 418-439). CRC Press, Boca Raton, Florida, USA.

Pham, Q.T., Wee, H.K., Kemp, R.M., and Lindsay, D.T. (1994). Determination of the enthalpy of foods by an adiabatic calorimeter. *J. Food Engr.* 21 : 137 – 156 .

Sahin, S. and Sumnu, S.G. (2006). Physical Properties of Foods. Springer Science+Business Media, New York, USA.

Singh, R.P. & Heldman, D.R. (2009). *Introduction to Food Engineering Fourth Edition* (pp. 501-539). Academic Press, USA

Smith , P. G. (2003). *Introduction to Food Process Engineering*. Kluwer Academic/Plenum Publishers , New York .

Toledo, R.T. (2007). *Fundamentals of Food Process Engineering 3rd Edition* (pp. 125-152). Springer Science+Business Media, New York, USA.

Watson , E.L. and Harper, J.C. (1988). *Elements of Food Engineering, 2nd ed.* Van Nostrand Reinhold , New York .

Yanniotis, S. (2008). *Solving Problems in Food Engineering*. Springer Science+Business Media, New York, USA.

3

Fluid Flow

Ramanpreet Kaur, U.S. Shivhare and Santanu Basu

1. Introduction

Movement of fluids from one location to another is an important unit operation in food processing. Various types of systems are used to convey fluids based on their flow behaviour. Understanding fluid flow behaviour is therefore important for designing of fluid flow systems. The fundamental mechanism involved in fluid flow is the momentum transfer occurring in moving fluids during various unit operations (fluid flow, aeration, drying, fluidization, filtration, centrifugation, sedimentation, mixing, etc.). This chapter deals with the basic concepts of fluid flow, viscosity, pressure measurement, conservation of mass and energy, boundary layer, momentum and velocity profile of fluid flowing in a pipe, flow metering devices, and pumps.

2. Fluid Flow

A fluid may be defined as a substance that does not permanently resist deformation and, hence, will change its shape upon application of shear stress. The rate of deformation is proportional to the applied shear stress. The fluids include air, carbon dioxide, vapors, water, steam, oils, milk, juice, syrup, puree, paste, etc. Flow of fluids can be divided into two branches: (i) fluid statics; and, (ii) fluid dynamics. Fluid statics deals with study of fluid characteristics when the fluid is at rest, whereas the fluid dynamics deals with the characteristic properties of it in motion.Understanding of the following terms related to the fluid behaviour is essential while dealing with the fluid statics and fluid dynamics.

Compressibility: The fractional change in the volume of fluid element per unit change in pressure is termed as compressibility and based onthese the fluids are classified as:

Compressible fluids: The fluids that undergo a decrease in volume with corresponding increase in density when pressure is applied are known as compressible fluids. Example: gases, vapors.

Incompressible fluids: These are the fluids where volume and density remains almost constant under the applied pressure. Example: liquids.

Uniform flow: When the velocity of a flowing fluid is same in magnitude as well as direction at all points, then the flow is considered to be uniform flow in nature.

Non-uniform flow: When the velocity of the flowing fluid is not same in magnitude as well as direction at any given point then the flow is termed as non-uniform flow.

Steady flow: If fluid properties(pressure, velocity, density) do not vary with time at any point in the flowing fluid, then this type of flowis termed as steady flow.

Unsteady flow: When the properties (pressure, velocity, density) of the fluid depend on time while the fluid is flowing, that flow is termed as unsteady flow.

Viscosity: The characteristic mouth-feel of various liquid food products like yoghurt, tomato ketchup, cream, ice cream, syrup, beverages is dependent on the important parameter, that is viscosity or consistency.

Viscosity is defined as the measure of resistance to a flow of fluid. The molecules of a fluid are in constant motion, the net velocity remains zero unless some force is applied to cause the fluid to flow in a particular direction. The viscosity of the fluid is related to the magnitude of the force needed to induce a flow at a certain velocity.

The fluid begins to move as soon as a force acts on it. A fluid may be visualized to be composed of several layers. Relative movement of one layer of the fluid over another is due to the force, commonly called shearing force, which is applied in a direction parallel to the surface over which it acts. A resistive force is offered by the fluid to the movement in the opposite direction to the shearing force, and it must also act in a direction parallel to the surface between the layers. This resistive force is a measure of an important property of fluid called viscosity.

The force that moves the fluid is the *shearing force* or *shear stress* and the velocity gradient along the flow is termed as *shear rate*. When shear stress is plotted against shear rate, simple fluids exhibit a linear relationship (Figure 1) and are termed as Newtonian fluids(water, milk, juice, most oils, honey, gases, and simple salt and sugar solutions).The proportionality constant μ (Pa.s), is known as viscosity, also called "absolute" or "dynamic" viscosity.

$$\tau = \mu\left(-\frac{dV}{dr}\right) \tag{1}$$

where, τ is the shear stress (Pa); and,dv/dr is the rate of shear (s^{-1}).Viscosity depends on the physico-chemical properties of the material and temperature. Viscosity of liquids decreases with increase in temperature while viscosity increases with temperature for most gases. Approximate value of viscosity of selected materials is given in Table 1.

Table 1: Approximate viscosity of selected fluids at room temperature.

Fluid	Approximate viscosity (Pa.s)
Air	10^{-5}
Water	10^{-3}
Olive oil	10^{-1}
Glycerol	10^{-0}
Liquid honey	10^{1}

If the relationship is non-linear, the fluids are termed as non-Newtonian fluids. Many foods like gums, pectin, suspensions, emulsions and concentrated solutions containing starch show Non-Newtonian behaviour. At low concentrations, these liquids behave as Newtonian fluids but as the concentration is increased, there is an increase in viscosity and they tend to act like non-Newtonian fluids (Rielly, 1997).Apparent viscosity is the term commonly used for non-Newtonian fluids. The non-Newtonian fluids can be classified as: (i) time independent; and, (ii) time dependent (Figure 2).

Various types of non-Newtonian fluids are discussed as follows:

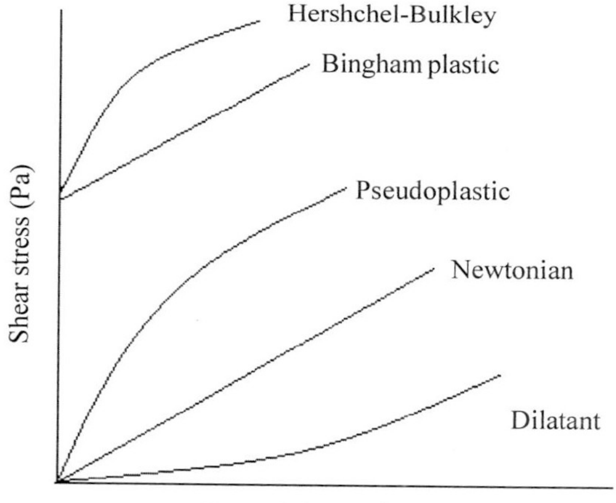

Fig. 1: Newtonian and Non-Newtonian fluids.

i) Shear-thinning (pseudoplastic) fluids:the apparent viscosity decreases with increase in shear rate in these type of fluids. The examples include emulsions, concentrated juices, purees, condensed milk, vegetable soups.

ii) Shear-thickening (dilatants) fluids: the apparent viscosity increases with the shear rate. This behaviour is found in liquid chocolate, starch-and cornflour- suspensions.

iii) Bingham plastic fluid:when there is no flow until a critical shear stress (yield stress) is reached and then the fluid exhibits a linear behaviour, these type of fluids are called as Bingham plastic fluids. Mayonnaise, toothpaste fall under this category.

iv) Herschel-Bulkleyfluids: Fluids exhibiting yield stress behaviour and apparent viscosity value changes with increasing shear rate (tomato ketchup, jelly, minced fish paste, raisin paste, etc.)

v) Thixotropic fluids: when structure breaks down and viscosity decreases with time. Mostly in creams, jelly etc.

vi) Rheopectic fluids: the structure builds up and viscosity increases with time (e.g. whipping cream).

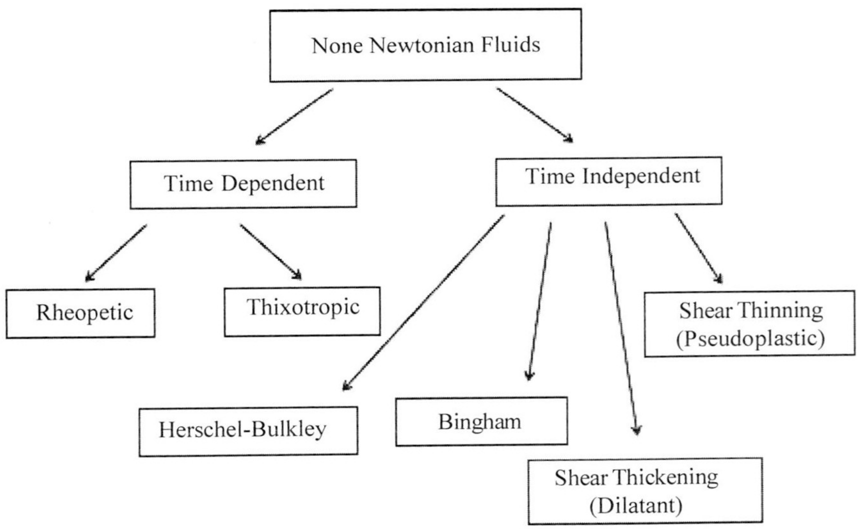

Fig. 2: Classification of non-Newtonian fluids.

The two most commonly used equations for characterising non-Newtonian fluids are the power law model (equation 2) and the Herschel-Bulkley(HB) model (equation 3). Power law is used to describe behaviour ofnon-Newtonian fluids without yield stress, while HB model describes behaviour of non-Newtonian fluids exhibiting yield stress.

$$\tau = K(\gamma)^n \tag{2}$$

$$\tau = \tau_o + K(\gamma)^n \tag{3}$$

where, K is the consistency index (Pa.s^n); n is the flow behaviour index (dimensionless); and, τ_o represents the yield stress (Pa). It is pertinent to mention here that HB model (equation 3) is general in nature and may be reduced to power law (τ_o=0) (equation 2) or Newton's law of viscosity (τ_o=0; n=1; K=μ) (equation 1).Values of the coefficients of HB model for Newtonian and non-Newtonian fluids are given in Table 2.

Table 2: Values of the coefficients of HB model

Fluid	K	n	σ_o	Example
Herschel-Bulkley	>0	$0 < n < \infty$	>0	Minced fish paste, raisin paste
Newtonian	>0	1	0	Water, fruit juice, honey, milk, vegetable oil
Shear-thinning	> 0	$0 < n < 1$	0	Apple sauce, banana puree, orange juice concentrate
Shear-thickening	> 0	$1 < n < \infty$	0	Some types of honey, 40% raw corn starch solution
Bingham plastic	> 0	1	>0	Tooth paste, tomato paste

2.1. Fluid Statics

When the fluid is at rest, it exerts pressure and that pressure is the static pressure or simply pressure exerted by it to the surrounding. When the fluid is at zero velocity,there is absence of shearing force and only normal stress prevails. Under the influence of gravitational force, mass of any material exerts a force on whatever supports it or holds it. The magnitude of this force is the product of mass of the material and acceleration due to gravity(equation 4).

$$F = mg = V\rho g \tag{4}$$

where, F represents force exerted (N); m is fluid mass (kg), V is fluid volume (m^3); and,ρ is fluid density (kg/m^3).

To maintain the equilibrium by the material, the force exerted by the mass must be resisted by some supporting medium. The fluid at any point must support the fluid above it. But as the fluid is at rest, there will be no shearing forces and the forces at any point in a fluid will exert equal force in all directions when at rest. The force per unit area in a fluid is known as fluid pressure and is exerted equally in all directions. This also defines the Pascal's law which states that the pressure exerted anywhere in a confined incompressible fluid is transmitted equally in all directions throughout the fluid such that the pressure variations remain the same.

Mathematically,

$$P = \rho\, g\, \Delta Z \tag{5}$$

where,P is pressure (Pa); and, ΔZ is the height of the fluid above the point of measurement (m)and can be measured as the difference in pressure at two points within a fluid column.

Consider a horizontal plane, in a fluid at depth Z below the surface as shown in Figure 3.

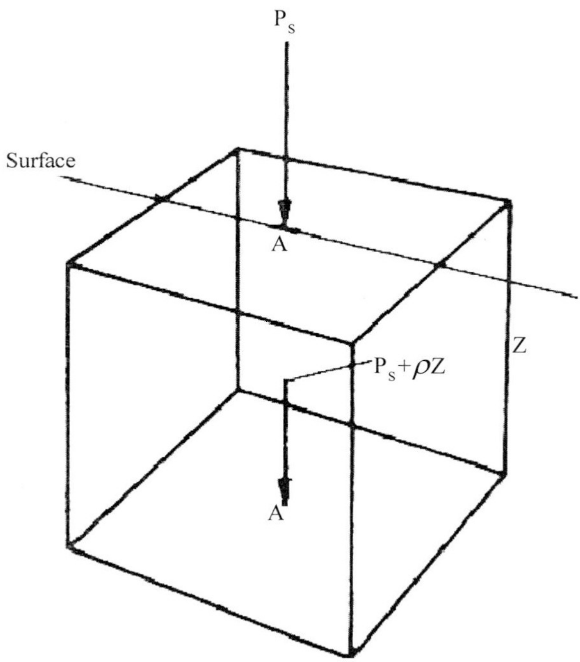

Fig. 3: Pressure in a fluid.

The volume of fluid above the area A will be AZ, and the weight of this mass of fluid will be AZρg. Further, the total force on the area A should also include any external force acting upon it. So, if the force on the surface is P_s per unit area, then

$$F = AP_s + AZ\rho g \tag{6}$$

where, F is the total force exerted on area A; and, P_s is the pressure above the surface of the fluid (it might be atmospheric pressure) (Earle, 1983).Since the total pressure P is the total force per unit area.

$$P = \frac{F}{A} = P_s + Z\rho g \tag{7}$$

Thus, the pressure of a fluid at any point is solely dependent on its density and height of the fluid along with any pressure that is present above the surface of the fluid.

Sometimes, the pressures are quoted as absolute pressures and this means the absolute value of the force per unit area exerted on a surface by the fluid. Thus, the absolute pressure is the difference between the pressure of a fluid at a certain point and the absolute zero of pressure or a perfect vacuum. Gauge pressure is the difference between the absolute pressure and the local atmospheric pressure. Local atmospheric pressure can vary depending upon the ambient temperature, altitude and local weather conditions. When the pressure is greater than atmospheric pressure,

$$P_{absolute} = P_{atmospheric} + P_{gauge} \tag{8a}$$

And, when the pressure is below atmospheric pressure, then,

$$P_{absolute} = P_{atmospheric} - P_{vacuum} \tag{8b}$$

Hence, various pressures can be calculated using equations (8a) and (8b) when the fluid is static. Pictorial representation of the various pressures is shown in Figure 4.

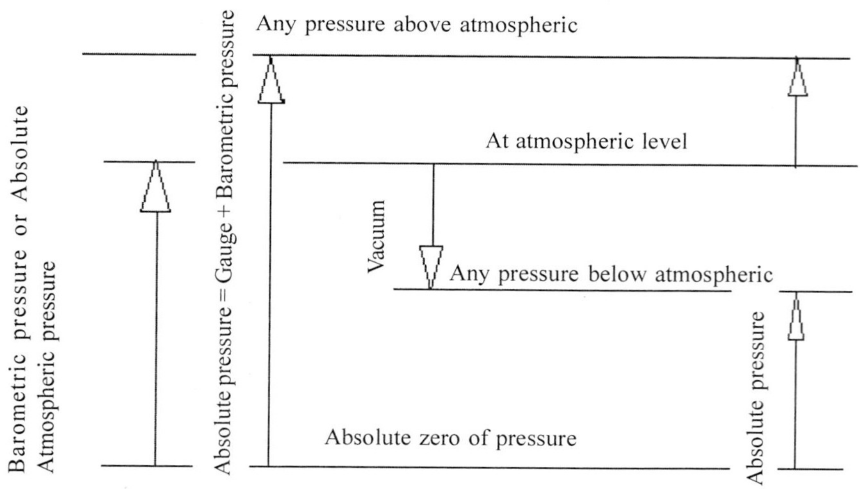

Fig. 4: Relationship between various pressure terms

2.1.1. Pressure Measurement

2.1.1.1. Manometers

The devices that are used for measuring the pressure at a certain point in a fluid with the help of balancing the column head, are known as manometers. Manometers can be Simple manometers or Differential manometers.

a) Simple Manometers

A glass tube when connected to a point where the pressure is to be measured at one end, and leaving the other end open to atmosphere, comprises a simple manometer. Simple manometers are of three types:

i) Piezometer

ii) U-tube Manometer

iii) Single Column Manometer

i) Piezometer

The simplestform of manometer to measure gauge pressure is Piezometer. In this type of manometer, one end remains connected to the point where pressure is to be measured while the other end remains open to the atmosphere (Figure 5).

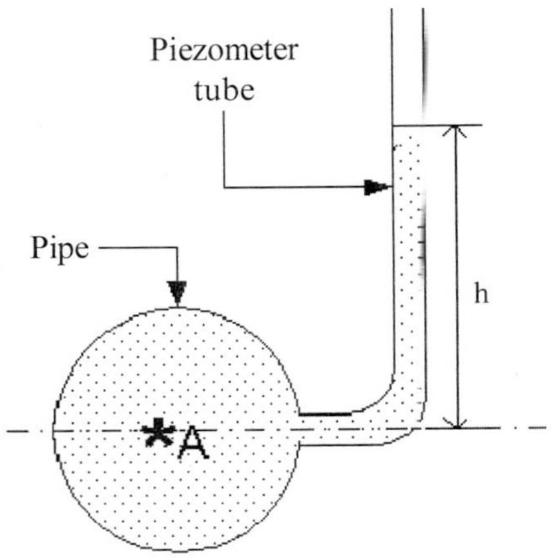

Fig. 5: Piezometer.

The rise of the liquid in the piezometer tube represents the pressure head at that point which is in alignment. If, at a point A, the height of the fluid is h in the tube,

Pressure at A = ρ.g.h (9)

ii) U - Tube Manometer

As the name implies, the U-tube manometer is consisted of U shaped glass tube in which one end is connected to the point where pressure is to be measured while the other end remains open to the atmosphere (Figure 6). Mercury or any other liquid whose specific gravity is greater than the specific gravity of the liquid whose pressure is to be measured, is used in the tube.

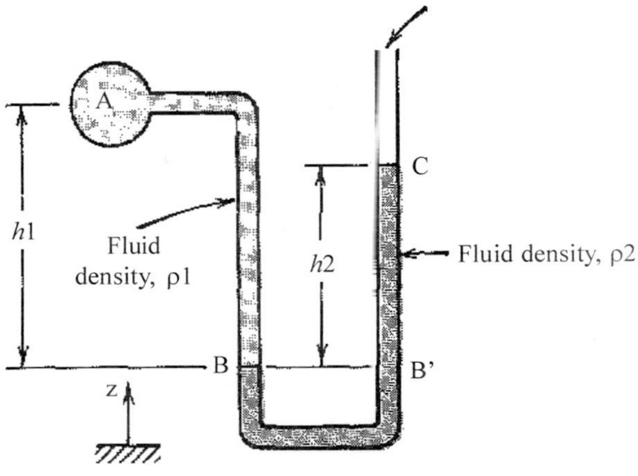

Fig. 6: U-tube manometer.

To measure the gauge pressure at A in having two liquids which are immiscible with different specific gravities, then taking the datum line as BB', and assuming the pressure isP at A, then,

Pressure at BB' in the left column = $P + \rho_1 . g. h_1$ (10)

Pressure at BB' in the right column = $\rho_2 . g. h_2$. (11)

where, ρ_1 is density of the lighter liquid; ρ_2 is density of the heavier liquid; h_1 is height of the lighter liquid above the datum line; and, h_2 is height of the heavier liquid above the datum line.

Equating the above two equations,

$$P + \rho_1 . g. h_1 = \rho_2 . g. h_2 \quad (12)$$

$$P = (\rho_2 . g. h_2 - \rho_1 . g. h_1) \quad (13)$$

For vacuum pressure,

Pressure above BB' in the left column = $\rho_2 . g. h_2 + \rho_1 . g. h_1 + P$ (14)

Pressure head in the right column above BB' = 0

Therefore,

$$\rho_2 . g. h_2 + \rho_1 . g. h_1 + P = 0 \quad (15)$$

$$P = -(\rho_2 . g. h_2 + \rho_1 . g. h_1) \quad (16)$$

iii) Single Column Manometer

Single column manometer is a modification to the U-tube manometer, in which a reservoir of a particular cross sectional area is present. Generally, the reservoir has cross sectional area 100 times as compared to that of the tube in U-tube manometers. There is a limb on the other end of the reservoir which helps in classifying the manometer as vertical single column manometer or inclined single column manometer depending upon the shape of the particular limb which is open at that end.

Vertical Single Column Manometer: A vertical single column manometer is shown in Figure 7. It consists of A_1B_1 datum line of one fluid in the reservoir and in the right limb of the manometer, when it is not connected to the pipe. If the reservoir is connected to the pipe, the heavy liquid in the reservoir will go downwards and will rise in the right limb due to the high pressure at A.

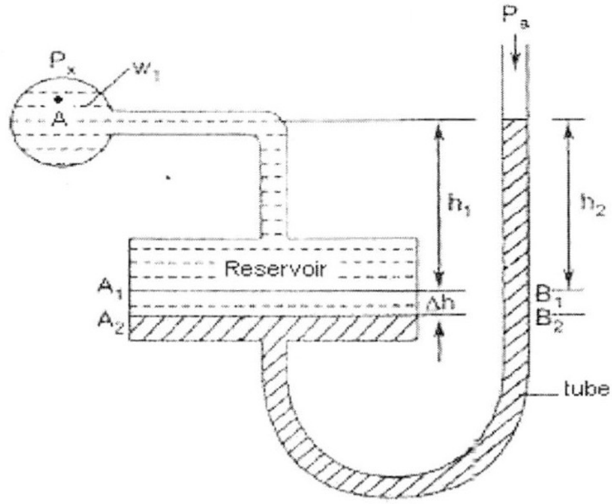

Fig. 7: Vertical Single Column Manometer.

Now, fall of heavy liquid in the reservoir will cause a rise in the right limb, thus,

$$A.\ \Delta h = a.\ h_2 \tag{17}$$

$$\text{and, } \Delta h = (a\ .\ h_2)\ /\ A \tag{18}$$

where, A is cross sectional area of the reservoir;Äh is the fall of heavy fluid in the reservoir; a is cross sectional area of the right limb; and, h_2 is the rise of heavy fluid in the right limb.

Considering the datum line shift as A_2B_2, the pressure in the right limb becomes

$$\rho_2.\ g.(\Delta h + h_2) \tag{19}$$

Pressure in the left limb above $A_2B_2 = \rho_1.\ g.\ (\Delta h + h_1\ ) + P_x$ (20)

where, P_x is pressure at A; h_1 is height of the centre of pipe above A_2B_2; ρ_1 is liquid density in the pipe; and, ρ_2 is density of liquid in the reservoir. Equating the pressures, we get,

$$\rho_2.\ g.\ (\Delta h + h_2) = \rho_1.\ g.\ (\Delta h + h_1) + P_x \tag{21}$$

$$P_x = \Delta h[\rho_2 g - \rho_1 g] + h_2\rho_2 g - h_1\rho_1 g \tag{22}$$

But, as $\Delta h = (ah_2)/\ A$, equation (22) becomes,

$$P_x = (ah_2)\ (\rho_2 g - \rho_1 g)/A + h_2\rho_2 g - h_1\rho_1 g \tag{23}$$

Since area of reservoir (A) is very large as compared to the area of the tube (a), the ratio a/A is very small, hence may be neglected. Thus, h_1 is the height of the centre of pipe above A_2B_2, ρ_1 is the density of the liquid in the pipe, ρ_2 is the density of liquid in reservoir.

$$P_x = h_2\rho_2 g - h_1\rho_1 g \tag{24}$$

Inclined Single Column Manometer:This manometer is more sensitive as the inclined tube attached to the reservoir, and the distance moved by the heavy liquid in the right limb will be more. Considering Figure 8, let L be the length of the heavy liquid moved in right limb from XX, θ be the inclination of the limb to the horizontal plane and h_2 be the vertical rise of the heavy liquid which will be equal to L sin θ.

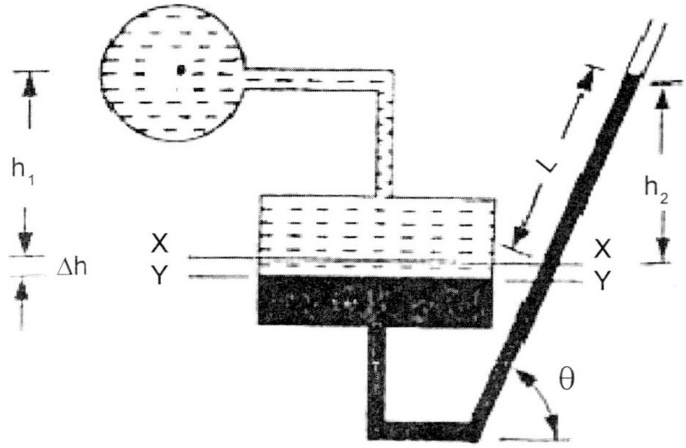

Fig. 8: Inclined Single Column Manometer.

Then, pressure at A

$$P_A = h_2\rho_2 g - h_1\rho_1 g \quad (25)$$

or,

$$P_A = L \sin\theta\, \rho_2 g - h_1\rho_1 g. \quad (26)$$

b) Differential Manometers

Differential manometers are usedto measure the difference in pressure when there is a difference of pressure between the two points in a pipe or in two different pipes. It consists of a U-tube, containing a heavy liquid, whose two ends are connected to the points, where the pressure is to be measured. There are two types of differential manometers.

i) U-Tube Differential Manometer

A U-tube differential manometer is shown in Figure 9, having two different levels containing liquids of different specific gravities. Let, the pressure at A be P_A and pressure at B be P_B; ρ_1 is density of liquid at A; ρ_2 is density of liquid at B; ρ_g is density of heavy liquid or mercury;h_1 is height of liquid in A from datum XX'; h_2 is distance of the centre of B from the mercury level in the right limb; and,h is the difference of mercury level in the tube.

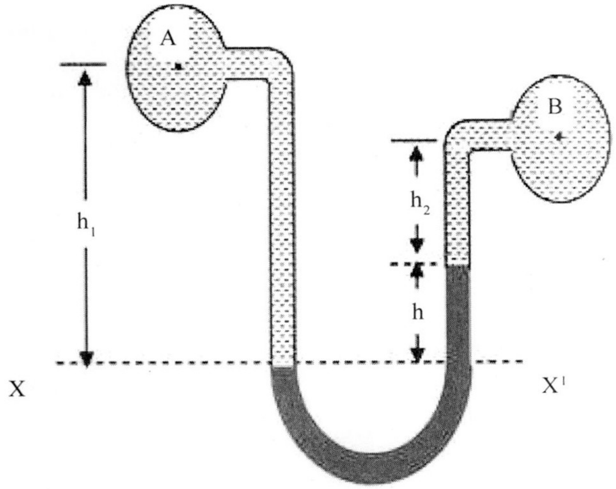

Fig. 9: U-tube differential manometer

Pressure in left limb above XX' $= \rho_1.g.h_1 + P_A$ (27)

Pressure in the right limb above XX' $= \rho_g .g.h + \rho_2. g. h_2 + P_B$ (28)

Equating the above pressure, we get,

$\rho_1. g. h_1 + P_A = \rho_g.g. h + \rho_2.g.h_2 + P_B$ (29)

$P_A - P_B = \rho_g.g.h + \rho_2.g.h_2 - \rho_1.g.h_1$ (30)

Since, $h = h_1 - h_2$

$P_A - P_B = \rho_g.g.(h_1 - h_2) + \rho_2.g.h_2 - \rho_1.g.h_1$ (31)

ii) Inverted U-Tube Differential Manometer

An inverted U-tube differential manometer basically consists of an inverted U-tube (Figure 10). The two ends of this inverted U-tube are connected to the two points where difference in pressure is to be measured. Generally, it is used to measure the light pressures. Considering the pressure at A be P_A and at B be P_B. The pressure at A is considered to be greater than pressure at B.

Pressure in the left limb below the reference line

$= P_A - \rho_1. g. h_1$ (32)

Pressure in the right limb below the reference line

$= P_B-(\rho_2. g. h_2) - (\rho_S. g. h)$ (33)

where, ρ_1 is density of liquid A; ρ_2 is density of liquid B; h_1 is height of liquid in the left limb below reference line; h_2 is height of liquid in right limb; h is difference between heavier and lighter liquid; and, ρ_S is density of the manometric fluid used to measure the difference between the two liquids.

Equating the two pressures, we get,

$$P_A - \rho_1 . g. h_1 = P_B - (\rho_2 . g. h_2) - (\rho_S . g. h) \quad (34)$$

$$P_A - P_B = \rho_1 . g. h_1 - \rho_2 . g. h_2 - \rho_S . g. h \quad (35)$$

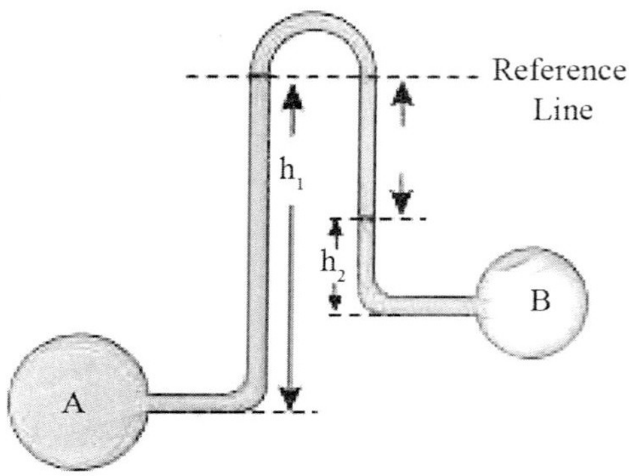

Fig. 10: Inverted U-tube differential manometer.

2.2. Fluid Dynamics

Fluid dynamics deals with the change in the flow behaviour when an external force is applied and the fluid sets in motion. The motion of fluids can be predicted in the same manner as the motion of solids is predicted using the fundamental laws of physics along with the physical properties of the fluids. Depending on the flow characteristics, the flow of fluid can be divided into four categories,

1. Steady-Uniform Flow: If the flow behaviour properties (pressure, velocity, density) do not change with time, then that flow is steady-uniform flow. Example: milk flowing through a pipe of constant diameter and at a constant velocity in a plant.

2. Steady- Non Uniform Flow: When the conditions or properties change from point to point in the stream but do not change with time, then a steady - non uniform flow behaviour exists. An example of this type of flow is when milk is flowing in a tapered pipe with a constant velocity at the inlet but the velocity will change along the length of the pipe towards the exit.

3. Unsteady-Uniform Flow: If the conditions are same at a certain point at specific time but show a change with the passage of time then, this type of flow will be unsteady uniform flow. If milk is being pumped with the help of a pump at a certain velocity in a constant diameter pipe, then unsteady uniform flow will be followed as velocity will decrease.

4. Unsteady-Non Uniform Flow: Every property when shows a change from point to point as well as with the passage of time, then flow is unsteady non uniform flow. Example of this would be waves in a confined channel or vessel.

2.2.1. Laminar and Turbulent Flow

When a fluid is flowing against a stationary solid surface, the fluid adheresto the wall at the actual point of contact and the adhesion is due to the force fields at the boundary which are also responsible for the interfacial tension between the fluid and the surface. Further, the fluid velocity at the interface is zero. The adjacent layers of the fluid therefore develop velocity gradient and this gradient plays an important role in predicting the flow behaviour in a pipe or over a flat surface.

If the fluid is flowing at a slow rate in a pipe, and there is presence of smooth streamlines and highly ordered motion being followed, then this type of flow is laminar flow. When these streamlines follow velocity fluctuations and the flow behaviour is highly unordered, at increased velocities, then the flow is turbulent in nature(McCabe et al., 2005). Sometimes, the flow is neither laminar,nor turbulent and may enter a transition state.

2.2.2.Reynold's Number

Osborne Reynolds (1842-1912), a British scientist and mathematician first developed a concept of differentiating the flow into laminar and turbulent flow using a simple apparatus as shown in Figure 11.

Reynold established a relationship to predict the flow behaviour using a dimensionless number known as Reynolds Number (Re), which is dependent on diameter (D) of the pipe in which fluid is flowing, with average velocity (v) density of fluid (ρ) and viscosity (μ). Based on numerous experiments, he proposed the relationship, which is the ratio of inertial to viscous forces.

$$\mathrm{Re} = \frac{Dv\rho}{\mu} \quad (36)$$

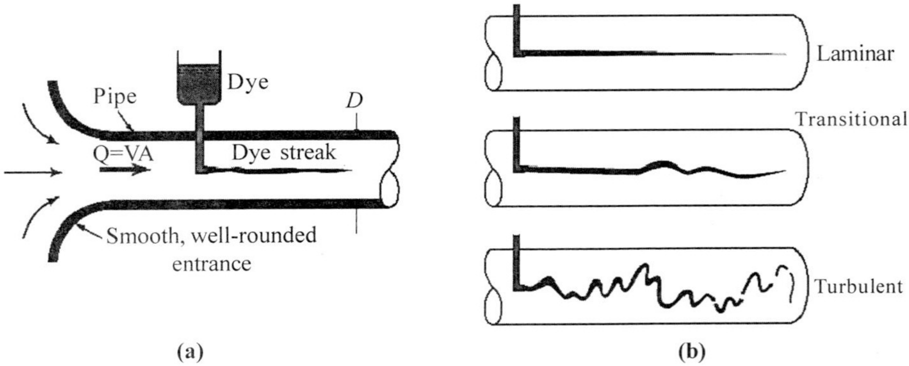

Fig. 11: **(a)** Reynolds experiment using water pipe to study laminar to turbulent, **(b)** Typical dye streak.

Laminar flow exists if the Reynolds number is less than approximately 2100. If the value of Reynolds number is greater than 4000 then, the flow will be turbulent and if the Re varies between 2100 and 4000, the flow is believed to be the transitional in nature (White, 1999).The non-Newtonian fluids may not follow equation (36), as the viscosity is varying in those fluids with the shear rate. A Critical Reynolds number at which transition to turbulent flow starts has been proposed, stated as follows in accordance with the power law number involving flow behaviour index (n):

$$Re_{Cr} = 2100\frac{\left(4^n + 2\right)\left(5^n + 3\right)}{3(3^n + 1)^2} \tag{37}$$

Equation (37) is generally used for pseudo plastic fluids ($n < 1$).

3. Conservation of Mass and Energy

3.1 Conservation of Mass-Equation of Continuity

Processing of liquid foodsmay involve heating, cooling, concentrating, mixing, etc. Transfer of liquid foods is possible through pipes and the flow characteristics in pipes are generally dependent on the basic law of nature, i.e. conservation of mass stating matter can not be destroyed nor created (can only transform to a different form of matter). To understand the basics of fluid flow, the principle of conservation of mass is used. Let us assume that a fluid is flowing upwards in a pipe, as shown in Figure 12.

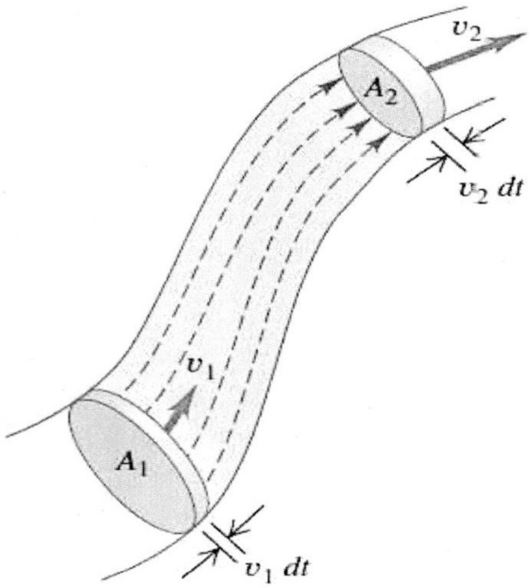

Fig. 12: Flow in a pipe.

The flow in the pipe is assumed to be steady and there is no accumulation of fluid within the pipe. In this scenario, the rate of fluid flow entering in the pipe must be equal to the rate of the exiting fluid according to the conservation of mass. Considering the cross-sectional area A (m^2), through which the fluid is flowing, covering a distance dL in dt time, then the volumetric flow rate (Q, m^3/s) can be expressed as

$$Q = A\frac{dL}{dt} = \text{v.A} \tag{38}$$

where, v is fluid velocity (m/s).

Denoting the mass flow rate (kg/s) as the product of density (ρ) and volumetric flow rate,

$$\dot{m} = \rho.\ Q = \rho.\text{v.A} \tag{39}$$

According to the conservation of mass we can write between two points having varying fluid velocity and cross sectional area, thus

$$\rho_1.\text{v}_1.\text{A}_1 = \rho_2.\ \text{v}_2.\ \text{A}_2 \tag{40}$$

which forms the continuity equation for a fluid flowing in a pipe having different cross sectional areas.

For incompressible fluids, $\rho_1 = \rho_2$, equation (40) yields

$$\text{v}_1.\text{A}_1 = \text{v}_2.\text{A}_2 \tag{41}$$

Hence, for an incompressible flow in a constant cross section, the velocity will remain constant. If other streams (branched streams) are entering in a conduit, the continuity equation will hold true by summing up the incoming fluid mass or volumetric flow rates to the sum of exiting mass and volumetric rates of the fluid (Shemmeri, 2012).

3.2. Conservationof Energy - Bernoulli's Equation

The law of conservation of energy deals with conservation of three forms of energy namely:(i) kinetic energy which is due to the motion of the fluid;(ii) potential energy that comes in existence with the positional elevation above the datum of the fluid; and, (iii) pressure energy which is due to the absolute pressure of the fluid at that point. The law states that the total energy of the fluid remains constant, however transformation from one form to another can occur.Assuming that the fluid is incompressible, moving at a constant flow rate ($A_1v_1 = A_2v_2 =$ constant) (Figure 13), the pressure will also remain constant as long as the diameter and elevation are constant, though diameter and elevation may change.

Total mechanical energies at both the points can be expressed as

Total mechanical energy of the volume element = Kinetic energy + Potential energy

$$E_1 = 1/2.\ \Delta\dot{m}_1 .v_1^2 + \Delta\dot{m}_1 .g.z_1 \tag{42}$$

where, $v_1$is velocity; Δ is mass flow rate; and,z_1 is the elevation.

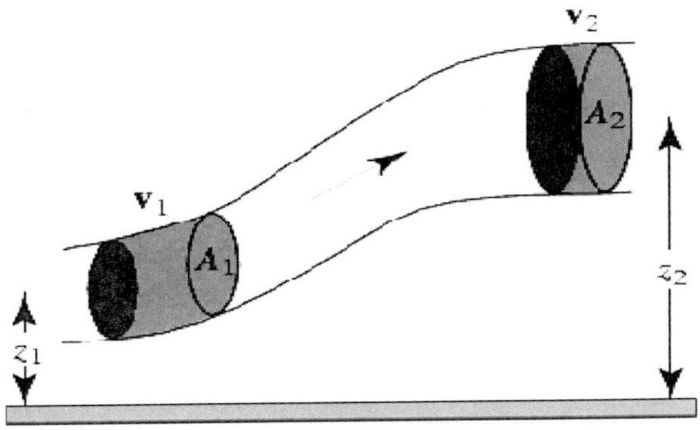

Fig. 13: Flow in a pipe for Bernoulli's equation.

Similarly the volume at element at the other point can be expressed as

$$E_2 = 1/2.\ \Delta.v_2^2 + \Delta.g.z_2. \tag{43}$$

Since, mass of the volume element remains constant during the flow,

$$\Delta \dot{m} = \Delta \dot{m}_1 = \Delta \dot{m}_2 \tag{44}$$

Obviously, there will be no change in the total mechanical energy at the two points for the steady flow ($v_1 = v_2 = v$); and,if there is no change in elevation ($z_1 = z_2 = z$). However, there will be a change in total mechanical energy of the fluid if either one or both of the values vary.

$$E_2 - E_1 = 1/2.\Delta.v_2^2 + \Delta \dot{m}.g.z_2 - 1/2.\ \Delta \dot{m}.\ v_1^2 - \Delta \dot{m}.g.z_1 \tag{45}$$

This difference in the mechanical energies arises from the net work done over the system or the flow. It will be equivalent to the energy or the net work done on the fluid along the length of the pipe.

Net work = Net work done on the fluid -Work done by the fluid , i.e.

$$W_{in} - W_{out} \tag{46}$$

Now, W_{in} = Force x Distance = (Pressure x Area) (Velocity x Time interval)

$= (P_1.A_1).(v_1.\Delta t)$ (Δt is the time to move an element length)

$$= P_1.\ (\rho.A_1.v_1.\Delta t)/\rho \tag{47}$$

But, $(\rho.A_1.v_1.\Delta t)$ is the mass of the volume element, so $W_{in} = P_1.\ \Delta/\rho$

Similarly, the work done by the system using the above formulations can be expressed as

$W_{out} = P_2.$

The net work done, $W_{in} - W_{out} = (P_1 - P_2).$ (48)

Hence, work done for different energies with different velocities and elevations can be written as;

$$E_2 - E_1 = \Delta \dot{m}.\ \{\frac{1}{2}(v_2^2 - v_1^2) + g.\ (z_2 - z_1)\} \tag{49}$$

and,

$$(P_1 - P_2).\frac{\Delta \dot{m}}{\rho} = \Delta \dot{m}.\ \{\frac{1}{2}.\ (v_2^2 - v_1^2) + g.\ (z_2 - z_1)\} \tag{50}$$

Fluid being incompressible, the mass terms will cancel and moving the density to the other side, equation (50) becomes

$$(P_1 - P_2) = \rho. \{\frac{1}{2} . (v_2^2 - v_1^2) + g. (z_2 - z_1)\}$$

$$= (\frac{1}{2} . \rho. v_2^2) - (\frac{1}{2} . \rho. v_1^2) + (\rho. g. z_2) - (\rho. g.z_1) \quad (51)$$

$$\text{or, } P_1 + (\frac{1}{2} . \rho. v_1^2) + (\rho. g.z_1) = P_2 + (\frac{1}{2} . \rho. v_2^2) + (\rho. g. z_2) \quad (52)$$

and,

$$P_1 + (\frac{1}{2} \rho\, v_1^2) + (\rho\, g\, z_1) = P_2 + (\frac{1}{2} \rho\, v_2^2) + (\rho\, g\, z_2) = \text{Constant} \quad (53)$$

which forms the Bernoulli's equation. Also, dividing the above equation by ρg,

$$(\frac{P_1}{\rho.g}) + (\frac{v_1^2}{2.g}) + (z_1) = (\frac{P_2}{\rho.g}) + (\frac{v_2^2}{2.g}) + (z_2) = \text{Constant} \quad (54)$$

where, $(P/\rho.g)$ is pressure head (also known as flow work); $(v^2/2.g)$ is velocity head; and, z is the elevation.

In real life situations, sometimes the flow will suffer loss of energy due to friction and presence of obstruction(s) in between the inlet and outlet points. Incorporating these losses, equation (54) becomes

$$(\frac{P_1}{\rho.g}) + (\frac{v_1^2}{2.g}) + (z_1) = (\frac{P_2}{\rho.g}) + (\frac{v_2^2}{2.g}) + (z_2) + h_L \quad (55)$$

where, h_L is the loss of energy.

Also, if the pump is used at the inlet for the upward movement of the fluid, there will be addition of that energy as well into the equation. Thus,

$$(\frac{P_1}{\rho.g}) + (\frac{v_1^2}{2.g}) + (z_1) + h_P = (\frac{P_2}{\rho.g}) + (\frac{v_2^2}{2.g}) + (z_2) + h_L \quad (56)$$

where, h_P is the head which is gained due to the pump. The Bernoulli's equation forms the basis of the various flow measurement devices which are needed in the process industry to regulate the flow behaviour. Equation(56) defines that the total head for a steady, inviscid and incompressible flow remains constant along the streamline (Seshadari and Patanakar, 1974).

4.Boundary Layer

The concept of boundary layer was developed by Prandtl in 1904 (White,1999).When fluid flows over a solid surface or inside a pipe, some part of the moving fluid is influenced by the presence of the solid boundary of the surface over which it is flowing.This part forms the boundary layer.The formation of boundary layer in a steady state flow of a fluid over a flat plate is shown in Figure 14.

Velocity of the fluid upstream of the leading edge at X = 0 of the plate is uniform across the entire fluid stream and has value v. Velocity of the fluid at solid-fluid interface is zero.The velocity (v_x) increases in Y-direction (farther from the surface) as the flow progresses along the plate in X-direction. The velocity v_x approaches asymptotically the undisturbed free stream velocity (v) of bulk of the stream. The dashed line L is drawn such that the velocity at that point is 99% of the bulk velocity (v) The region between the dotted line and the plate constitutes the boundary layer.

The flow is initially laminar as the velocities near the surface are small thus forming a boundary layer in laminar region. Farther away from the surface of the plate, the fluid velocity though less than the undisturbed velocity, may be large and causes turbulence in the flow via transition zone and thus, forming the boundary layer in turbulent region. Onset of turbulence is sometimes characterized by a sudden rapid increase in the thickness of the boundary layer. The transition zone is developed between fully developed turbulent flow and laminar flow region, sometimes also referred to *buffer layer*.When the boundary layer formed is turbulent in nature, a thin viscous sublayer is persistent next to the plate. The presence of this viscous sublayer gives rise to viscous shear and

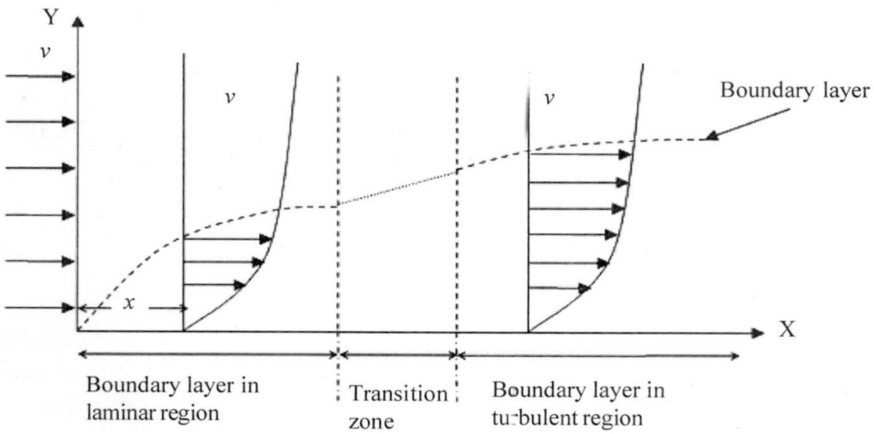

Fig. 14: Boundary layer formation over a flat plate.

the drag thus formed due to this shear in the boundary layer is called *skin friction or skin drag* and it is the only drag present for flow past a flat plate (Geankoplis, 1993).

4.1. Boundary Layer Formation in Pipe

Consider a straight pipe in which the fluid is entering at uniform velocity. The boundary layer will begin to form at the entrance region of the pipe and will thicken when crosses the entrance region as shown in the Figure 15.

Initially the fluid when enters the pipe, it forms an annular boundary layer between the edges and the centre point of the pipe and, will move in a rod like manner as the flow moves ahead at constant velocity. The velocity of the boundary will show an increasing trend from zero at the wall to the constant velocity existing at the midpoint of the flow or at the core of the pipe. Further, when the fluid moves ahead, the boundary layer occupies the entire region of the pipe and the velocity distribution reaches its final form and remains unvarying throughout the tube. This peak is when reached without any disturbance in velocity throughout the tube, the flow will be *fully developed flow.* The length of the pipe from entrance to the location when the fully developed flow is initiated is termed as *transition length* (L_e) (McCabe et al., 2005). The transition length is expressed as

$$L_e = 0.05\,(R_e)\,(D) \tag{57}$$

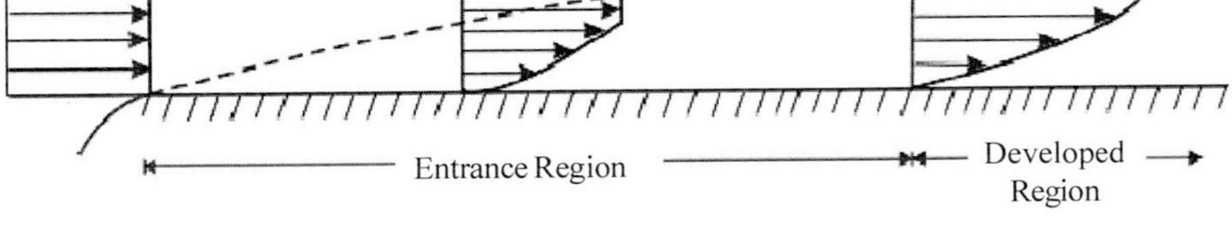

Fig. 15: Boundary layer development of fluid flowing in a pipe.

4.2. Boundary Layer Separation and Formation of Wakes

Velocity gradient develops while fluid flows past an object, but it fades away as the as soon as the distance is covered and the boundary layers from both the sides (layers formed above and below the object) intermingle and disappear. The fluid again starts moving at a uniform velocity at which the flow initiated.

However, in blunt obstacles or spheres (Figure 16), after reaching the maximum point when the flow should move downwards according to the shape, the momentum prevents the fluid from making a sharp turn along the surface of the body and thus, separating the boundary layer frcm the surface. Once fluid leaves the surface, there is backwater zone of fluid with decelerated velocity which promotes formation of *eddies* in that region, known as *wake* (White, 1999). The eddies in the wake remain in motion due to shear stresses between the wake and the separated fluid.Consider that a fluid is flowing against an object (Figure 16). There is a formation of boundary layer along the surface of the object. After reaching the point of maximum diameter of the object, the boundary layer tends to separate as the fluid flows ahead and is termed as *point of separation.* The separation of the boundary layer occurs due to the velocity gradient that persists in the flow.

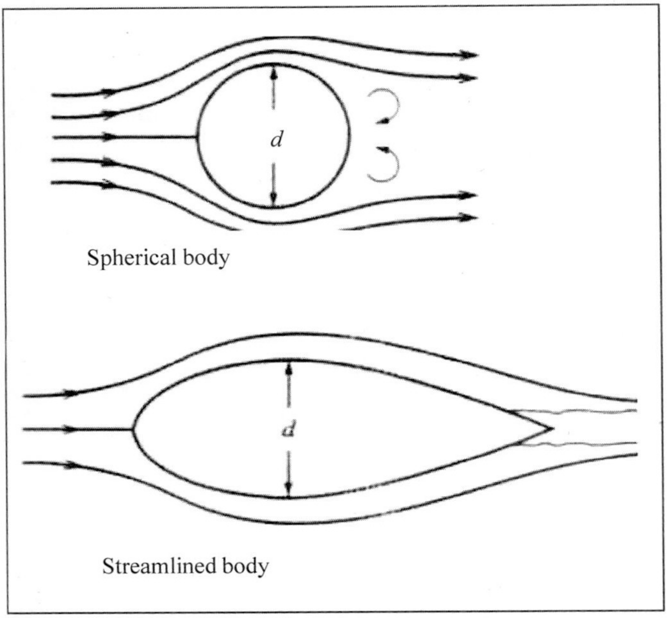

Fig. 16: Formation of wakes.

5. Momentum and Velocity Profiles of Flows in Pipe

Generally, for the actual view of what's happening inside a pipe, we apply the momentum balance over an enclosed conduit, along with the concept of viscosity, we are in a position to get the expression for the velocity profile and the pressure drop that is happening inside the pipe while the fluid is flowing. According to the law of momentum, in an overall system, we know that,

Sum of forces acting on a control volume = rate of momentum out of control volume - rate of momentum into control volume + rate of accumulation of momentum in a control volume.

The velocity profiles remain constant along the axis in the direction of the flow for a fully developed flow in a pipe. In one dimensional, steady state and laminar flow, it will be constant in x-direction. Consider a pipe of inside radius r, thickness Δr and length Δx (Figure 17).The equation for conservation of momentum having no accumulation at steady state is expressed as

Sum of forces acting on a control volume = Rate of momentum at outlet - rate of momentum at inlet

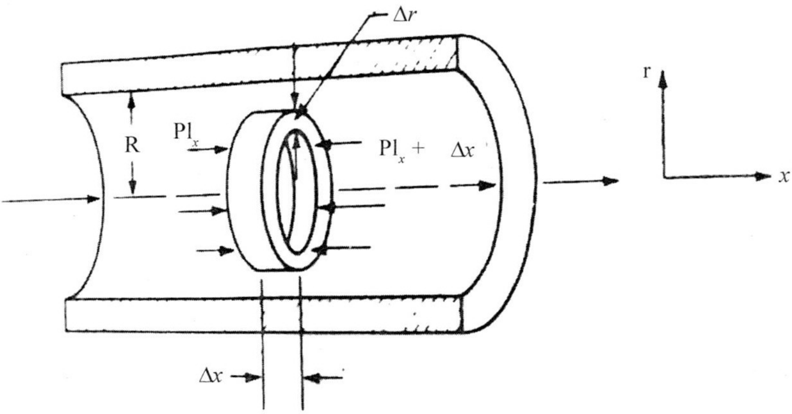

Fig. 17: Control volume for shell momentum balance on a fluid flowing in a pipe.

That is, $pA|_x - pA|_{x+\Delta x} = p\,(2\pi\Delta r)|_x - p\,(2\pi\Delta r)\,|_{x+\Delta x}$ (58)

The shear force acting over the surface at the radius r is the product of shear stress (τ_{rx})and the area (2π r Δx). This may also act as the rate of momentum flow entering into the pipe and thus making net flux as,

Net flux = $(\tau_{rx}\ 2\pi\ r\ \Delta x\ )|_{r+\Delta r} - (\tau_{rx}\ 2\pi\ r\ \Delta x)\ |_r$ (59)

Since the flow is fully developed, the net convective momentum flux across the annular space x and x + Δx will be zero, thus making v_x at x is equal to v_x at x + Δx.

Equating equations (58) and (59), we get,

$[(r\ \tau_{rx})|_{r+\Delta r} - (r\ \tau_{rx})|_r\ ]\ /\ \Delta r\ =\ [r\,(\,p|_x - p|_{x+\Delta x})]\ /\ (\Delta x)$ (60)

The pressure gradient will be constant in a fully developed flow. Thus, $(\Delta p/\Delta x)$ can be written as $(\Delta p/L)$, where Δp is the pressure drop and L will be the length of pipe. Considering Δr approaching zero, we have,

$$\frac{d(r\tau_{rx})}{dr} = \left(\frac{\Delta p}{L}\right) r \tag{61}$$

Separating the variables, we will get,

$$_{rx} = \left(\frac{\Delta p}{L}\right)\left(\frac{r}{2}\right) + \frac{C_1}{r} \tag{62}$$

Also for a definite momentum flux, constant of integration must be zero at $r = 0$, thus,

$$\tau_{rx} = \left(\frac{\Delta p}{2L}\right) r = \frac{r(p_o - p_L)}{(2L)} \tag{63}$$

That is, the momentum flux varies linearly with the radius with maximum value prevailing at the wall$(r = R)$ (Figure 18). Substituting, Newton's law of viscosity inequation, we will get,

$$_{rx} = -\mu \frac{dv_x}{dr} \tag{64}$$

and, the differential equation of for the velocity becomes,

$$\frac{dv_x}{dr} = \frac{\left[-(p_o - p_L)r\right]}{2\mu L} \tag{65}$$

Using the boundary condition at the wall, $v_x = 0$ at $r = R$, the equation of velocity distribution becomes,

$$v_x = \frac{(p_o - p_L)R^2[1 - (r/R)^2]}{(4\mu L)} \tag{66}$$

Equation (66) reveals that the velocity distribution is parabolic in nature (Figure 18).

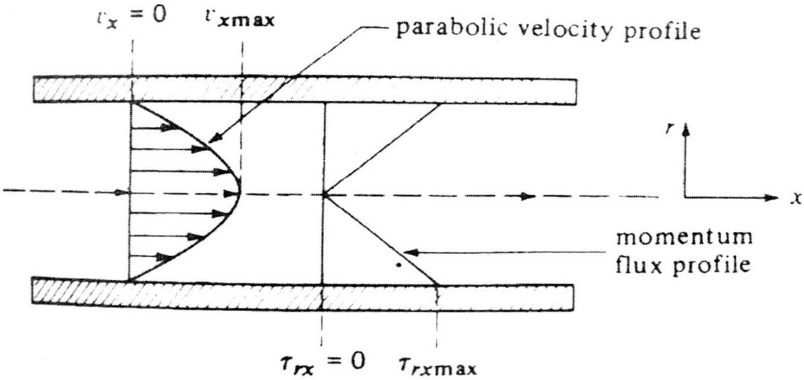

Fig. 18: Velocity and momentum flux profiles in pipe for laminar flow

Average velocity can be computed by summing up all the velocities over the cross section and dividing by the cross sectional area as

$$v_{av} = \frac{1}{A}\iint v\,dA \tag{67}$$

Here, taking dA = r dθ d and A = π R²,

$$v_{x.av} = \frac{1}{A}\int\int_{A} v(x)\,dA = 1/(\pi R^2)\int_0^R v(x)\,2\pi r\,dr \tag{68}$$

Combining and equating above two equations we will get,

$$v_{x.av} = \frac{(p_o - p_L)R^2}{8\mu L} = \frac{(p_o - p_L)D^2}{32\mu L} \tag{69}$$

Equation (69) is known as Hagen-Poiseuille equation, which relates the pressure drop and average velocity for the laminar flow in the pipe.

Maximum velocity can be obtainedwhen r=0, thus from the equation (66) we get,

$$v_{x,\,max} = \frac{(\Delta p)R^2}{4\mu L}, \tag{70}$$

occurring at r = 0.Here,$\Delta p = (p_O - p_L)$.

Combining equations (69) and (70),

$$v_{x,av} = \frac{V_{x,max}}{2} \tag{71}$$

For turbulent flow, the average velocity is 0.8 times the maximum velocity.

5.1. Pressure Drop and Loss due to Friction

Using the momentum shell balance in a pipe , the Hagen Poiseuille's equation and writing the differential momentum balance equation using equation (64) and (69) as:

$$\Delta p_f = (p_1 - p_2)_f = \frac{\left[32\mu\, v(L_2 - L_1)\right]}{D^2} \tag{72}$$

where, p_1 is the upstream pressure, p_2 is the downstream pressure, v is the average velocity in tube, D is the inside diameter through which the fluid is flowing, $(L_2 - L_1)$ is the length of the tube for laminar flow.

In practice, the pressure loss for every type of fully developed flow that can be laminar or turbulent, or be in circular or non-circular pipes, or be over smooth or rough surfaces and in horizontal or inclined pipes, can be given as,

$$\text{Pressure loss, } \Delta p_L = f\left(\frac{L}{D}\right)\left(\frac{\rho V^2}{2}\right) \tag{73}$$

where, $\rho v^2/2$ is the dynamic pressure and f is the Darcy friction factor, and it can also be given in terms of shear stress (τ_w) as,

$$f = \frac{(8\tau_W)}{\rho v^2} \tag{74}$$

which is also known as Darcy-Weisbach friction factor. Henry Darcy (1803-1858) and Julius Weisbach (1806-1871) provided the important contribution in the development of the friction factor. There is also an another parameter, which is a friction coefficient also known as Fanning friction factor, which is defined as, $C_f = 2\,\tau_W/(\rho V^2) = f/4$

For, a fully developed flow which is laminar in nature, flowing through a circular pipe can be given as,

$$f = \frac{64\mu}{\rho v D} = \frac{64}{Re} \tag{75}$$

Equation (76) is the relation between friction factor and Reynolds number. It further reveals thatthe friction factor is directly dependent on Reynolds number and is independent of roughness of the pipe surface.

In piping systems, the head loss can be given as by simply dividing the pressure loss equation by ρg and the pressure losses are commonly expressed as in terms of the equivalent fluid column height.

$$\text{Head loss, } h_L = \frac{\Delta p_L}{\rho g} = \frac{fLv^2}{2Dg} \tag{76}$$

The head loss may be caused by viscosity or can be directly related to the wall shear stress and represents the additional height that the fluid might need to be raised by pumpin order to overcome the frictional losses in the pipe. For non-circular pipes the friction factor may vary depending upon the cross section. Also, if the non-circular pipes are present, the diameter may be replaced by hydraulic diameter of the cross section which is given as, $D_h = 4A_c/p$, where, D_h is the hydraulic diameter, A_c is the cross sectional area of the pipe and p is the wetted parameter.

In case of turbulent flows, the friction factor in a fully developed flow is dependent on the Reynolds number and the relative roughness, ε/D, which is the ratio of mean height of roughness and the pipe diameter. Theoretically it becomes very difficult to predict the pressure loss and the head loss in a turbulent flow, and the solution is available in chart form by Lewis F. Moody (1880-1953), also known as Moody's chart (Figure 19), which presents the relationship between Darcy friction factor of pipe flow with Reynolds number and roughness parameter over a wide range. It is commonly used chart in engineering and can also be used for non-circular pipes by replacing the diameter with the hydraulic diameter.

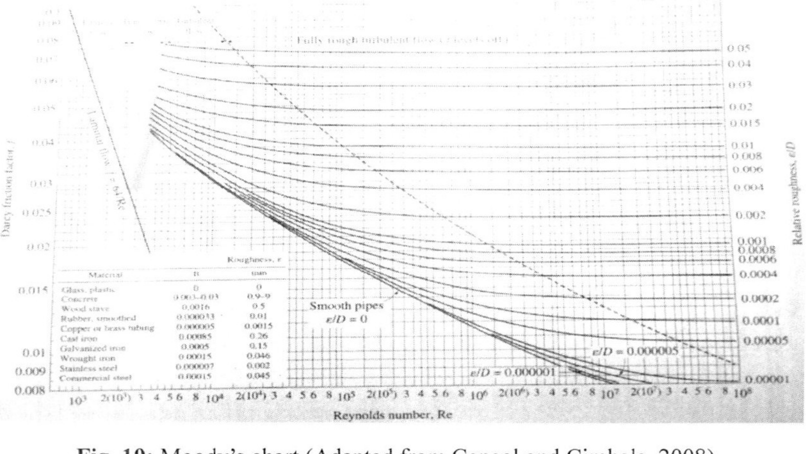

Fig. 19: Moody's chart (Adapted from Cengal and Cimbala, 2008).

5.2. Frictional Loss

The frictional energy loss may take place due to various factors when a fluid is flowing in a pipe. Broadly, the friction losses can be major friction loss and minor friction loss. The major friction loss is the loss due to sudden pressure drop in a viscous flow.

$$E_{major} = \frac{\Delta P}{\rho} = \frac{2fv^2 L}{D} \tag{77}$$

The minor losses (E_{minor}) may be present due to the various components fitted in the piping system. These can be contractions(when fluid entering from a reservoir to pipe), expansions (when fluid is entering from a pipe into the reservoir), valves, elbows and tees.

The frictional loss due to contraction can be given as,

$$E(\text{contraction}) = \frac{\Delta P}{\rho} = \frac{C_{fc}.v^2}{2}, \tag{78}$$

where,\

$= 0.4\ [\ 1.25 - (A_2/A_1)];$ for $A_2/A_1 < 0.715$

$= 0.4\ [1 - (A_2/A_1)];$ for $A_2/A_1 > 0.715$(79)

Here, is the coefficient of contraction and A_1 is the area of the reservoir and A_2 is the area of the pipe.

Similarly, the expansion loss can be given as,

$$E(\text{expansion}) = \frac{\Delta P}{\rho} = \frac{C_{fe}.v^2}{2}, \tag{80}$$

where, the coefficient of expansion $C_{fe} = (1-(A_1/A_2)^2)$

If certain pipe fittings are involved in the piping system like valves, tees and elbows then the frictional loss can be given as,

$$E(\text{fittings}) = \frac{\Delta P}{\rho} = \frac{C_{ff} v^2}{2}. \tag{81}$$

Loss due to these fittings are given in Table 3 below which can be used and may added up if more than one fitting is involved in the system. Hence, this is how various losses can be predicted depending upon the frictional losses.

Table 3: Frictional losses due to valves and fittings (Perry and Green, 1999).

Type of fittings	Additional frictional loss.
45° Elbow, std.	0.35
45° Elbow, long radius.	0.2
90° Elbow, std.	0.75
Tee, std.	0.4
Gate valve,open	0.17
Gate valve,3/4 open	0.9
Gate valve,1/2 open	4.5
Gate valve,1/4 open	24
Globe valve, std.	6
Globe valve, 1/2 open	9.5

6. Flow Metering Devices

The fluid flow measuring devicesare based on Bernoulli's equation. There are three basic devices that are used for the measurement of flowing fluid namely, venturimeter, orificemeter and pitot tube.

6.1. Venturimeter

The rate of flow of fluid flowing through a pipe can be measured using a venturimeter consisting of a short converging part, a throat and a diverging part (Figure 20). It can also be classified as one of the application of the Bernoulli's equation.

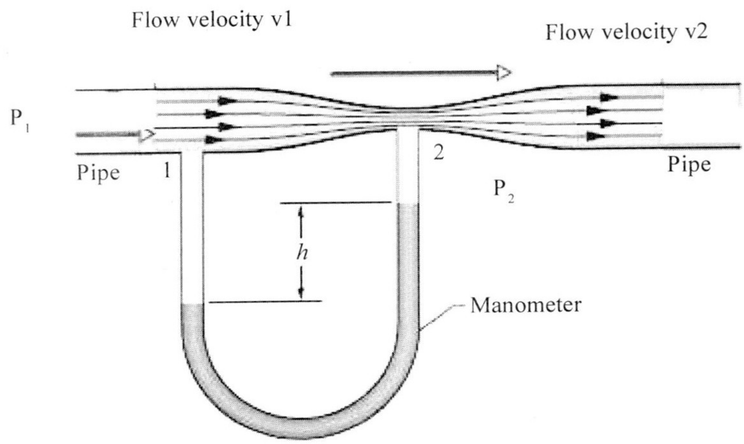

Fig. 20: Venturimeter.

Let, d_1 is the inlet diameter at section1, d_2 is the inlet diameter at section 2, P_1 is the pressure at section 1 and P_2 at section 2, v_1 and v_2 are the respective velocities of the fluid at section 1 and 2. Applying Bernoulli's equation at section 1 and 2, we get,

$$\frac{P_1}{\rho.g}+\frac{v_1^2}{2g}Z_1=\frac{P_2}{\rho.g}+\frac{v_2^2}{2g}Z_2 \qquad (82)$$

Being a horizontal pipe, Z_1=Z_2. That is

$$\frac{P_1}{\rho.g}+\frac{v_1^2}{2g}=\frac{P_2}{\rho.g}+\frac{v_2^2}{2g}$$

$$\frac{P_1-P_2}{\rho.g}=\frac{v_2^2}{2g}-\frac{v_1^2}{2g} \qquad (83)$$

But, $(P_1$-$P_2)/\rho.g$ is the difference in the pressure heads at section 1 and 2 and is equal to h;

$$h=\frac{v_2^2}{2g}-\frac{v_1^2}{2g} \qquad (84)$$

Applying continuity equation which takes area into account as $A_1v_1 = A_2v_2$, and which leads to on rearranging as, $v_1 = (A_2v_2)/A_1$. Substituting this value in equation (85), we get,

$$h=\frac{v_2^2}{2g}-\frac{\left[(A_2v_2)/A_1\right]^2}{2g} \qquad (85)$$

$$h = \frac{v_2^2}{2g}\ [1 - A_2^2/A_1^2] \qquad (86)$$

$$v_2^2 = 2gh\ [\frac{A_1^2}{A_1^2-A_2^2}] \qquad (87)$$

Therefore, $v_2 = (A_{1.}\sqrt{2gh})/\sqrt{A_1^2-A_2^2}$

Now the theoretical discharge would be, $Q = A_2V_2$, and substituting the above value of V_2, the discharge will thus become,

$$Q = (A_1A_2\ \sqrt{2gh}.)\ /\ \sqrt{\left[A_1^2-A_2^2\right]} \qquad (88)$$

Actual discharge is less than the theoretical discharge, and involves a coefficient of venturimeter (C_v) which is less than 1.

$$Q\ (\text{actual}) = C_v.\ (A_1A_2\ \sqrt{2gh}\ .)\ /\ \sqrt{\sqrt{([A_1^2 - A_2^2])}} \tag{89}$$

6.2. Orificemeter

An orificemeter (Figure 21) is used to measure flow rates in a pipe for restricting flow or for reducing pressure and works on the same principle as venturimeter. An orifice is basically a circular sharp edged hole in a circular plate, placed inside a pipe. In general, the orifice diameter is half of the diameter of the pipe in which it is placed, although it varies from 0.4 to 0.8 times the pipe diameter. The velocity that is obtained is assumed to have no loss in the flow, and considered to be maximum velocity at the constriction (Bean, 1971).A differential manometer is connected at the pressure taps, which helps in measuring the discharge from the equipment.

Let, P_1 be the pressure at the inlet and P_2 at the outlet, v_1 the velocity at the inlet, v_2 the velocity at the outlet, A_1 area of pipe at inlet. Applying Bernoulli's equation at the inlet and outlet of the orifice, we get,

$$\frac{P_1}{\rho.g} + \frac{v_1^2}{2g} + Z_1 = \frac{P_2}{\rho.g} + \frac{v_2^2}{2g} + Z_2 \tag{90}$$

$$\text{or,}\quad \left[\frac{P_1}{\rho.g} + Z_1\right] - \left[\frac{P_2}{\rho.g} + Z_2\right] = \frac{v_2^2}{2g} - \frac{v_1^2}{2g}$$

But, [$P_1/\rho.g + Z_1$] - [$P_2/\rho.g + Z_2$] act as differential head and will be equal to h, thus,

$$h = \frac{v_2^2}{2g} - \frac{v_1^2}{2g} \tag{91}$$

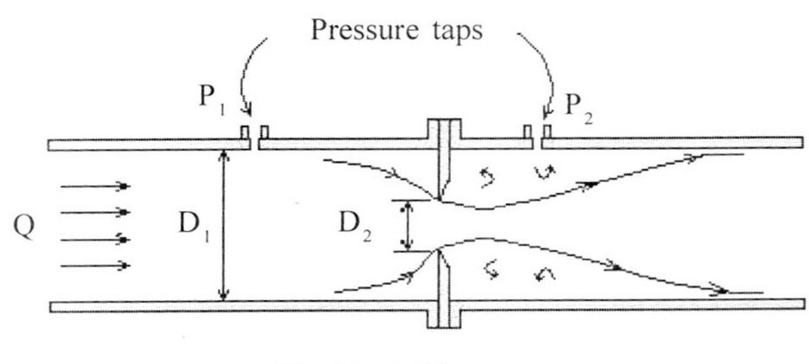

Fig. 21: Orificemeter.

or, $2gh = v_2^2 - v_1^2$

$$\text{or, } v_2 = \sqrt{2gh + v_1^2} \tag{92}$$

As the fluid cross the orifice, it enters vena-contracta and A_2 is the area at the vena-contracta. If A_0 is the area of orifice, then, we have,

$$C_c = A_2 / A_0 \tag{93}$$

where, C_c is the coefficient of contraction; and, $A_2 = C_c.\, A_0$

Since,, $A_1 v_1 = A_2 v_2$

$$v_1 = (A_0.C_c.v_2) / A_1 \tag{94}$$

Substituting equation (95) in equation (93),

$$v_2 = \sqrt{(2gh + \left[(A_0.C_c.v_2)/A1\right]^2}$$

or, $v_2^2 = 2gh + (A_0/A_1)^2.\, C_c^2.v_2^2$

$$\text{Therefore, } v_2 = \frac{\sqrt{2gh}}{\sqrt{1-\left(\frac{A0}{A1}\right)^2 Cc^2}} \tag{95}$$

$Q = v_2.A_2 = v_2.A_0C_c$, thus,

$$Q = A_0C_c.\frac{\sqrt{2gh}}{\sqrt{1-\left(\frac{A0}{A1}\right)^2 Cc^2}} \tag{96}$$

$$Q = C_d \,.\, A_0.\, A_1 \frac{\sqrt{2gh}}{\sqrt{\left(A1^2 - A0^2\right)}} \tag{97}$$

where, C_d is the coefficient of discharge from orificemeter. This coefficient of discharge is comparatively very small than that of venturimeter.

6.3. Pitot tube

These tubes are named after Henri de Pitot (1695-1771), and are widely used for the measurement of flow rates. A Pitot tube measures stagnation pressure and other circumferential static pressures. Henry Darcy (1803-1858) further developed many features of the instrument, which involved the small openings.

This tube is a most reliable device as it has no moving parts, inexpensive, and is simple to use (Cengal and Cimbala, 2008). Further, there is generation of small pressures which does not disturb the flow. A pitot tube consist of a glass tube, bent at right angles (Figure 22). One end is directed in the upstream direction placed at 90° angle while the other end remains open in the reservoir (or where the pressure to be measured). The liquid rises up in the tube as the conversion of kinetic energy into pressure energy takes place and the velocity is measured by the rise of liquid in the tube. Consider two points 1 and 2 at the same level in such a way that point 2 is just at the inlet of the pitot tube and point 1 is far away from the tube.

Let, P_1 be the intensity of pressure at point 1 and P_2 at point 2; v_1 be the velocity of flow at 1 and v_2 be the velocity at 2 which is zero; H is the depth of tube in the liquid; and, h is the rise of the liquid in the tube above the surface.

Applying Bernoulli's equation at both the points,

$$\frac{P_1}{\rho.g}+\frac{v_1^2}{2g}+Z_1=\frac{P_2}{\rho.g}+\frac{v_2^2}{2g}+Z_2$$

As, $Z_1 = Z_2$ being the points on the same line and v_2 being zero, and $P_1/\rho.g = H$ (pressure head at 1) and $P_2/\rho.g = H + h$ (i.e. pressure head at 2), we get,

$$H + v_1^2/2g = (H + h) \tag{98}$$

$$v_1 = \sqrt{2gh} \tag{99}$$

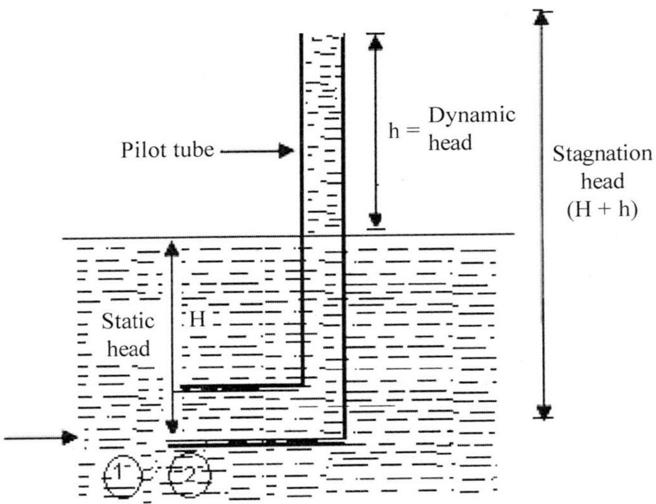

Fig. 22: Pitottube.

wherev_1 is the theoretical velocity. The actual velocity can be calculated by introducing coefficient of Pitot tube (Cp),

$$v_1 = Cp.\ \sqrt{2gh} \tag{100}$$

7. Pumps

Pump plays a significant role in transportation of liquid and semi liquid food materials in the food industry. Pump can be defined as the device which raises or transfers liquids at the expense of power input, or it may be defined as a machine designed to elevate, deliver and move various liquids, or in a broad term as a unit that helps in transferring the mechanical energy of a motor or an engine into potential and kinetic energy of a liquid. Generally, there are two broad categories of pumps depending upon their operation and principle namely, Dynamic pressure pumps and Positive displacement pumps which are further categorised as shown in Figure 23.

The basic criteria for the selection of pump involves the characteristics of the liquid to be handled, that is, pressure and capacity of the liquid to be handled; properties like viscosity, temperature, corrosiveness of the flowing liquid; speed of rotation and power required; maintenance cost of the plant, etc. Each pump has its unique characteristics which allow us the selection of the pump to be used. The basic terms which are associated with the use of pumps are discussed below.

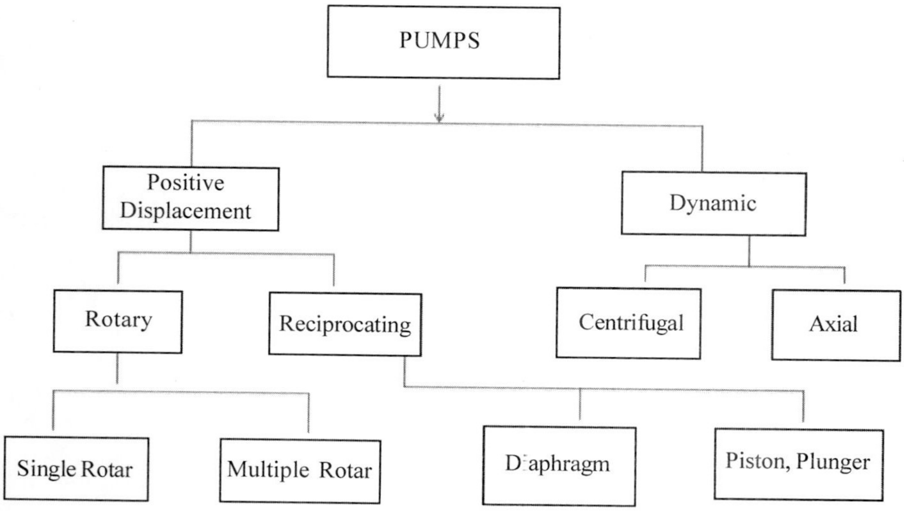

Fig. 23: Classification of pumps.

1. Head: To maintain a required rate of flow in the pipe or transporting medium, the pressure which is needed to make liquid flow is called head. Generally, the head losses are of two types, static head loss and dynamic head loss. The difference in the height of the supply and the target reservoir or equipment in the plant, is basically the static head. Whereas, the dynamic head, also known as friction loss head, may caused in the pipes, involving bends, valves, fittings having various friction factors.

2. Power: The power required by the pump drawn from an external source (P_B) can be calculated from the work done by the pump (W_P) which is the ratio of the change in heads and efficiency of the pump. Symbolically,

$$P_B = \dot{m} . W_P = (\dot{m} . \Delta H)/\eta \quad (101)$$

where, is the mass flow rate of the liquid; ç is efficiency of pump; and,ÄH is the difference between the suction and the discharge heads. The power delivered to the fluid (P_f) is the product of mass flow rate of the liquid and the head developed by the pump,

$$P_f = \dot{m} . \Delta H \quad (102)$$

ΔH can be calculated using the Bernoulli's equation as well in terms of head,

$$\Delta H = [(P_1/\rho.g) + (v_1^2/2.g) + (z_1)] = [(P_2/\rho.g) + (v_2^2/2.g) + (z_2)] \quad (103)$$

$$\Delta H = (P_2 - P_1)/\rho g + (v_2^2 - v_1^2)/2g + (z_2 - z_1) + E_{major}/g + E_{minor}/g \quad (104)$$

$$\Delta H = (P_2 - P_1)/\rho g + (v_2^2 - v_1^2)/2g + (z_2 - z_1) + 2fv^2L/Dg$$

$$+ C_{fc} v^2/2g + C_{fe} v^2/2g + C_{ff} v^2/2g \quad (105)$$

where, E_{major} are the losses due to valves, fittings etc. and E_{minor} are the losses due to bents, T-s and elbows as well as contractions and expansions in the flow.

3. Suction lift and cavitation in the pump: Sometimes, when the pressure at the suction head (suction pressure) is slightly greater than the vapour pressure of the liquid, the liquid may form vapours and erode the pump and causes the condition of cavitation. This cavitation reduces the pump capacity and severely affect the efficiency by the formation of vapours inside the pump. Also, if the suction pressure is less than the vapour pressure, no liquid will be drawn into the pump due to the vaporisation in the suction line. To overcome this situation, a net positive suction head (NPSH) must be provided by exceeding the pump pressure than the vapour pressure by a certain value. Mathematically,

$$NPSH = P_a + P_s - P_{vp} - H_{fs} \quad (106)$$

where, NPSH is the net positive suction head available; P_a is the absolute pressure on the liquid; P_s is the static suction pressure; P_{vp} is the vapour pressure of the fluid; and, H_{fs} is the friction loss in the suction pipe (McCabe et al.,2005).

Centrifugal pump is used for a small pressure differential change causing a relatively large change in flow. On the other hand, positive displacement pump delivers a constant quantity regardless of the pressure variations. To maintain a constant head/ pressure without taking into account the capacity or the discharge, centrifugal pumps are preferred among most of the pumps giving high output and efficiency (Kumar, 2001).

7.1. Centrifugal Pumps

Pumps when used for liquids, may impart kinetic energy, increase pressure or may provide circulating action depending upon the use. Centrifugal pumps are commonly used by the processing industries and are available in sizes of 0.004 to 380 m^3/min and may discharge pressures from head of having a few meters to 5000 kPa(Geankoplis,1993). Schematic diagram of a centrifugal pump is shown in Figure 24. The liquid enters the pump and reaches a high speed rotary element impeller, through a suction connection concentric with the axis of the impeller. In the impeller, the liquid spreads out radially and then enter the channels located between the vanes and reach the periphery of the impeller. For the proper functioning of the pump, there should be no cavitation in the vanes and that area must completely be filled with liquid. The liquid leaving the periphery of the impeller is then collected in the volute which is a spiral casing and then, leaves the pump. In the impeller section, there is a high velocity head supplied to the fluid, which further transforms into the pressure head as it passes through the volute chamber towards the discharge, through a tangential connection.

In an ideal case of the centrifugal pumps, there will be a specific discharge at a particular rate with a specific developed head. But, this ideal condition is difficult to achieve due to the presence of friction and other departures from perfection. Generally, the performance is given by the means of the curves, known as characteristic curves (Figure 25), by manufacturers and is basically with water taken as the pumping fluid by the pumps. It is a curve plotted among various parameters of the pump like, discharge head, efficiency, capacity, and flow rate.If the characteristic curves are unavailable for a specific pump or operating conditions, the centrifugal pump performance can be obtained with the help of certain affinity laws. These are the similarity relationships between (i) capacity (Q) being directly proportional to the impeller rotation speed (N); (ii) head (H) varying as the square of the impeller rotation speed; and, (iii) brake horsepower (BHP) varying as the cube of the impeller rotation speed. Brake horsepower is the power input to the pump with the help of the motor installed along with the pump.

Most pumps are usually rated on the basis of capacity and head at the point of peak efficiency.

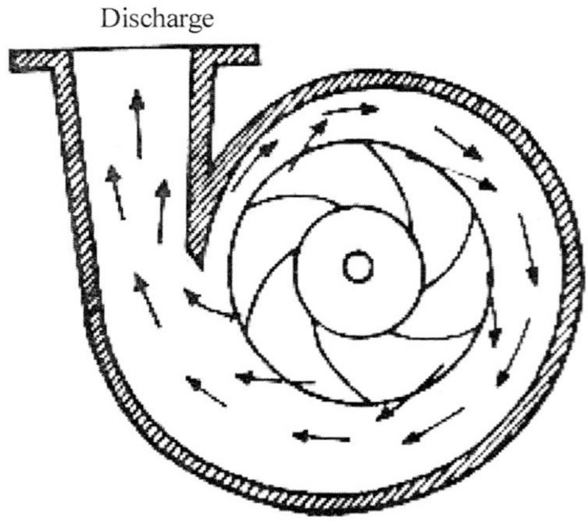

Volute centrifugal pump cross section

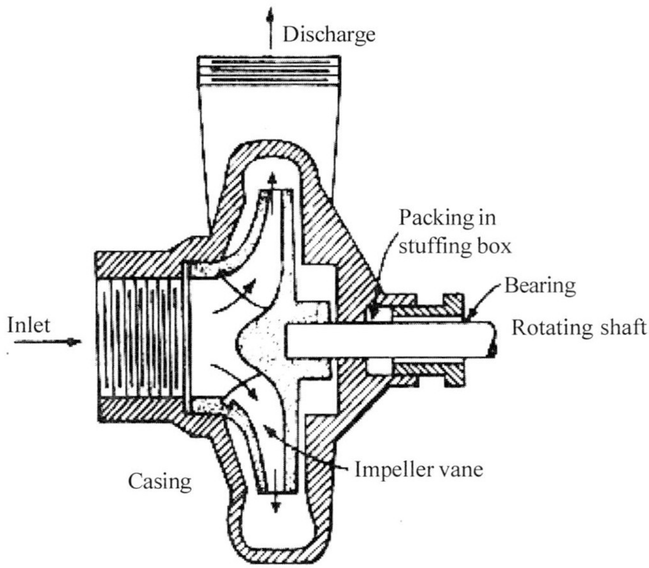

Fig. 24: A centrifugal pump.

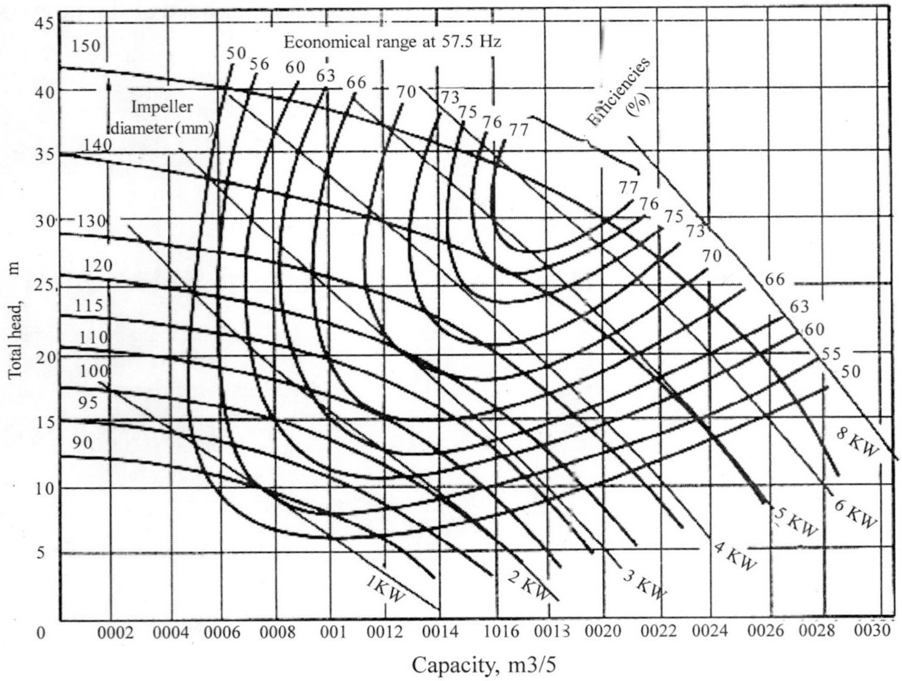

Fig. 25: Characteristic curves for centrifugal pumps. Adapted from Coulson and Richardson (1973).

8. Conclusion

A fluid is something that flows and a fluid while flowing show different characteristic patterns depending upon various conditions surrounding it. These different patterns helped us to develop certain pressure measuring devices and incorporating their applications in various processing conditions in the industries. Fluids not only act as an important field of study in food or chemical or biochemical industries, but also in aerodynamics and other natural science areas. In consideration to food industry, the shelf life, quality characteristics of the end product completely depend upon the rate at which mixing or incorporation of fluids is done. Thermal processing of the raw material also influence the flow characteristics. All these parameters can be monitored if we have proper control of fluid flow and parameters basic knowledge of flow behaviour in different processing and storage conditions. This will further help us to develop new efficient methods for manufacturing of processed foods.

References

Bean H.S. (1971) , Fluid Meters : Their theory and applications, 6th edition, New York : American Society of Mechanical Engineers.

Seshadari C V and Patanakar S V (1974), Elements of fluid mechanics, Book, Prentice hall of india private limited, New Delhi. ISBN- 0-87692-009-1.

Neddermann R M, Horton D J and Davies S T (1980), The flow of granular materials round obstacles. Powder Technol, 25 (pg.215-223)

Earle R L (1983), Unit operations in Food processing, Second edition, Massey University, New Zealand, Pergamon Press

Lewis M.J (1990), Physical properties of foods and food processing systems, Woodhead publishers. Cambridge

Gekas, V G (1992), Transport phenomena of foods and biological materials. Boca raton, F L : CRC press.

Geankoplis C J (1993), Transport processes and unit operations, 3rd edition. Printed by Prentice-Hall Inc, Englewood Cliffs, N.J., USA

Steffe J F (1996), Rheological methods in Food Process Engineering, 2nd edition. ISBN 0-9632036-1-4

Rielly C D (1997), Book chapter on Food Rheology in Chemical engineering for the food industry by Fryer P J, Pyle D L and Rielly C D. Published by Chapmann and Hill

White F M (1999), Fluid Mechanics, 7th edition. Published by Mc-Graw Hills.

Perry R H and Green D W (1999), Perry's Chemical Engineers Handbook. Published by Mc-Graw Hills.

Kumar D S (2001), Fluid Mechanics and Fluid Power Engineering. Published by SK Kataria and Sons, New Delhi.

McCabe W L, Smith J and Hariott P (2005), Unit operations of chemical engineering, Mc-Graw Hill Publisher, 7th edition.

Cengal Y A and Cimbala J M (2008), Fluid Mechanics Fundamentals and Applications (In SI Units), Third reprint, Tata McGraw Hill.

Shemmeri T T (2012), Engineering Fluid Mechanics, published by Bookboon.com (Venus Publishing ApS). ISBN 978-87-403-0114-4.

4

Heat Transfer

Debabandya Mohapatra

1. Introduction

Thermal energy is transferred in the form heat from one body to another through conduction, convection and radiation or in combination of these modes. Some of the examples of heat transfer that occur during various agricultural processing is drying and dehydration, sterilization, pasteurization, evaporation, distillation, to name a few.

The basic equation of conservation of heat can be written as:

Rate of heat in + rate of heat generation= rate of heat out+ rate of accumulation of heat

1.1. Basic Mechanism of Heat Transfer

Conduction:- In conduction, heat is conducted through solids, liquids and gases. The heat is conducted by the transfer of energy from one part of to the other through the adjacent molecules, under influence of the temperature gradient. Heat usually transfers from the greater energy level to lower energy level. In conduction heat is transferred by the by the free electrons as it is observed in case of metal rod getting heated up at one end by the fire and the heat can be experienced at the other end (Fig 1).

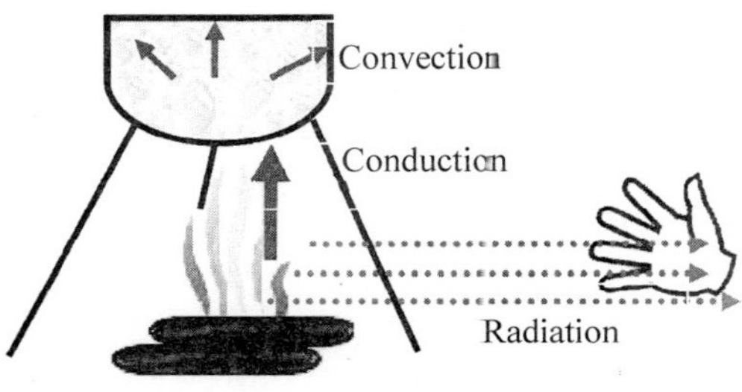

Fig. 1: Modes of heat transfer

Convection:- In convection, transfer of thermal energy occurs due to the movement of molecules from one part of the material to another. The example of convective heat transfer is the warm air which is heated by fire experienced at a distance, cooking of food while stirred, boiling of water etc (Fig 1).

Radiation:- In radiation transfer of heat energy through empty space. No medium is necessary, for it is transferred through electromagnetic waves. All object with a temperature above absolute zero radiate energy. Solids and liquids tend to absorb radiant energy being transferred through it. The most important example of solar energy reaching the earth's surface and heating things up.

1.2 Fourier's Law of Heat Conduction

The rate of heat transfer depends on the driving force which could be the temperature differential or thermal energy differential and the resistance to heat flow, so the generalized form of the conduction heat can be written as:

$$Rate\ of\ heat\ transfer = \frac{Driving\ force}{Resistance\ to\ heat\ flow}$$

For the heat flow to occur, the driving force must overcome the resistance. The equation can be written in the form of total heat energy involved:

$$\frac{q_x}{A} = -K\frac{dT}{dx}$$

Where, q_x (Watt) is the rate of transfer of heat in the x direction through a distance dx (m), A is the surface area (m^2) normal to the direction of heat flow, T is the temperature in Kelvin and k is thermal conductivity in W/ m-K. q_x/A is called the heat flux and dT/dx is known as temperature gradient along x-direction (Fig. 2).

1.2.1 Thermal Conductivity

Thermal conductivity is defined as the amount of heat flows through unit thickness of material over a unit area per unit time for temperature difference.

$$Q = -k\frac{Adt}{x}\times t$$

Ability to conduct heat: $q = -kA\frac{dt}{dx}$

The temperature difference (dT) is between T_1 and T_2,

$$q = -kA\ \frac{dT}{dx}$$

where t= Time, dT= temperature difference, dx= thickness

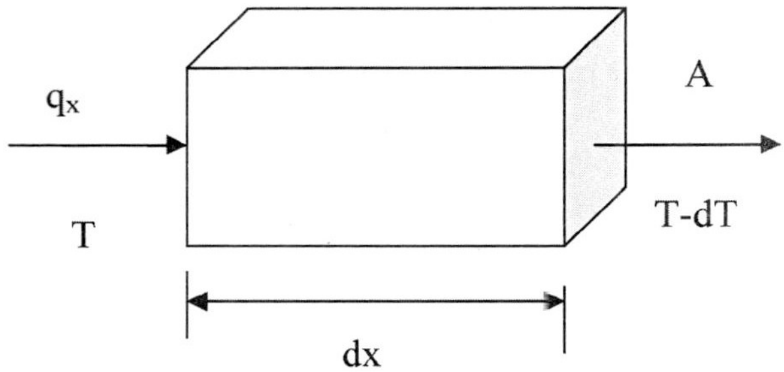

Fig. 2: Heat transfer through conduction in x direction

$$\frac{dQ}{d\theta} = -kA\frac{dT}{dx}$$ A= Surface area, k= thermal conductivity.

The final equation for

$$\text{'k'} = \frac{kcal - m}{m - hrk} \text{ or } {}^{W}\!/_{mk}$$

The thermal conductivity of gases, liquids and solids vary significantly from each other. The mechanism of conduction in gas is due to the random molecular motion and energy transfer occurs due to collision of molecules. Similar mechanism works in case of liquids. However, in case of solids the mechanism of conduction of two types i.e. conduction of heat through free elections, as in case of metallic solids, and conduction of heat is through the transmission of vibrational energy between two adjacent atoms.

1.2.2 Conduction Through a Flat Slab or Wall

In one dimensional steady state condition, the heat flow from one end to another end through a crosssectional area of thickness dx. The temperature of the outside wall is T1 and that of the lower end is T2. Heat transfer from higher to lower so (dT) accounts for the negativity (So final temperature will have lower value then the initial) to negate that (-) sign.

$$\frac{dT}{\left(\frac{dx}{kA}\right)} = \frac{(T_1 - T_2)}{R} = \frac{driving\ force}{resistance}$$

Example 1: Calculate the total heat transfer rate through a rectangular wall (10m×25 m) of a room, composed of a brick wall of 15 cm thick, where the outside temperature is 45°C and inside room temperature is 25°C. The thermal conductivity of brick is 0.6 W/m-K.

Solution:

k=0.6W/m-K

$A=10\times25=250\ m^2$

dx=15 cm= 0.15 m

dT=45-25=20

$$q = Ak\frac{dT}{dx} = 250\times0.6\times\frac{20}{0.15} = 20000W = 20kW$$

1.2.3 Conduction Through a Hollow Cylinder

In many processing industry fluid flows through pipes, and heat is transferred through the walls of thick walled cylinders, which may or may not be insulated. Let us consider a hollow cylinder having inside diameter r_1 and outside diameter r_2 and length L (Fig 3). The outside temperature is T_1 and inside temperature is T_2.

$$\frac{q}{A} = -k\frac{dT}{dr}$$

The cross sectional area of through which heat flows in 2πrL

Now, substituting equation —— into equation —and rearranging

$$\frac{q}{2\pi rL} = -k\frac{dT}{dr}$$

$$\therefore \frac{q}{2\pi L}\int_{r1}^{r2}\frac{dr}{r} = -k\int_{T1}^{T2} dT$$

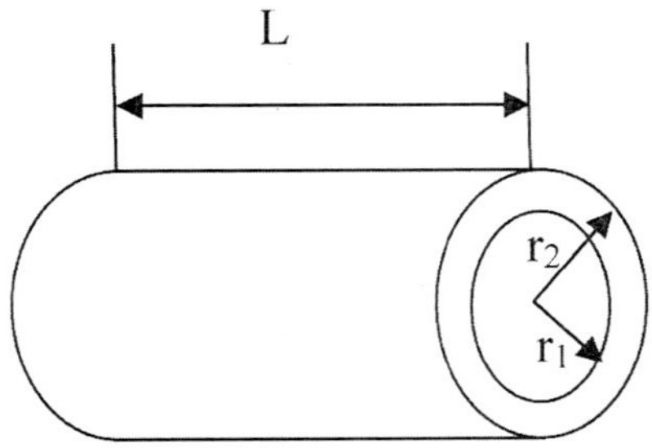

Fig. 3: Heat conduction in a hollow cylinder

$$q = k\frac{2\pi L}{\ln\left(r_2/r_1\right)}(T_1 - T_2)$$

Multiplying the numerator and denominator by (r_2-r_1), we get

$$q = kA_{lm}\frac{(T_1 - T_2)}{(r_2 - r_1)} = \frac{(T_1 - T_2)}{(r_2 - r_1)/(kA_{lm})} = \frac{(T_1 - T_2)}{R}$$

Where,

$$A_{lm} = \frac{(2\pi L r_2) - (2\pi L r_1)}{\ln\left(2\pi L r_2/2\pi L r_1\right)} = \frac{A_2 - A_1}{\ln\left(\frac{A_2}{A_1}\right)} = \log mean\ Area$$

$$R = \frac{(r_2) - (r_1)}{kA_{lm}} = \frac{\ln\left(\frac{r_2}{r_1}\right)}{2\pi k L}$$

1.2.4. Conduction Through a Hollow Sphere

In case of one dimensional heat flow through a hollow sphere, the cross sectional area normal to the heat flow can be written as $4\pi r^2$

Hence,

$$q = \frac{4\pi k(T_1 - T_2)}{(1/r_1 - 1/r_2)} = \frac{(T_1 - T_2)}{(1/r_1 - 1/r_2)/(4\pi k)} = \frac{(T_1 - T_2)}{R}$$

1.2.5. Conduction Through Solids in Series

a) Plane walls in series

In cases where there is more than one wall, the temperature profile will follow as shown in fig 4.

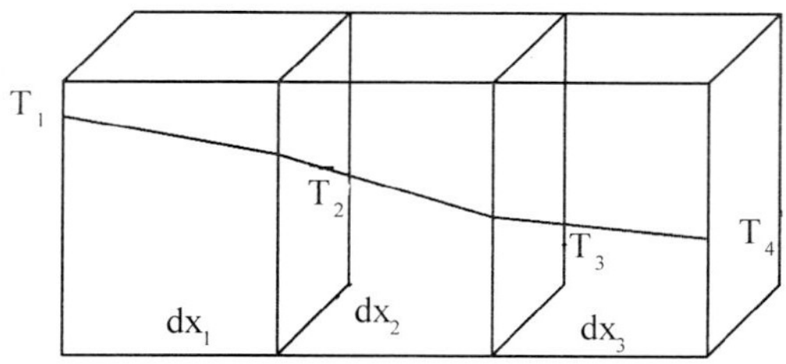

Fig. 4: heat flow through multi layer wall

So q will be

$$q = \frac{k_1 A_1 (T_1 - T_2)}{dx_1} = \frac{k_2 A_2 (T_2 - T_3)}{dx_2} = \frac{k_3 A_3 (T_3 - T_4)}{dx_3}$$

Since the cross sectional area through which the heat flows is same for all the blocks the equation can be rewritten as

$$q = \frac{k_1 A (T_1 - T_2)}{dx_1} = \frac{k_2 A (T_2 - T_3)}{dx_2} = \frac{k_3 A (T_3 - T_4)}{dx_3}$$

Solving each equation for dT

$$T_1 - T_2 = q\frac{dx_1}{Ak_1}; T_2 - T_3 = q\frac{dx_2}{Ak_2}; T_3 - T_4 = q\frac{dx_3}{Ak_3}$$

Adding the equations for T1-T2, T2-T3, T3-T4, the final equation becomes

$$q = \frac{T_1 - T_4}{dx_1/Ak_1 + dx_2/Ak_2 + dx_3/Ak_3} = \frac{T_1 - T_4}{R_1 + R_2 + R_3}$$

Where, $R = \frac{dx}{kA}$

b) Conduction through multilayer cylinders

In processing operation, heat sometimes flows through multilayer pipes such as insulated pipes carrying steam or chilled water. Hence the temperature at each of the intersection of the material will be T1, T2, T3, T4, and the temperature across the section A, B and C will be T1-T2, T2-T3 and T3- T4.

At steady state the heat transfer rate will be same across different layers and it can be expressed as.

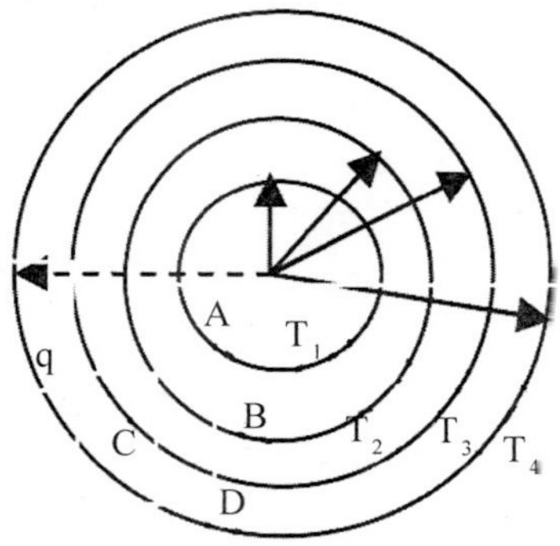

Fig. 5: Radial heat flow through multilayer cylinders

$$q = \frac{T_1 - T_2}{(r_2 - r_1)/(k_A A_{Alm})} = \frac{T_2 - T_3}{(r_3 - r_2)/(k_B A_{Blm})} = \frac{T_3 - T_4}{(r_4 - r_3)/(k_C A_{Clm})}$$

Where

$$A_{Alm} = \frac{A_2 - A_1}{\ln\left(\frac{A_2}{A_1}\right)};\ A_{Blm} = \frac{A_3 - A_2}{\ln\left(\frac{A_3}{A_2}\right)};\ A_{Clm} = \frac{A_4 - A_3}{\ln\left(\frac{A_4}{A_3}\right)}$$

Combining the equations we get

$$q = \frac{T_1 - T_4}{(r_2 - r_1)/(k_A A_{Alm}) + (r_3 - r_2)/(k_B A_{Blm}) + (r_4 - r_3)/(k_C A_{Clm})} = \frac{T_1 - T_4}{R_A + R_B + R_C} = \frac{T_1 - T_4}{\sum R}$$

1.3 Convective Heat Transfer Coefficient

what The convective heat transfer occurs through a liquid or gaseous medium. The rate of heat transfer can be expressed as follows:

$$q = hA(T_s - T_f)$$

q (Watt) is the rate of transfer of heat, A is the surface area (m^2), Ts and Tf are the average bulk temperature of the solid surface and average bulk temperature of the fluid in and h is the heat transfer coefficient in W/ m^2-K. The convective heat transfer coefficient is a function of system geometry, fluid properties, flow velocity and temperature difference. When a fluid flows by a surface, adjacent to the wall there is almost a stationary layer of fluid which gives maximum resistance to the heat transfer and it is known as film coefficient or convective heat transfer coefficient.

1.3.1 Combined Convection and Conduction and Overall Heat Transfer Coefficient

In some cases, there is fluid at both sides of solid surface such as in case of heat exchangers, where cold and hot fluids are in the both sides of the metal pipes. If the temperature of hot fluid is T1 and that of cold fluid is T4 and the temperature at the intermediate nodes are T2 and T3, and the heat transfer coefficient of hot fluid and cold fluids are ho and hi, respectively then the equation for heat transfer rate can be written as:

$$q = h_i A(T_1 - T_2) = \frac{k_A A(T_2 - T_3)}{\Delta x_A} = h_o A(T_3 - T_4)$$

$$q = \frac{(T_1 - T_4)}{\frac{1}{h_i A} + \frac{\Delta x_A}{k_A A} + \frac{1}{h_o A}} = \frac{(T_1 - T_4)}{\sum R}$$

The overall heat transfer coefficient by combined conduction and convection is termed as overall heat transfer coefficient U

$$q = UA\Delta T_{overall}$$

$$U = \frac{1}{\frac{1}{h_i} + \frac{\Delta x_A}{k_A} + \frac{1}{h_o}} \frac{W}{m^2.K}$$

The heat transfer from a fluid outside to a fluid inside the tube (fig 6b), the equation can be written as

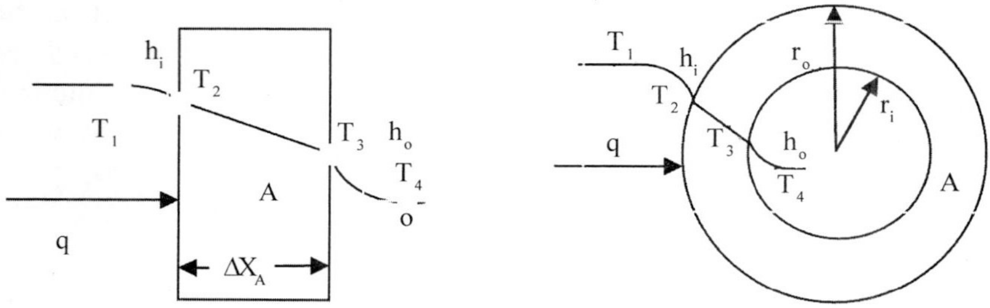

Fig. 6 a: heat flow through a plane wall (b) cylindrical wall

$$q = \frac{T_1 - T_4}{\frac{1}{h_i A} + \frac{(r_o - r_i)}{k_A A_{Alm}} + \frac{1}{h_o A}} = \frac{T_1 - T_4}{\sum R}$$

Where Ai, the inside area of the metal tube and A_{Alm} is the log mean area of the metal tube and is Ao outside area of the tube

The overall heat transfer coefficient for the cylinder will be based on inside and outside area of the cylinder

$$q = U_i A_i (T_1 - T_4) = U_o A_o (T_1 - T_4) = \frac{T_1 - T_4}{\sum R}$$

$$U_i = \frac{1}{\frac{1}{h_i} + \frac{(r_o - r_i) A_i}{k_A A_{Alm}} + \frac{A_i}{h_o A_o}}$$

$$U_o = \frac{1}{\frac{A_o}{A_i h_i} + \frac{(r_o - r_i) A_o}{k_A A_{Alm}} + \frac{1}{h_o}}$$

Example 2

Hot water at 95 C is flowing inside a steel pipe, having OD 25 mm and ID 20 mm. The pipe is insulated with 15 mm insulation on outside. The convective heat transfer coefficient for the inside hot water surface is 200 W/ m2 K and the outside 0.3 W/ m2 K. the thermal conductivity of the metal is 45 W/m-K and 0.026 W/m-K for the insulation. Calculate the heat loss to the surrounding from a 5 m pipe, if the surrounding air temperature is 35 C

Solution

The inside radius of the steel pipe (r_i)= 10/1000= 0.01 m

The outside diameter of the steel pipe (r_1)=12.5/1000=0.0125 m

The r_o= 0.0125+0.015= 0.0275 m

For 5 m length pipe

$A_i=2\pi\ Lr_i= 2\pi\ 5\ 0.01=0.3141 m^2$

$A_1=2\pi\ Lr1_i=2\pi 5\ 0.0125= 0.3927\ m^2$

$A_o=2\pi\ Lr_o = 2\pi 5\ 0.0275=0.8639\ m^2$

The log mean areas of steel pipe and the insulated area are

$A_{Alm}=A_1-A_i/\ln(A_1/A_i) = (0.3927-0.314)/\ln(0.3927/0.314) = 0.0787/\ln(1.2506) = 0.0787/.02236=0.3519$ m2

$A_{Blm}= A_o-A_1/\ln(A_o/A_1)= (0.8639-0.3927)/\ln(0.8639/0.3927) = 0.4712/\ln(2.199) =0.4712/0.7884 =0.5977\ m^2$

$R_i= 1/h_iA_i=1/200\ 0.3141=0.0159$

$R_A= (r_1-r_i)/k_AA_{Alm}= 0.0125-0.010/(45\ 0.3519)=0.0001578$

$R_B=(r_0-r_1)/k_BA_{Blm}=(0.0275-0.0125)/(0.026\ 0.5977)=0.9652$

$R_o=1/h_oA_o=1/0.3\ 0.8639=3.8595$

$Q=(Ti-To)/\ R_i+ R_A+ R_B+ R_o = (95-35)/(0.0159+.0001578+0.9652+3.8595)=$ 12.39 W

1.4 Radiation Heat Transfer

When thermal radiation falls upon any body, part of it is reflected, part is absorbed and part is transmitted. Most of the biological bodies absorb some part and reflect some part, thus for biological materials

$= \rho+\alpha=1$

Where, ρ is reflectivity, and α is absorptivity

A black body on the other hand absorbs all radiant energy and reflect none, hence $\alpha=1$ for black body and $\rho=0$.

1.4.1 Radiation from a Body

Stefan Boltzmann's law states that the total energy radiated per unit surface area of a black body is unit time in proportional to the fourth power of the thermodynamic temperature.

$Q = A\sigma T^4$

Where, q is heat flow in watt, A is surface area of the body, in m^2, σ is Stefan Boltzmann's constant=5.676 10^{-8} W/m^2K^4

1.4.2 Thermal Emissivity(ε)

It is a ratio of total emissive power of a blackbody to the same temperature.

This thermal constant is associated with the rate of heat transfer by radiation from a hot body to a cold body.

$q = AF_aF_e\ \sigma\ (T_1^4 - T_2^4)$ ε= 1 black body F_aF_e<1 real object

A= surface area

F_a = Angle factor, representing the angle that one body sees the other body.

F_e = Emissivity factor

σ = Stefan Boltzmann constant

T_1 &T_2 = Absolute temperature of body 1 and 2, respectively

While dealing with grey body Kirchhoff's law of thermal radiation is applied emissivity equals to absorptivity (for an object in thermal equilibrium) so that the body does not absorb all incident light will also emit less radiation then ideal blackbody.

When a small gray object of area A_1 m^2 at T_1 is enclosed in a high temperature (T_2) medium, there will be heat transfer from the small gray body to the surrounding and similarly there will be the small body will absorb some energy from surrounding. So the net absorption will be

$q = A_1\ \varepsilon_1\ \sigma\ T_1^4 - A_1\ \alpha_{12}\ \sigma\ T_2^4 = A_1\ \sigma\ (\varepsilon_1 T_1^4 - \alpha_{.2}\ T_2^4)$

Example 3

A metal tube having with an OD of 2.5 cm and 35 cm length, at a surface temperature of 30°C is heated by a furnace having temperature of 900°C. the emissivity of the is metal tube at 30°C is 0.3 and at 900°C is 0.6. Calculate the heat transfer to the body by radiation

Solution

Area of tube=πDL= $\pi \times 0.025 \times 0.35 = 0.02475 m^2$

Since the temperature of furnace is very large compared to the small body the emissivity of higher temperature will be considered

$$q = A_1\,\sigma\,(\varepsilon_1 T_1^4 - \alpha_{12}\,T_2^4) = 0.02475 \times 5.676 \times 10^{-8} (0.6) \times (1173^4 - 303^4) = -1588 \text{ W}$$

pl add 3-4 small numerical with application in food industry add determination of 'h' by dimensionless numbers

5

Psychometry

Vijay S Sharanagat , Vinkel Kumar Arora

1. Introduction

Psychrometric is the study thermal and physical properties of gas–vapour mixtures. Ambient air is the mixture of dry air and water vapour and some pollutants. The moisture and pollutant content vary from place to place. The amount of pollutant and water vapour decrease with increase in the altitude. The air is completely dry above 10 km altitude. In many unit operations moist air is required and properties of air changes with temperature and pressure. The pollutants has to be removed before using moist air as a process air either by filtration or circulation. So it is necessary to determine the amount of water present in the air and effect of different unit operation on thermal properties and change in the heat and moisture content of mixture.The International Joint Committee of psychrometric data have fixed the composition of dry by mole fraction as: Oxygen: 0.2095, Nitrogen: 0.7803, Argon: 0.0093, Carbon dioxide: 0.0003. The following are the properties of the dry air

Molecular weight	28.966
Gas constant R	287.035kJ/kgK
Specific heat at constant pressure, Cp	1.005 kJ/kgK
Specific heat at constant volume, Cv	0.791 kJ/kgK

In moist air the composition of the dry air is fixed but the water vapour present may vary from zero to maximum. When the air is possessing the maximum amount of moisture which it can hold than the moist air is called as saturated air. The moist air has a molecular mass of 18.015 and its gas constant is 461.52 J/kg.K.

Properties of air-water vapour mixture changes during heating or cooling, adding or removal of moisture and when one air steam mixed with another for design and analysis of various food processing and storage systems.Changes in properties of air-water vapour mixture during different unit operation can be determined with mathematical calculation. As these calculations are time consuming and same can be determined by chat containing the thermal and physical properties of air at different temperatures called as psychrometric chart. The psychrometric chart is the graphical representation of physico-thermal properties of air –vapour mixture.

Gibbs Dalton law of total pressure "the total pressure exerted by the homogeneous mixture gases is the sum of the partial pressure of individual gas". It is assumed that the gases are a perfect homogeneous mixture and do not interact with other gases.

$$Pt = P_1 + P_2 + P_3 - - - + P_n$$

Where Pt = total pressure; P1- partial pressure of 1st gas air, P2= partial pressure of 2nd gas and so on.

When Dalton's law of partial pressure is applied to moist air the total pressure is equal to the summation of partial pressure of dry air and partial pressure of water vapour.

$$Pt = P_a + P_v$$

Where Pt = total pressure; Pa- partial pressure of dry air, Pv= partial pressure of water vapour.

2. Psychrometric Pproperties

Different terms are used to express the physical and thermal properties of air-water mixture are defined and discuss here.

2.1 Dry Bulb Temperature (DBT)

The dry-bulb temperature (DBT) is the temperature of air measured by a thermometer freely exposed to the air. DBT is the temperature that is usually thought of as air temperature, and it is the true thermodynamic temperature. It is measured by a glass thermometer.

2.2 Wet Bulb Temperature (WBT)

It is the temperature of air measured by the thermometer when the bulb of thermometer is covered by wet cloth or cotton wick. When the stream of unsaturated air at constant temperature and humidity passes through a wet surface, then evaporation of water from the wetted surface tends to reduce the temperature of water. When the water become cooler then air, sensible heat transferred from air to water. When, heat loss from the water by evaporation is exactly equal to sensible heat passing from the air, steady state will be reached and temperature of water become constant.

2.3 Dew Point Temperature (DPT)

The dew-point temperature is the temperature at which the air can no longer hold all of the water vapour which is mixed with it, and some of the water vapour must condense into liquid. The dew point is always lower than (or equal to) the air temperature. At this point where the dew point temperature equals the air temperature, the relative humidity is 100%. At constant pressure when the unsaturated moist air is cooled the temperature at which the air begins to condense is called as Dew Point temperature. It can be expressed by the equation

$$DBT = \frac{4030(DBT + 235)}{4030 - (DBT + 235)\ln \ln \phi} - 235$$

Where ϕ = *relative humidity,*

Saturation temperature: for a given pressure the temperature at which the phase change occurs is called as saturation temperature. so the Dew Point temperature is a saturation temperature at a given vapour pressure, it can also be obtained steam table for a given vapour pressure.

2.4 Saturation Vapour Pressure (Psat)

For a given dry bulb temperature of a water vapour, the partial pressure of the water vapour is called as saturation vapour pressure. The American Society of Heating, Refrigeration and Air conditioning Engineers (ASHRAE) have suggested an equation for temperature range of zero to hundred.

$$\ln(P_{sat}) = \frac{C_1}{T} + C_2 + C_3T + C_4T^2 + C_5T^3 + C_6 \ln(T)$$

Where,

psat = saturated vapor pressure of water in kPa

T = temperature in K

The regression coefficients c1 to c6 are given by

$c_1 = -5.80022006E + 03$, $c_2 = -5.516256E + 00$

$c_3 = -4.8640239E - 02$, $c_4 = 4.1764768E - 05$,

$c_5 = -1.4452093E - 08$, $c_6 = 6.5459673E + 00$

2.5 Saturation state

If the water vapour added to dry air alimit will be reached when the air will be saturated and cannot hold more water vapour. If more water vapour is added drop of water remain in suspension stage and make air moist and fogy. Such drops are the condensedpartials of water vapour.

$$P = P_a + P_w$$

$If\ P_w = P_{saturated}$ at particular temperature

$$P = P_a + P_{sat}$$

$$P_{sat} = 610.78xe^{\left\{\frac{t}{(t+238.3)X}17.2694\right\}}$$

Where

$P = Total\ pressur\ of\ mixture\ of\ air\ and\ water\ rvapour$

$P_a = presseure\ of\ air$

$P_w = pressure\ of\ water\ vapur$

$t = Temperature\ in\ Degree\ Celsius$

1.2 Absolute Humidity/Humidity Ratio/ Specific Humidity (H)

Absolute humidity is defined as the ratio of mass of water vapour per unit mass of dry air in a mixture of air and water vapour. Absolute humidity will depend upon partial pressure of water vapour in the air and total pressure and can be expressed by –

$$H = \frac{m_w}{m_a} = \frac{V/v_w}{V/v_a} = \frac{v_a}{v_w}$$

$$H = \frac{18.016P_w}{28.96(P - P_w)} = 0.622\frac{P_w}{(P - P_w)}$$

Where,

H = Absolute Humidity/Humidity ratio/ Specific humidity (kg/kg)

P_w = partial pressure of water vapour in air

P= Total pressure

2.7 Degree of Saturation (μ)

The ratio of actual specific humidity to the specific humidity of saturated air at a given temperature is called as degree of saturation

$$\mu = \frac{H}{H_s} = \frac{0.622\dfrac{P_w}{(P-P_w)}}{0.622\dfrac{P_s}{(P-P_s)}} = \frac{P_w}{P_s}\left[\frac{1-Ps/P}{1-Pw/P}\right]$$

2.8 Relative Humidity (RH or ϕ)

Relative humidity is the ratio of partial pressure of water vapour in the air to partial pressure of water vapour in the saturated air at same temperature.

$$Relative\ humidity = \frac{P_w}{P_s} x100$$

Humidity reflects to the amount of moisture (water vapour) in the surrounding air. Relative humidity is a measure of the amount of moisture in the air compared with the amount of moisture the air can hold.

Relative humidity is expressed as a percentage of how much moisture the air could possibly hold at the temperature, it happens when you measure it. The wetter or damper you feel, the higher is the relative humidity. If you feel the air is dry around you, the relative humidity is low.We are very sensitive to humidity. Sweating keeps our body cool and maintains its current temperature.If the air is at 90% relative humidity, sweat will not evaporate in to the air. As a result, we feel much hotter than the actual temperature when the relative humidity is high.If the relative humidity is low we can feel much cooler than the actual temperature because our sweat evaporates easily, cooling the body. The relative humidity can also be expressed in terms of degree of saturation as:

$$\phi = \frac{\mu}{1-(1-\mu)Ps/P}$$

2.9 Humid heat (S)

Humid heat is the energy required to raise the temperature of one kg of dry air and its accompanying water vapour to one degree. It is represented in kcal/kg/°C and calculated by –

$S = 0.24 + 0.45H$

2.10 Enthalpy (h)

Enthalpy of air–water mixture is the total heat content of one kg of dry air and its accompanying water vapour. It is represented in kcal /kg.

h = ha + $W.\ hw$

For dry air above 0°C the enthalpy of dry air is expressed as:

$$ha = Cpa.t = 1.005t\ {}^{kJ}\!/_{kgK}$$

$$hw = hg + Cp(t-0) = 2500 + 1.88t$$

$$h = 1.005t + W(2500 + 1.88t)$$

If air–water mixture is at the datum temperature of 0°C and 1 atmpressure then the enthalpy of mixture at t °C is -

h = 0.24(t-0)+H(lemda+045(t-0))

h = (0.24 + 0.45 H)t + *lemda* H

2.11 Humid volume (V)

Humid volume is the total volume of one kg of dry air and its accompanying water vapour. It is represented in m³/kg.

$$V = \frac{22.4}{29}\left(\frac{t+273}{273}\right) + \frac{22.4}{18}H\left(\frac{t+273}{273}\right)$$

$$= \left(\frac{22.4}{273}\right)(t+273)\left(\frac{1}{29} + \frac{H}{18}\right)$$

$$V = (0.00283 + 0.00456H)(t+273)$$

2.12 Saturated volume

Saturated volume is the volume of air-water mixture and volume of water vapour required to saturate at a given temperature.

3. Psychrometric chart

Usually a psychrometric chart (Fig. 1) is prepared at 1atm pressure. psychrometric chat is used to find out the seven properties dry bulb temperature, wet bulb temperature, dew point temperature, absolute humidity, relative humidity, humid volume and enthalpy of air-vapour mixture. All of the above properties of air-vapour can be determine by using psychrometric chart provided any two are known.

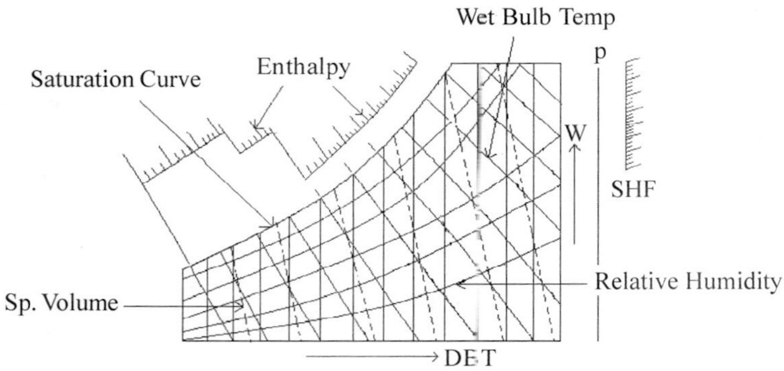

Fig. 1: Psychrometric Chart

- Dry bulb temperature is read on X-axis, Wet bulb temperature is read on X-axis,Dew point temperature is read on X-axis.Specific humidity is along y-axis, Sensible heat factor is along y-axis,There are constant RH lines as 10 %, 20 %, 30 %,40 %, 50 %, 60 % etc. and 100 % is on extreme left., Constant enthalpy lines, Constant specific volume lines In order to use the chart, any two properties must be given.Other properties are found from the chart after locating a point from the given condition of air on the chart.
- Point is marked with red is at DBT of 95^{0}F and WBT of 70^{0}F. Other properties are found from the Chart.
- Find specific humidity on y axis by moving horizontally.
- Find Dew point temperature by moving horizontally towards left up to the saturated curve.
- Find enthalpy by moving along constant enthalpy from the red point.
- Find RH by drawing a curve parallel to nearing curve of RH through the red point. Find the specific volume of air.
- The DBT, WBT and DPT are equal when RH =100%

Psychrometric Processes

Damp air undergoes number of processes in air conditioning. Some of the important psychrometric processes are discussed and plotted on a psychrometric chart.

3.1 Sensible Cooling

Sensible cooling means the cooling of moist air at a constant specific humidity. Fig. 2 shows the sensible cooling. The process is represented by a horizontal line 1-2

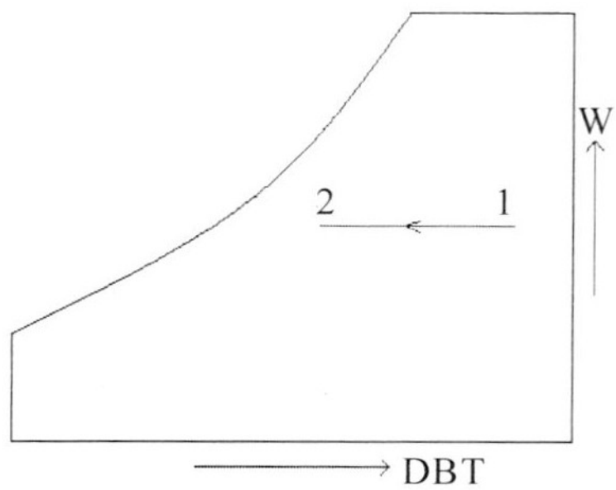

Fig. 2: Sensible Cooling

From energy balance

$mh_1 = mh_2 + Q$

$Q = m(h_1 - h_2)$

where m, h_1, h_2 and Q are mass, enthalpy at the inlet, enthalpy at the exit and heat lost by the moist air, respectively.

3.2 Sensible Heating

Heating of moist air at a constant specific humidity is called sensible heating.Fig. 3 displays the sensible heating.

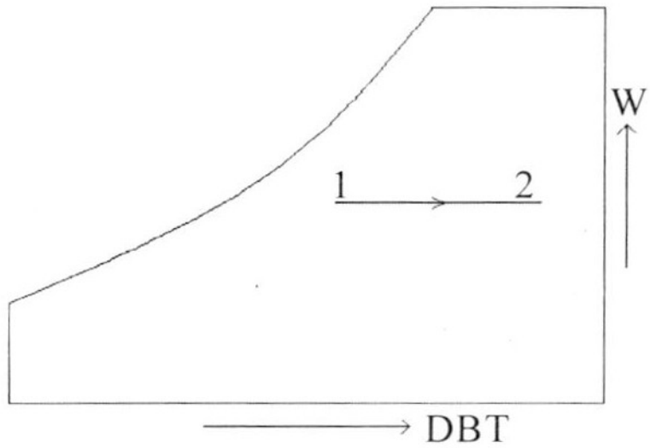

Fig. 3: Sensible heating

From energy balance

$mh_1 + Q = mh_2$

$Q = m (h_2 - h_1)$

where m, h_1, h_2 and Q are the mass of moist air, enthalpy at the inlet, enthalpy at the exit and heat supplied to the moist air during the process. Heating can be accomplished by passing steam in a coil kept in a stream of moist air

3.3 Humidification

The addition of water vapour to the moist air at the constant dry bulb temperature is called humidification. Figure 4 shows the humidification process.

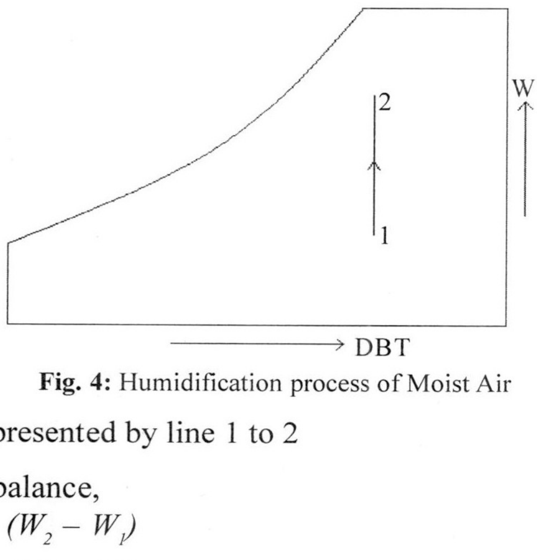

Fig. 4: Humidification process of Moist Air

The process is represented by line 1 to 2

From the energy balance,

$mh_1 = mh_2 + hfg (W_2 - W_1)$

$m (h_1 - h_2) = hfg (W_2 - W_1)$

where h_{fg} is the enthalpy of the moist air

3.4 Dehumidification

The removal of water vapour from the moist air at a constant dry bulb temperature is called the dehumidification. Figure 5 displays the dehumidification process.

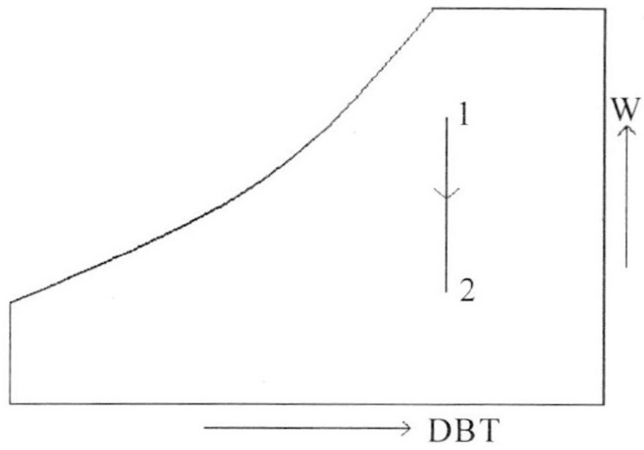

Fig. 5: Dehumidification process of moist air

The process is represented by line 1to 2

From the energy balance

$$mh_1 = mh_2 + hfg\ (W_1 - W_2)$$

$$\text{or } m\ (h_1 - h_2) = hfg\ (W_1 - W_2)$$

where different terms have the usual meaning

3.5 Cooling and Dehumidification of Moist Air

Cooling and dehumidification of moist air is required in summer air conditioning. Figure 6 represents the cooling and dehumidification of moist air.

Fig. 6: Cooling and dehumidification process on a psychrometric chart

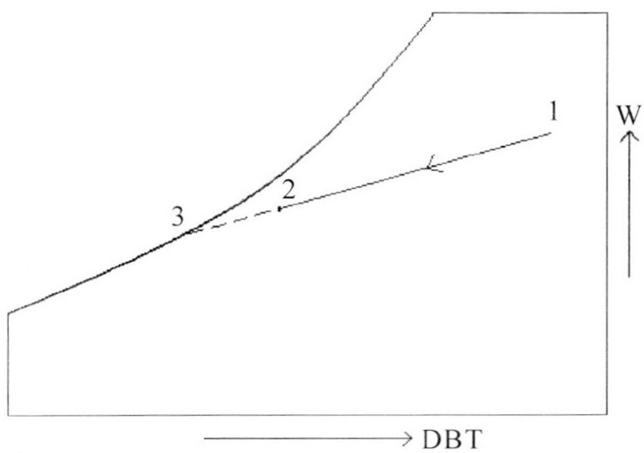

The moist air is passed over a coil in which a cold fluid passes. The process is shown by line 1-2. As moist air passes from state 1 to state 2, its temperature and specific humidity both decrease. Therefore, cooling and dehumidification take place. For the water vapour of moist air to condense, the temperature of the coil must be kept at a temperature below dew point temperature of the moist temperature. Air is entering the coil at state 1 and leaves it at state 2. Under ideal conditions, moist air will leave the coil at state 3. The condensed water is removed at temperature t_2.

From energy balance,

$$m_a h_1 = m_a h_2 + m_w hf_2 + Q_{12}$$
$$Q_{12} = m_a (h_1 - h_2) - m_w hf_2$$

where m_a = Mass of air

h_1 = Enthalpy of moist air at state 1

h_2 = Enthalpy of moist air at state 2

Q_{12} = Total heat removed

m_w = Mass of water vapour condensed

h_{f2} = Enthalpy of water at state 2

Mass of water vapour condensed

$$m_w = m_a (w_1 - w_2)$$

3.6 Heating and Humidification of Moist Air

Heating and dehumidification of moist air is needed in winter air conditioning. Fig. 7 shows the heating and humidification process.

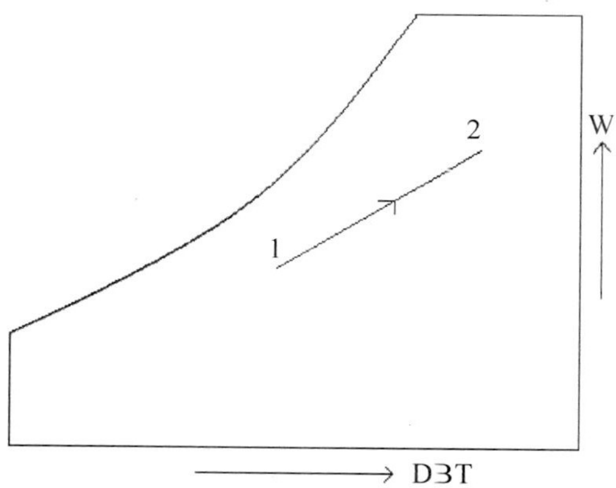

Fig. 7: Heating and humidification process on a psychrometric chart

Generally, steam is injected into the moist air to decrease its specific humidity and temperature. The process is represented by line 1-2 on the psychrometric chart. As moist air changes from state 1 to 2, its specific humidity and dry bulb temperature both increase resulting in heating of humidification.

From energy balance, we obtain

$$m_a h_1 + m_s h_g + Q_{12} = m_a h_2$$

where

m_a = Mass of dry air entering

m_s = Mass of steam introduced into the air

h_1 = Enthalpy of moist air at the entrance

h_2 = Enthalpy of moist air at the exit

h_g = Enthalpy of steam introduced into the moist air

Q_{12} = Heat added to the moist air during the process

$$Q_{12} = m_a \left(h_2 - h_1\right) - m_s h_g$$

From mass balance

$$m_s = m_a \left(W_2 - W_1\right)$$

3.7 Adiabatic Mixing of Two Streams of Moist Air

In summer air conditioning, generally a stream of room air is mixed with a stream of fresh air and the mixture is passed through the air conditioning unit to get the desired state of the mixture. Fig. 8 shows the mixing process on a psychrometric chart.

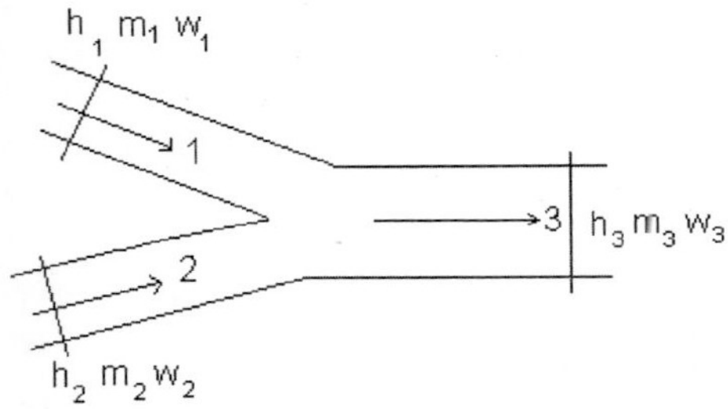

(a) Two Streams of air 1 and 2 and mixed with stream 3

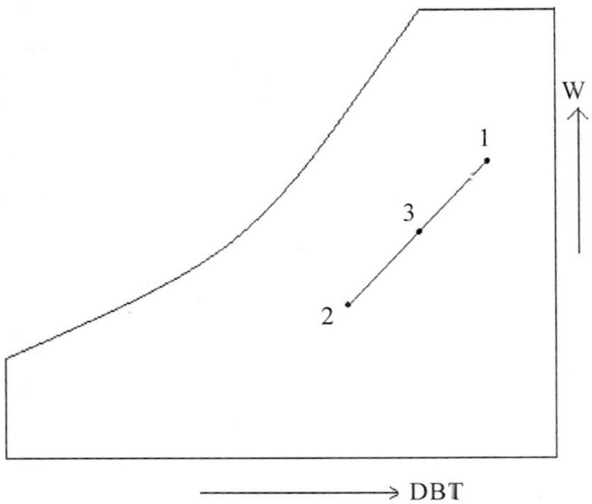

(b) Adiabatic mixing on psychrometric chart

Fig. 8 Adiabatic mixing of two streams of moist air

Moist air stream 1 mixes with the moist air stream 2 and the mixed stream is shown by state 3.

From energy balance, we obtain

$$m_1h_1 + m_2h_2 = m_3h_3$$

From mass balance of moist air

$$m_1 + m_2 = m_3$$

From mass balance water vapour, we get

$$m_1w_1 + m_2w_2 = m_3w_3$$

From above three equations, we get

$$\frac{m_1}{m_2} = \frac{h_3 - h_2}{h_1 - h_3} = \frac{w_3 - w_2}{w_1 - w_3}$$

3.8 Evaporative Cooling

In evaporative cooling water is introduced to the moist atmospheric air. In 'this process, the wet bulb temperature remains constant throughout the process. Evaporative cooling is nothing but cooling and humidification of moist air. The process is shown by line 1-2 on a Psychrometric chart.

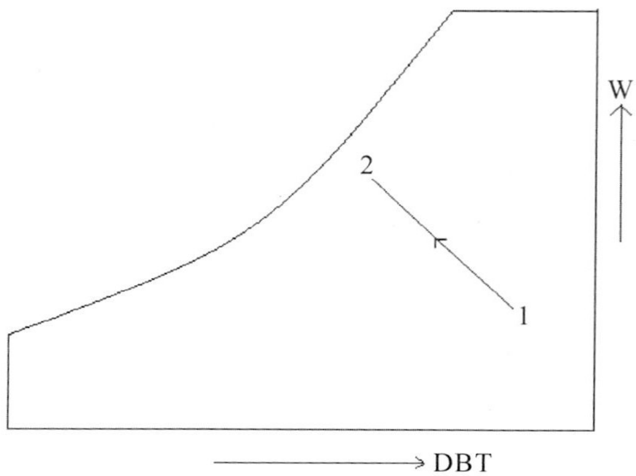

Fig. 9: Evaporation process on a psychrometric chart

But bulb temperature at state 1 and at state 2 are same. From energy balance,

$$mh_1 + m_w h_f = mh_2$$

where h_1 = Enthalpy of moist air at the inlet

h_2 = Enthalpy of moist air at the exit

m_w = Mass of water introduced into the moist air

h_f = Enthalpy of water introduced into the moist air

m = mass of moist air per kg of dry air

$$h_2 = h_1 + m_w / m h_f$$

m_w / m h_f is very small. It can be said that evaporation process is a constant enthalpy process.

4. Different processes on psychrometry

There are different methods of achieving comfort conditions (Heating, Cooling, Humidity Control, Ventilation & Filtration) in domestic and industrial air conditioning system. Some of psychrometirc process involved to change the psychrometric properties of air for obtaining desired comfort condition in a closed space. Air conditioning is the combination of different common psychrometric process are sensible heating or cooling of air, humidification and dehumidification of air, mixing to two air steams of air or combination of various process. Analysing the psychrometric properties and psychrometric processes by using the psychrometric chart is very easy and time saving. In the next few paragraphs we shall see some of the most commonly employed psychrometric processes in and how they are represented on the psychrometric chart.

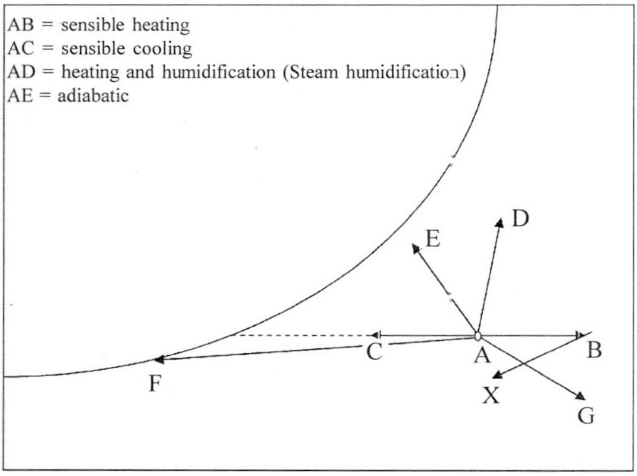

Fig. 10: Different processes on Psychrometric chart

4.1 Adiabatic humidification

Adiabatic humidification is a process whereby water is picked up by air and the heat required to vaporize the added water comes from the sensible heat loss that results from a reduction of the air temperature. The path traced by the temperature and humidity of the air is represented by D in Fig. 10, and this path parallels the wet bulb lines. The wet bulb temperature of the air remains constant, humidity and % RH increase, and T_{db} decreases. This is the process that occurs when air is passed over or through a bed of wet solids in drying or when air is passed through water sprays in a water cooling tower.

4.2 Mixing of air steam

When air streams are mixing the properties of the resulting stream can be determined by mass balances or graphically on the psychrometric chart. Mixture properties will between the properties of two air. The actual condition will vary in proportion to the relative masses of the two air streams.

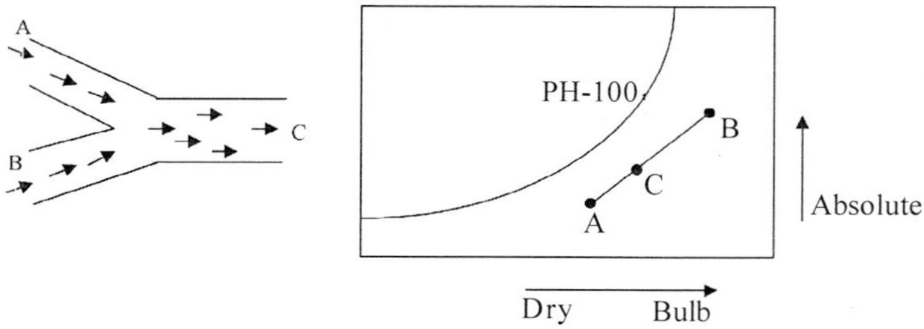

If air steam A, B and C having mass flow rate M_A, M_B and M_C, temp T_A, T_B and T_C, Weight of water SH_A, SH_B and SH_C, total heat contain H_A, H_B and $H_C$and $T_A > T_B$then

1. Heat loss by A= Heat gain by B

$$M_A C_P (T_A - T_C) = M_B C_P (T_C - T_B)$$

$$\frac{M_A}{M_B} = \frac{T_C - T_B}{T_A - T_C}$$

2. *Weight of water in A + weight of water in B = weight of waterin C*

$$SH_A M_A + SH_B M_B = SH_C M_C$$

$$\frac{M_A}{M_B} = \frac{SH_C - SH_B}{SH_A - SH_C}$$

3. Heat contain in A + Heat contain in B=Heat contain in C

$$H_A M_A + H_B M_B = H_C M_C$$

$$\frac{M_A}{M_B} = \frac{H_C - H_B}{H_A - H_C}$$

Problems on Psychrometry

Example 1. Air has a dew point of 40 °C and a RH of 50%. Determine

(a) The absolute humidity

(b) The dry bulb temperature

(c) The wet bulb temperature

Sol:

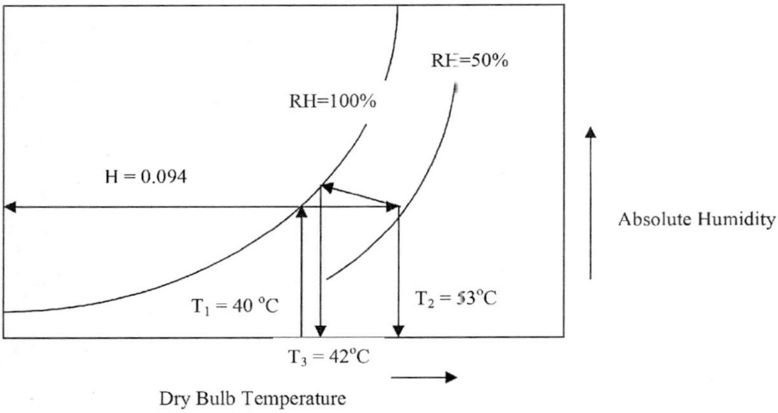

Fig. 11

Following points to be followed to solve the problem

1. Draw a vertical line through T_1 = 40 °C in a psychrometric chart.
2. At the intersection with the line representing 100% RH draw a horizontal line and read the absolute humidity represented by this horizontal line at the abscissa. Therefore, H = 0.094 kg water / kg dry air (Fig. 11).
3. Extend the horizontal line representing the humidity until it intersects the diagonal line representing 50% RH. Draw a vertical line through this intersection Fig.11 and read the dry bulb temperature representing the vertical line at the abscissa. The dry bulb temp (T_2) = 53 °C
4. From the point represented by H = 0.094 and T_2 = 53 °C, draw a line that parallels the wet bulb lines. Project this line to its intersection with the 100% RH line. Draw a vertical line at this intersection, project to the abscissa, and read the wet bulb temperature at the abscissa. T_3 = 42 °C.

Example 2: Air in a house has a wet bulb temperature of 54 °C and a dry bulb temperature of 71 °C.Determine

(1) The humidity,

(2) The dew point,

(3) The percentage of relative humidity.

Sol:-

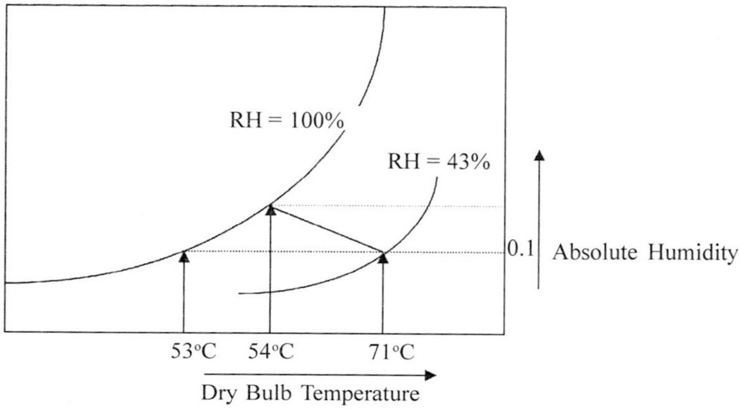

Fig. 12

Following points to be followed to solve the problem

1. Draw vertical line through T_1 = 54 °C and the line representing 100% RH. Draw a line that parallels the wet bulb lines.

2. From the intersection of this drawn Wet bulb line with the vertical line representing T = 71 °C, draw a horizontal line and project to the abscissa. Read the humidity represented by this horizontal line (Fig. 12). H = 0.1 kg water / kg dry air.

3. From the intersection of the horizontal line representing H = 0.1 with 100% RH line, draw a vertical line and connect to the abscissa. T_{dp} = 53 °C.

4. From the intersection of the line representing T = 71 °C and that representing H = 0.1, interpolate between the diagonal lines representing 40% and 50% RH and estimate the % RH. % RH = 43 % (Fig. 12).

Example 3: Atmospheric air at dry bulb temperature of 40°C and 50% RH. Determine –

(1) Humid volume

(2) Enthalpy

(3) Weight of air containing in room measuring 5m x 4m x 3m

(4) Amount of water vapour present

Sol:-Given

DBT=40°C, RH=50%

$$RH = \frac{P_W}{P_{sat.at\,40^o C}}$$

$$0.50 = \frac{P_W}{0.0752}$$

$$P_W = 0.0376$$

$$S_h = \frac{0.662 P_W}{P_a}$$

$$S_h = \frac{0.662 P_W}{1.0332 - P_W}$$

P_W=0.0234 kg of water vapour per kg of dry air

$H.V. = (0.00283 + 0.00456\ S_h)T$

$$H.V. = 0.9193\, m^3/kg\ of dry air$$

$$Density = \frac{1}{H.V.}$$

$$Density = 1.0877\, kg/m^3$$

Mass of air present in 5m x 4m x 3m = density x volume

= 1.0877 x 60

= 65.26 kg

Mass of water vapour = Mass of airx S_h

Mass of water vapour = 65.26 x 0.023

= 1.5 kg

$h = (0.24 + 0.45\ S_h)T + 597\ S_h$

$h = 23.745 \, {}^{kcal}\!/_{kg} \, f dry air$

For dew point temperature of air

$P_{sat(27^0 C)} = 0.0363$

$P_{sat(28^0 C)} = 0.0386$

One degree rise in temperature increases 0.0023 pressures

or

0.0023 pr increases one degree temperature

So DPT= 27.5652 °C

Mixing of air steam

Example 4 – An air steam at 60°C at relative humidity 10% with mass flow rate of 250 kg/min mixed with another air steam having mass floe rate of 100 kg/min at 25°C and relative humidity of 80%. Determine the temperature and relative humidity of mixed air.

Solution-

Given

T_A=60°C, Rh=10%, M_A=250kg/min

T_B=25°C, Rh=80%, M_B=100 kg/min

Heat loss by A= Heat gain by B

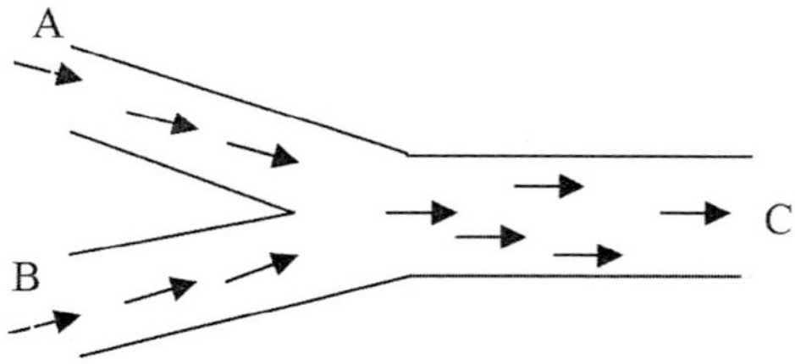

$$M_A C_P (T_A - T_C) = M_B C_P (T_C - T_B$$

$$\frac{M_A}{M_B} = \frac{T_C - T_B}{T_A - T_C}$$

$$\frac{250}{100} = \frac{T_C - 25}{60 - T_C}$$

$$T_c = 50^o C$$

Weight of water in A + weight of water in B = weight of waterin C

$$SH_A M_A + SH_B M_B = SH_C M_C$$

$$\frac{M_A}{M_B} = \frac{SH_C - SH_B}{SH_A - SH_C}$$

$$\frac{250}{100} = \frac{SH_C - 0.01595}{0.01247 - SH_C}$$

$$SH_C = 0.01346$$

$$SH_c = \frac{0.662 P_W}{1.0332 - P_W}$$

$$0.01346 = \frac{0.662 P_W}{1.0332 - P_W}$$

$$P_W = 0.02188$$

$$RH = \frac{P_W}{P_{sat.at 50^o C}}$$

$$RH = \frac{0.02188}{0.1258}$$

$$RH = 17.39\%$$

6

Thermodynamics

Vinkel Kumar Arora

1. Introduction

Thermodynamics is science of energy and may be defined as "*The branch of science which deals with the Interoction between the heat and work is referred as Thermodynamics*".

'Thermo' means heat and 'dynamics' means motion. It deals with energy interaction when work is converted into heat energy or vice versa. Precisely, thermodynamics may be defined as a science which deals with energy transformation from one form to another and dealing with relationship among the properties of the system. In power plants, internal combustion engines, boilers, nozzles, diffusers etc. are the example where heat is converted into work. In compressor, turbine, cooling of engines by water, application of brakes in a vehicle etc. are the examples where work is converted into heat.

2. Approaches of Thermodynamics

2.1 Macroscopic approach

If the state of matter is considered as a whole, without considering the events occurring at the molecular level it referred as macroscopic approach. Lesser computational requirements and fewer properties are required to describe the state of the system in this approach.

2.2 Microscopic approach

If the state of matter is considered to be consisting of large small number of particles viz. molecules and the events occurring at the molecular level are considered for investigation, it is referred as microscopic approach. This approach is also known as statistical thermodynamics

3 Thermodynamic system

The specified part of the universe or the definite quantity of matter or volume in the space upon which investigation is focused is called as thermodynamic system. The open system, closed system, and isolated system are three types of thermodynamic systems.

3.1 Open system

A system which can exchange both heat and mass with the surrounding is called as open system. Any flow process is characterized by the open system. Hot water placed in an open cup, air compressor, boiler, evaporator are the examples of the open system.

3.2 Closed system

A system which can exchange only heat but no mass with the surrounding is called as closed system. Any non-flow process is characterized by the closed system. Hot water placed in a closed steel vessel, cylinder piston assembly compressing a gas adiabatically are the examples of the closed system.

3.3 Isolated system

A system which can exchange neither heat but nor mass with the surrounding is called as isolated system. Hot water placed in a thermos-vaccuum flask, the universe are the examples of an isolated system.

4. Other systems

4.1 Adiabatic system: A system in which only work transfer takes place across the boundary but no heat transfer $(\partial Q = 0)$ takes place is called an adiabatic system.

A phase may be defined as the quantity of the matter having homogeneous chemical composition and physical structure throughout its structure.

4.2 Homogeneous system: A system which comprises of a single phase, eg. moist air

4.3 Heterogeneous system: A system which comprises of two or more phase eg. ice and water

5. Thermodynamic property

The properties of the system that describes the thermal and energy state of the system are termed as thermodynamic properties e.g. internal energy, enthalpy, entropy, temperature etc.

Thermodynamic properties are of two type's namely intensive and extensive property.

5.1 Intensive property

The property which is independent of the mass of the system is termed as Intensive property. Temperature, pressure, density, specific volume etc. are the examples of intensive property.

5.2 Extensive property

The property which is dependent on the mass of the system is termed as Extensive property. Volume, mass, kinetic energy, potential energy etc. are the examples of extensive properties. The extensive property per unit mass is termed as intensive property or specific property.

If we divide total volume, V, by mass, m, we get specific volume. Specific volume is an intensive property.

$$v = \frac{V}{m} = \frac{1}{\rho}$$

where

$v = specific\ volume(m^3/kg), V = total\ volume(m^3),$

$m = mass(kg)\ and\ \rho = density(kg/m^3)$

5.3 Characteristics of a thermodynamic property

1. It has a fixed value at each system state.
2. It is a point function.
3. It is exact differential i.e. if h is a thermodynamic property with dh depicting its differential change, then the integral from state 1 to state 2 will be given as

$$\int_1^2 dh = (h_2 - h_1)$$

State

The point at which the thermodynamic properties attains a constant value is called as the state of the system.

Path of change of state

A series of infinitesimal changes that takes place in a system from initial to final state is termed as the path of change of state.

Process

A completely defined path of change of state from initial to final state is called as a process

The different types of the processes are:

1. Reversible process- which can be reversed without affecting the surrounding.
2. Irreversible process-which cannot be reversed without affecting the surrounding.
3. Flow process-in which the fluid flow takes places (open system).
4. Non-flow process-in which the no fluid flow takes places (closed system).
5. Free expansion process- work done against vacuum.
6. Quasi- static process- almost stationary, assumed to be a reversible process.
7. Cyclic process- in which the initial and final state are same.

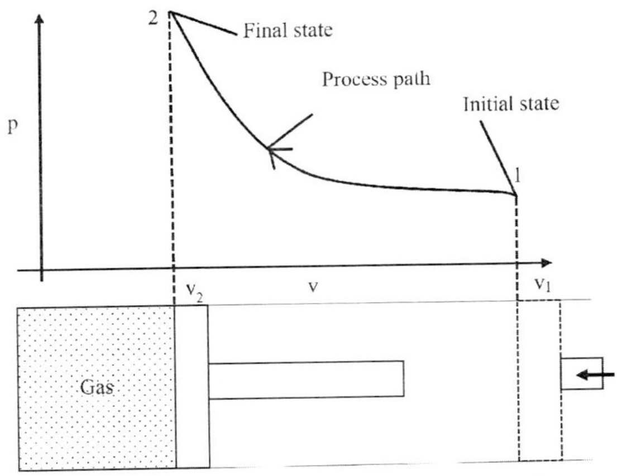

Fig. 6.1: P-V diagram for compression of gas

6. Thermodynamic Equilibrium

A system is said to be in thermodynamic equilibrium if the properties of system are same at all the points and do not change with time. The important properties are temperature, pressure and chemical composition. The other types of the equilibrium and their conditions are:

6.1 Thermal equilibrium: A system is said to be in thermal equilibrium, if the temperature of the system is same at all the points and does vary with time.

6.2 Mechanical equilibrium: A system is said to be in mechanical equilibrium, if there exist no unbalanced force within the system and between the system and the surrounding.

6.3 Chemical equilibrium: A system is said to be in chemical equilibrium, if the chemical concentration of the system is same at all the points of the system and does not change with time.

If the three types of equilibriums i.e. thermal, mechanical and chemical equilibrium are achieved by the system then it is under the thermodynamic equilibrium. When system goes from one state to another, and deviation from equilibrium is very small or infinitesimal, it is said to be in quasi-equilibrium.

7. Temperature, Equity of Temperature and Zeroth Law of Thermodynamics

7.1 Temperature

Temperature is a thermodynamic property which will be obtained from 2nd law of thermodynamics. As regards sensory perception, the temperature may be considered as a measure of hotness and coldness. At the same time, it is not reliable. If a wooden piece and metallic piece are kept in the sun, they will have same temperature. If we touch both pieces with our fingers, it will be realized that the metallic piece is hotter than the wooden piece. So, we cannot rely upon sensory perception. From microscopic point of view temperature is proportional to stored energy of molecules or average kinetic energy of molecules. Temperature scale in history

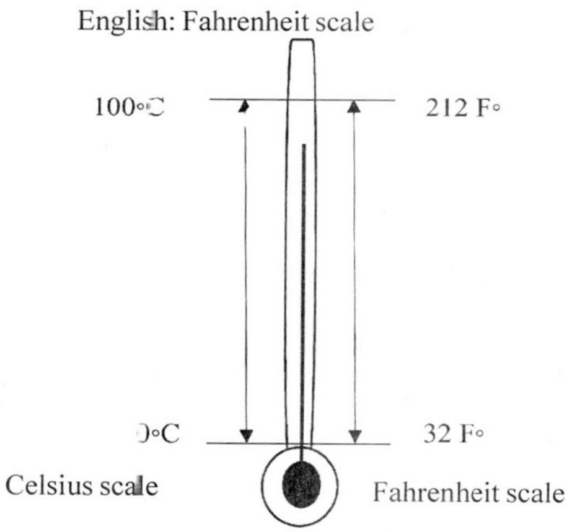

Fig. 6.2: Temperature scale

7.2 Zeroth Law of Thermodynamics (ZLOT)

"Zeroth law of thermodynamics states that if a body A is in thermal equilibrium with body B and separately in thermal equilibrium with body C than body B and C are in thermal equilibrium with each other". Fig. 1.16 shows three bodies A, B and C which are brought in contact with each other. When thermal equilibrium is attained, both the systems have the equity of temperature or are at the same temperatures.

Zeroth law of thermodynamics can be utilized for the measurement of temperature. If we consider thermometer a third system, and if the thermometer is in thermal equilibrium with another system, the temperature of the system will be equal to the temperature given by the thermometer.

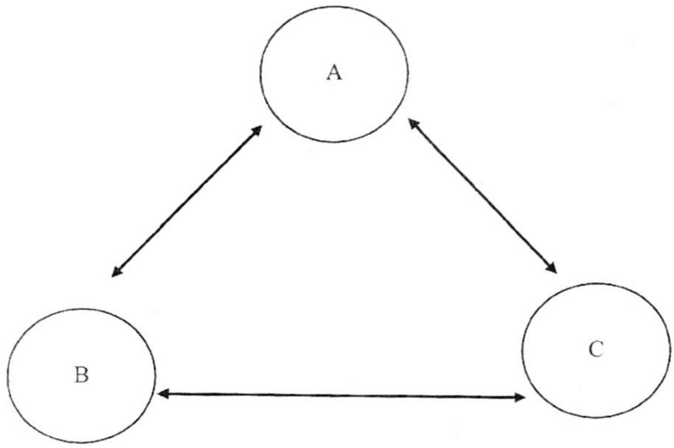

Fig. 6.3: Three systems in thermal equilibrium

8. Thermodynamic work and Heat

8.1 Work

In mechanics, it is defined as the product of force and distance moved in the direction of applied force. This definition can be used to find the work to lift a weight, to stretch a spring or a wire. But in thermodynamics "*Work is said to be done by a system on the surroundings if sole effect external to the system could be the raising of a weight*".

8..1.1 Salient feature of work

1. Work is energy in transit and takes place across the system boundary when the process is executed.

2. Work is a boundary phenomenon or energy in transition.
3. Work is a path function and hence inexact differential.
4. Work is represented by area under the pressure volume plot.
5. Work has unit as Nm or Joule.
6. Work rate is called as power and represented as

$$\dot{dW} = \frac{dW}{dt} = P$$

Unit for work rate (watt)

$$\dot{dW} = \frac{dW}{dt} = \frac{Nm}{s} = \frac{J}{s} = W$$

8.1.2 Sign and convention for work

1. The work done by the system on the surrounding is arbitrary considered as positive work. In all the expansion processes, work is done by the system hence positive.
2. The work done by the surrounding on the system is arbitrary considered as negative work. In all the compression processes, work is done by the system hence negative.
3. During a cyclic process, when the system does positive work and surrounding does as equal amount of negative work then the work net is zero.

8.2 Heat

The energy flow which takes place from system to the surrounding or vice-versa by virtue of temperature difference is referred as heat. Naturally, heat flows takes place from higher temperature to lower temperature. In thermal equilibrium no heat energy interaction takes place.

8.2.1 Sign convention for heat

1. The heat supplied by the system to the surrounding is taken as negative (-ve) i.e. heat transferred is negative when surrounding is at lower temperature.
2. The heat supplied to the system from the surrounding is taken as positive (+ve) i.e. heat transferred is positive when system is at lower temperature.

3. Heat transfer, Q, is not a property and not an exact differential. Therefore, it is denoted by äQ and not by dQ $\int_1^2 \delta Q$ is denoted by Q_{12} and not by $Q_1 - Q_2$. So to conclude both work and heat are transient phenomena, boundary phenomena and path functions.
4. Area under T-S diagram represent heat.
5. The heat rate is represented as:

$$\dot{dQ} = \frac{dQ}{dt}$$

Unit for work rate (watt)

$$\dot{dQ} = \frac{dQ}{dt} = \frac{J}{s} = W$$

8.2.2 Sensible and latent heat

8.2.2.1 Sensible heat: The heat that can be sensed by temperature change but the phase remains the same.

$$dQ = mC_p dT$$

Where; m =mass (kg), Cp= Specific heat at constant pressure (kJ/kg K), dT= Temperature difference (K).

Specific heat at constant pressure (C_p): At constant pressure, the amount of heat required to raise the temperature of the unit mass of the substance through unit degree in temperature, is called as specific heat at constant pressure.

Specific heat at constant volume (C_v): At constant volume, the amount of heat required to raise the temperature of the unit mass of the substance through unit degree in temperature, is called as specific heat at constant pressure.

Heat transfer and Work done for various quasi-static processes

The heat transfer and work done for various quasi-static processes in a tabular form considering the law governing the process, heat transfer and work done by considering different values of n in the equation:

$$PV^n = C$$

Table 6.1 Heat transfer and Work done for various quasi-static processes

Sr. No	Process	Governing equation	Value of n in $PV^n=C$	Heat transfer (dQ)	Work done (dW)
1.	Isobaric process	p = C	n = 0	$dQ = mC_p dT$	$dW = \int_1^2 p\,dv =$
		$\frac{v_1}{T_1} = \frac{v_2}{T_2}$			$p(v_2 - v_1)$
2.	Isochoric process	v = C	$n = \infty$	$dQ = mC_v dT$	0
		$\frac{p_1}{T_1} = \frac{p_2}{T_2}$			
3.	Isothermal process	T = C	n = 1	$dQ = p_1 v_1 \ln \frac{v_2}{v_1}$	$dW = p_1 v_1 \ln \frac{v_2}{v_1}$
		$p_1 v_1 = p_2 v_2 = c$			

4.	Polytropic process	$pv^n = c$	n = n	$Q = \frac{r-n}{n-n}(\frac{R(T1-T2}{n-1}$	$dW = \frac{p_1 v_1 - p_2 v_2}{n-1}$
		$p_1 v_1^n = p_2 v_2^n = c$		$= \frac{r-n}{n-1}(w)$	$= \frac{R(T_1 - T_2)}{n-1}$
5.	Adiabatic process	$pv^\gamma = c$	n = γ	0	$dW = \frac{p_1 v_1 - p_2 v_2}{\gamma - 1}$
		$p_1 v_1^\gamma = p_2 v_2^\gamma = c$			

4. First Law of Thermodynamics (FLOT) for a cyclic process

Joule conducted number of experiments involving different types of heat and work interaction using a paddle wheel and insulated chamber and he observed that work was propotional to the heat energy. He expresses this interaction as

$$Q \alpha\ W \text{ or } Q = \frac{W}{J}$$

Where J is the Joule's equivalent or mechanical equivalent of heat.

He stated that "First law of thermodynamics for a cycle states that the net heat interaction is equal to net work interaction during a cycle". He proposed an equation

$$\oint dQ_{net} = \oint dW_{net}$$

4.1 Energy

Energy is defined as the capacity to do work. Energy is an abstract quantity and cannot be perceived with eyes. First law of thermodynamics is law of conservation of energy. Generally it states that "*Energy can neither be created nor destroyed but it can be converted from one form to another form keep the total energy constant*". In thermodynamics, kinetic energy, potential energy, work, heat, internal energy etc. are dealt with. Potential energy is the energy possessed by a system by virtue of its elevation from a given datum.

It is given by:

$$P.E = mgz$$

Where m is mass in kg, g is acceleration due to gravity m/sec^2 and z elevation of the system from the datum in m. The kinetic energy is the energy possessed by a system by virtue of its velocity.

It is given by:

$$K.E = \frac{1}{2} mv^2$$

Where m is the mas in kg and v is the velocity in kg/sec.

5. First Law of Thermodynamics for a closed system executing a process

Fig.6.4. shows a system undergoing two cycles 1-a-2-b-1 and 1-a-2-c-1

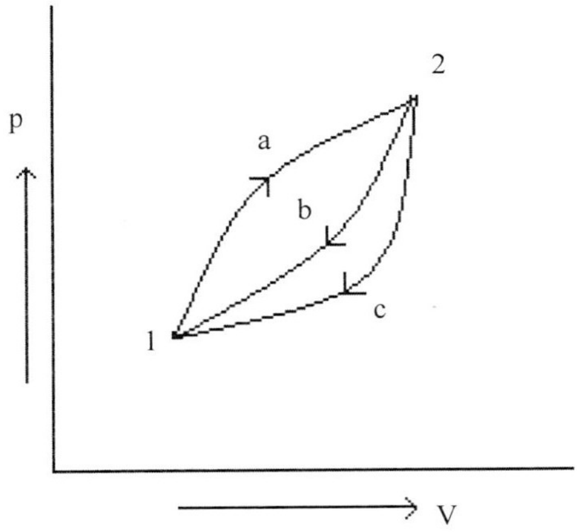

Fig. 6.4 P-V diagram for two cycles

Cycle 1-a-2-b-1 consists of two processes a and b. Applying first law to the cycle

1-a-2-b-1, we get:

$$\int_{a1}^{a2} \delta Q + \int_{2b}^{1b} \delta Q = \int_{a1}^{a2} \delta W + \int_{2b}^{1b} \delta W \qquad (1.9)$$

Again applying first law of thermodynamics to the cycle 1-a-2-c-1, we obtain

$$\int_{a1}^{a2} \delta Q + \int_{2c}^{1c} \delta Q = \int_{a1}^{a2} \delta W + \int_{2c}^{1c} \delta W \qquad (1.10)$$

Subtracting Eq. (1.10) from Eq.(1.9), we get

$$\int_{2b}^{1b} \delta Q - \int_{2c}^{1c} \delta Q = \int_{2b}^{1b} \delta W - \int_{2c}^{1c} \delta W$$

$$\int_{2b}^{1b} (\delta Q - \delta W) = \int_{2c}^{1c} (\delta Q - \delta W)$$

It is observed that ($\delta Q - \delta W$) is same whether we follow path b or c. If we take other path d or e, same conclusions will be drawn. So it can be concluded that the quantity $\int_1^2 \delta Q - \delta W$ is independent of path followed. It only depends on two states, hence a property. $\delta Q - \delta W$ is denoted by dE. E is called total stored energy.

$$\delta Q - \delta W = dE$$

Characteristics of total internal Energy and Stored Energy

1. Energy is a property derived from first law of thermodynamics when applied to a closed system executing a process.
2. E represents total energy i.e. sum of microscopic energy and macroscopic energy, which a system possess at a given state. Macroscopic types include kinetic and potential energy. Microscopic energy includes Kinetic energy, chemical energy, nuclear energy, vibrational energy and electronic energy of molecules. This microscopic energy is called internal energy. So, total energy is sum of kinetic energy, potential energy, work of all types, heat and internal energy.

$$E_{total} = E_{micro} + E_{macro}$$

$$E = U + mgz + \frac{1}{2}mv^2 + Flow\ work$$

For a Non-flow process (closed system)

$$E_{total} = E_{micro}$$

$$E = U$$

For a flow process (open system)

$$E_{total} = E_{micro} + E_{macro}$$

$$E = U + mgz + \frac{1}{2}mv^2 + Flow\ work$$

3. For the universe or for an isolated system, $\delta Q = 0$ and $\delta W = 0$.

11. Perpetual Motion Machine of First Kind, PMMFK

A system or machine which produces work continuously without involving any expenditure of energy is called perpetual motion machine of first kind. Such machine is not feasible as it violates the first law of thermodynamics.

12. Enthalpy

Enthalpy may be expressed as the sum of internal energy and product of pressure p and volume V.

$$H = U + pV$$

where U is specific internal energy, p is the pressure and V is the total volume.

Characteristics of enthalpy

1. H is an extensive property and h is an intensive property.
2. If heat is supplied at a constant pressure process from initial to final state , change in enthalpy is equal to heat supplied.
3. At 0°C, enthalpy is taken to be zero.
4. The enthalpy remains constant in throttling process.

6. FLOT for an open system (flow Process)

In majority of the application a certain mass of fluid having certain thermodynamics properties enter the control volume (cv) and leaves the control volume. For simplicity the flow processes are considered as steady flow. Assumption for steady flow:

1. The mass entering the control volume is equal to the mass leaving the control volume.
2. The energy entering the control volume is equal to the energy leaving the control volume.
3. The thermodynamic properties don't vary with respect to time.

If the above conditions are not met than the flow is called as transient flow and depending upon the variation in thermodynamic property in different direction it can be 1D- transient, 2D- transient and 3D- transient flow.

Steam turbine and petrol engine are the examples of the open system. In a steam turbine, steam enters at high pressure and leaves at low pressure.

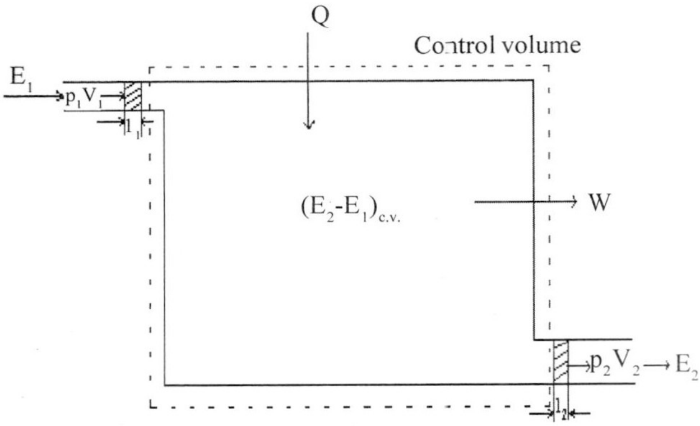

Fig. 6.5: First Law for an Open System

Steady flow energy equation, Assuming

$m_1 = m_2 = m,\ E1 = E2$

$$dQ + p_1V_1 + m\left[gZ_1 + \frac{1}{2}C_1^2 + u1\right] = dW + p_2V_2 + m\left[gZ_2 + \frac{1}{2}C_2^2 + u2\right]$$

Where

dQ= Heat supplied or rejected by the system

dW= Work done by the system or on the system

p_1V_1, p_2V_2= flow work at inlet and oulet respectively

u_1,u_2 = specific internal energy at inlet and outlet respectively

C_1, C_2 = Inlet and outlet flow velocity respectively

Z_1, Z_2= Inlet and outlet height from the datum surface respectively

Steady flow energy equation, Assuming 1kg mass:

$$dQ + p_1V_1 + \left[gZ_1 + \frac{1}{2}C_1^2 + u1\right] = dW + p_2V_2 + \left[gZ_2 + \frac{1}{2}C_2^2 + u2\right]$$

$$dQ + h_1 + gZ_1 + \frac{1}{2}C_1^2 = dW + h_2 + gZ_2 + \frac{1}{2}C_2^2$$

Normally the kinetic and potential energy are so small that they are neglected so the Steady flow energy equation is represented in terms of enthalpy only as:

$$dQ + h_1 = dW + h_2$$

Table 6.2: SFEE for various flow devices

Flow device	Process taking place	Assumptions	Steady flow energy equation	Outcome
Steam Turbine	Expansion	Kinetic and potential energy are so small that they are neglected. $C_1=C_2$; $Z_1=Z_2$ No heat is lost during the flow i.e. dQ=0	$dW = h_1 - h_2$ Positive work	Enthalpy decreases during the flow
Compressor	Compression	Kinetic and potential energy are so small that they are neglected. $C_1=C_2$; $Z_1=Z_2$ No heat is lost during the flow i.e. dQ=0	$dW = h_2 - h_1$ Negative work	Enthalpy increases during the flow
Boiler or steam generator	Heat supplied	Kinetic and potential energy are so small that they are neglected. $C_1=C_2$; $Z_1=Z_2$ No work is done during the flow i.e. dW=0	$dQ = h_2 - h_1$	Enthalpy increases during the flow
Steam Condenser	Heat rejected	Kinetic and potential energy are so small that they are neglected. $C_1=C_2$; $Z_1=Z_2$ No work is done during the flow i.e. dW=0	$dQ = h_1 - h_2$	Enthalpy decreases during the flow
Nozzle and	Pressure drop velocity	Potential energy are so small that they	$\frac{C_2^2 - C_1^2}{2} = h_1 - h_2$	Enthalpy increases

diffusor	increases	are neglected. $Z_1=Z_2$ No work is done during the flow i.e. dW=0 No heat is lost during the flow i.e. dQ=0		during the flow
Pump	Work is done on the system to lift water from lower to higher level	Potential energy are so small that they are neglected. $Z_1=Z_2$ No work is done during the flow i.e. dW=0 No heat is lost during the flow i.e. d Q=0	$dW = (h_2 - h_1) + g(Z_2 - Z_1)$ Negative work	Enthalpy increases during the flow

14. Application of First Law of Thermodynamics applied to a control volume or Applications of Steady Flow Equation

Example 1.

5 kg of air is contained in a container. The temperature of air is raised from 30^{0}C to 50^{0}C by heating the air. The heat supplied to the air is 100 kJ. Specific heat of air at constant volume is 0.718 kJ/kg^{0}C. Find the work done by the air and its change in internal energy.

Solution

Change in internal energy of air is given by $= m\, c_v\, (t_2 - t_1)$

$$= 5 \times 0.718\, (50 - 30) = 71.8 \text{ KJ}$$

From first law of thermodynamics,

$$Q_{12} = U_2 - U_1 + W_{12}$$

$$100 = 71.8 + W_{12}$$

$$W_{12} = 28.2 \text{ kJ}$$

So work is done by the air.

Example 2

A tank containing gas is stirred by a rotating paddle. The power given to the paddle is 2.5 kW. Find change in internal energy of the gas when (a) Tank is insulated and (b) heat lost from the tank to the surroundings is 1 kW.

Solution

(a) From first law of thermodynamics for a process,

$$Q_{12} = U_2 - U_1 + W_{12}$$

$$0 = U_2 - U_1 + (-2.5)$$

or $U_2 - U_1 = +2.5$ kW

So, temperature of the gas will increase.

(b) If the tank is not insulated, $Q_{12} = -1$ kW

$$Q_{12} = U_2 - U_1 + W_{12}$$

$$-1 = U_2 - U_1 - 2.5$$

$$U_2 - U_1 = 1.5 \text{ kW.}$$

Temperature of gas will increase but slightly less than the previous case.

14.1 Limitations of First Law of Thermodynamics

1. First law of thermodynamics don't provide any information regarding the possibility of heat flow from lower temperature to higher temperature.
2. According to first law of thermodynamics work can be converted into heat and heat into work. But heat is a low grade of energy and work is a high grade of energy and complete conversion of low grade energy into high grade energy is not possible.

15. Heat Engine, Heat Pump and Refrigerator

A heat engine is a device which while working in a cycle converts the heat into work. It exchanges the heat with two thermal energy reservoir i.e. high temperature source and low temperature sink. Fig. 6.6 shows a heat engine.

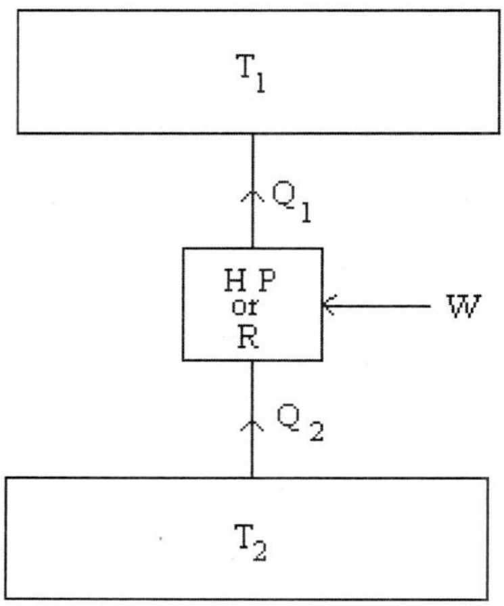

Fig. 6.6: Heat Engine

The engine receives heat Q_1 from the source and rejects heat Q_2 to the sink. The work done by the engine is W. The efficiency is given by:

$$\eta = \frac{W}{Q_1}$$

From 1st law $W = Q_1 - Q_2$, so

$$\eta = \frac{Q_1 - Q_2}{Q_1}$$

A heat pump is a system which operates in a cycle and transfers heat from low temperature system to high temperature system when work is done on it. Fig. 6.7 represents a heat pump or a refrigerator.

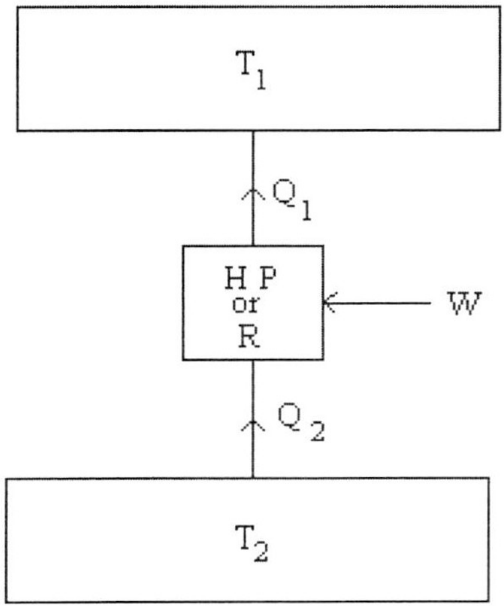

Fig. 6.7: Heat pump

Heat pump extracts heat Q_2 from a low temperature system and transfer heat Q_1 to the higher temperature system. Work is also done on the system. For heat pump, coefficient of performance, COP is defined as the ratio of heat transferred to the high temperature system to work done on it or

$$C.O.P = \frac{Q_1}{W}$$

From first law of thermodynamics,

$W = Q_1 - Q_2$, so

$$COP_p = \frac{Q_1}{Q_2 - Q_1}$$

Refrigerator extracts heat Q_2 from a low temperature system and transfer heat Q_1 to the higher temperature system. Work is also done on the system. For heat pump, coefficient of performance, COP is defined as the ratio of heat extracted from a low temperature sink to work done on it or for refrigerator, coefficient of performance is defined as the ratio of heat extraction from cold system or sink to work done on it.

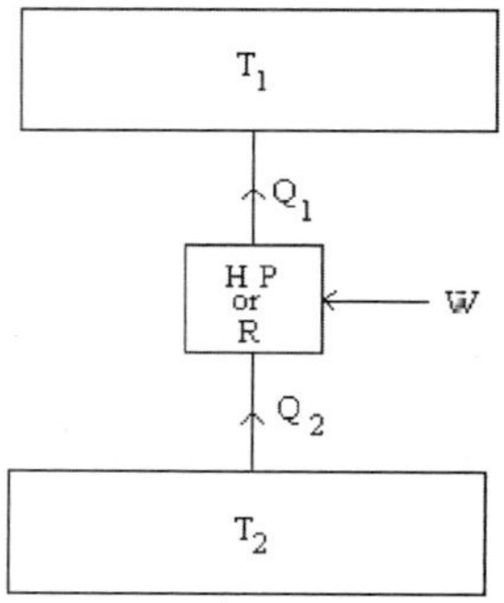

Fig. 6.8: Refrigerator

$$C.O.P = \frac{Q_2}{W}$$

$$COP_r = \frac{Q_2}{Q_2 - Q_1}$$

$$COP_p = 1 + COP_r$$

15.1 Statements of Second Law of Thermodynamics

All low gradecannot be converted into high grade energy. This observation is included in Kelvin-Planck statement.

15.1.1. Kelvin-Planck Statement

It is impossible for a heat engine working in a cycle and produces work net by exchanging the heat with a single thermal energy reservoir. It state that no heat engine can be 100% efficient, so heat has to be rejected to sink.

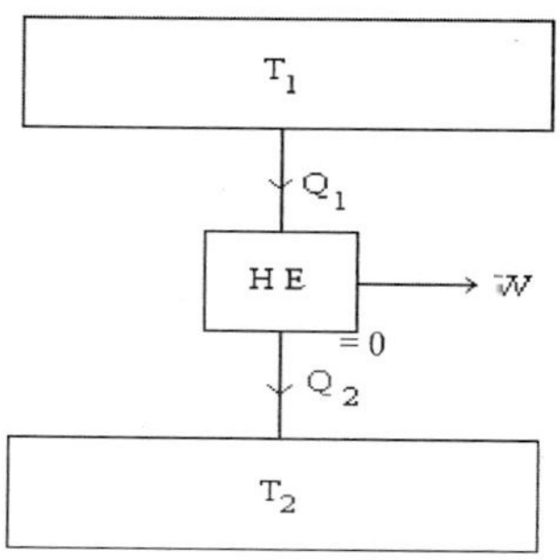

Fig. 6.9: PMM of Second kind

15.1.2. Claucius Statement

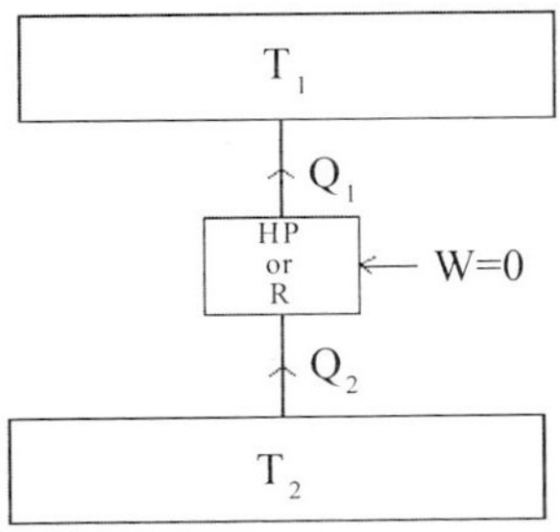

Fig. 6.10: PMM of Second kind

It states that it is impossible to construct a device which operates in a cycle and produces no other effect than transferring heat from a lower temperature system to higher temperature system.

15.2 Perpetual Motion Machine of Second Kind, PMMSK

A system which operates in a cycle and converts all heat into work and exchanges heat with a single reservoir is called perpetual motion machine of second kind, PMMSK. Such system cannot exist as it is violation of Kelvin-Planck's statement.

15.3. Reversible Process and Carnot Engine Efficiency

Reversible process is a process which once having taken place, can be reversed and doing so leaves no change in either the system or the surroundings. A process which is not reversible is called irreversible process. If a wire is pulled down by a force, its length increases. If the force is removed it regains its original length and state also. This process is a reversible.

Friction, heat transfer and free expansion make a process irreversible. A reversible process is also quasistatic process, but a quasistatic process may not be reversible.

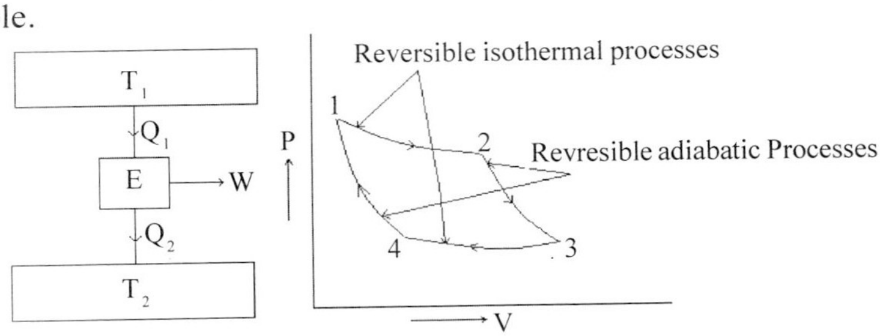

(a) Carnot Engine (b) p-V diagram of a Carnot Engine

Fig. 6.11. Carnot cycle and heat engine

Efficiency of the Carnot cycle is the ratio of work done to heat supplied.

$$\eta = \frac{W}{Q_1}$$

From 1st law of thermodynamics,

$W = Q_1 - Q_2$ so

$$\eta = \frac{Q_1 - Q_2}{Q_1} \quad (1.29)$$

During isothermal process 1-2, heat supplied Q_1 is given by:

$Q_1 = W_{1\,2} = p_1V_1 \ln V_2 / V_1$

Heat lost in the condenser, Q_2, during the isothermal process is given by:

$Q_2 = W_{3\,4} = p_3V_3 \ln V_3 / V_4$

Putting values of Q_1 and Q_2 in Eq. (1.29) and simplifying, efficiency is given by:

$$\eta = \frac{T_1 - T_2}{T_1} \quad (1.30)$$

Example 3

The condenser of a domestic refrigerator rejects 0.5 kJ/s heat to the room air. Heat extracted from the contents of the refrigerator is 0.15 kJ/s. What is power supplied to the motor of the compressor?

Solution

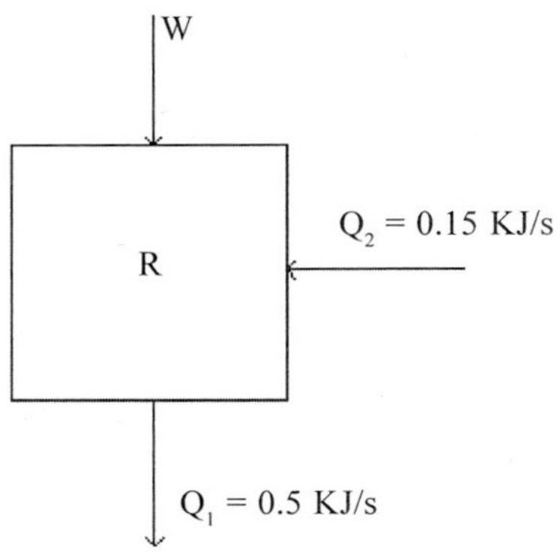

Fig. 2: Energy balance of example 3

Applying energy balance to the refrigerator as shown in Fig. 2, we obtain

$W + Q_2 = Q_1$

or $W = Q_1 - Q_2$

$= 0.5 - 0.15 = 0.35$ kW

16. Claucius theorem

From the Carnot's expression for efficiency of a heat engine and equality of temperature it may be mathematically written as:

$$\eta = \frac{Q_1 - Q_2}{Q_1} = \frac{T_1 - T_2}{T_1}$$

$$\frac{Q_1}{T_1} - \frac{Q_2}{T_2} = 0$$

Considering heat entering a system positive and heat leaving a system negative above equation can be re-written as:

$$\frac{Q_1}{T_1} + \frac{Q_2}{T_2} = 0$$

$$\Sigma \frac{dQ}{T} = 0$$

- $\frac{dQ}{T} = 0$

Cyclic integral of is zero, therefore is a property and denoted by, where is the change in entropy. This equation is referred as Claucius theorem.

17. Clausius inequality

- $\frac{dQ}{T} \leq 0$ it is known as clausius inequality and provides the criteria for

reversibility of a cycle.

The compiled list of equation for a process and cycle whether reversible or irreversible.

Table 6.1 Clausius theorem & inequality equations

Process	Equation applicable	
Reversible	$\frac{dQ}{T} = dS$	
Irreversible	$\frac{dQ}{T} \leq dS$	
Cycle	**Equation applicable**	
Reversible	• $\frac{dQ}{T} = 0$	Clausius theorem
Irreversible	• $\frac{dQ}{T} \leq 0$	Clausius inequality
	• $\frac{dQ}{T} > 0$	Not possible

18. Entropy

It may be defined as the property which shows the possibility of conversion of heat into work.

$$dQ = T\Delta S$$

$$\frac{dQ}{T} = \Delta S$$

When heat is supplied at higher temperature the changes in entropy is small and higher possibility of conversion of heat into work is there while when heat is supplied at lower temperature the changes in entropy is large and lower possibility of conversion of heat into work. Hence this property is a state depended hence is a point function.

18.1 Characteristics of entropy

1. S is an extensive property and s is an intensive property.
2. At 0^0 K, entropy is zero.
3. The entropy remains constant in a reversible adiabatic process or isentropic process.

4. For sensible heating at constant pressure, where temperature changes from T_1 to T_2 the entropy change can be given as:

$$dS = \frac{dQ}{T} = \int_{T1}^{T2} \frac{mC_p dT}{T} = mC_p \ln \frac{T_2}{T_1}$$

5. For latent heating or cooling the entropy changes can be determined as:

$$ds = \frac{LH_{melting/vapourization}}{T}$$

18.2 Principle of entropy

From the table above the entropy change for an irreversible process can be mathematically given as:

$$\frac{dQ}{T} \leq dS$$

For an isolated system neither heat nor mass exchanges takes place with the surrounding

$$dQ = 0$$

Hence

$$0 \leq dS$$

For an irreversible process executed in an isolated system the entropy change will always be greater than zero

$$0 \leq S_2 - S_1$$

or in other word the "*entropy of an isolated system executing an irreversible process will always increase*".

Universe is considered to be an isolated system and the processes executed in the universe are irreversible processes.

$$0 \leq dS_{univ}$$

The entropy of the universe will always increase, this is the principle of entropy. The entropy of the universe always increases and become maximum at the equilibrium state.

$$dS_{\max} \ @ \ when; \frac{dS}{dx} = 0$$

19. Combining first and second law equations

Table 6.2: Equation for FLOT & SLOT combined

Equation	Applicable for	Governing law
$dQ - dW = \Delta E$	Process reversible or irreversible and Flow or Non-flow system	First law of thermod ynamics
$dQ - dW = \Delta U$	Process reversible or irreversible and Non-flow system	First law of thermodynamics
$dQ - \int pdV = \Delta U$	Process reversible or irreversible and where displacement work is done	First law of thermodynamics
$dQ = TdS$	Reversible process	Second law of thermodynamics
$TdS - \int pdV = \Delta U$	Process reversible or irreversible and Non-flow system and where displacement work is done	Combined equation
$TdS = dH - Vdp$	Process reversible or irreversible and flow & Non-flow system. As no path function is present.	Combined equation

20. Third Law of Thermodynamics

$$S = K.lnW$$

where K= Boltzmans constant

when W is equal to unity than S=0. It occurs only at 0K. This equation is known as Nernst-Simon equation for third law of thermodynamics. It states that entropy of a pure substance is zero at the absolute zero temperature. It helps to calculate absolute value of entropy.

7

Refrigeration

Vinkel Kumar Arora

1. Introduction

Refrigeration may be defined as the process of achieving and maintaining a temperature below that of the surroundings. The aim being to cool some product or space to the required temperature. Refrigerated system or a room system is maintained at a temperature which is lower than the atmospheric temperature. Due to higher temperature of atmosphere, heat enters the system. To keep the temperature of the system constant, heat must be removed continuously from the system. Heat removed must be equal to the heat transferred to the system. One of the most important applications of refrigeration has been the preservation of perishable food products by storing them at low temperatures. Refrigeration systems are also used extensively for providing thermal comfort to human beings by means of air conditioning.

Air conditioning refers to the treatment of air, so as to simultaneously control of

- Temperature (Heating/cooling)
- Moisture content (Humidification/dehumidification)
- Cleanliness (Odor, filtration, ventilation)
- Circulation (Air distribution)

as required by occupants, a process, or products in the space.

2. Applications

There are enormous applications of refrigeration and air conditioning in industries and in our daily life. Refrigeration is essential for the storage and transportation of vegetables, fruits, dairy products, fruit juices and beverages. Air conditioning units are installed in cinema halls, hospitals, hotels, industries, offices and even class rooms. Industrial air conditioning is done in textile industries to reduce wastage, printing plants to improve quality of printing, photographic product to excellent prints and in computer rooms to function favorably. Refrigeration has wide applications to get reliable results from instruments, manufacturing and to preserve human tissues. Refrigeration is also utilized to heat air in blast furnaces, to produce rocket fuels to treat metals and to manufacture synthetic rubber. The most common application are as below:

1) Industrial Applications

- Processing of food products.
- Processing of farm crops.
- Processing of textiles, printing work, photographic materials
- Cooling of concrete for dams.
- Treatment of air for the blast furnace.
- Processing of tobacco, petroleum and other chemical products.

2) Preservation of Perishable Goods

- Manufacturing of ice.
- Freezing or chilling, storage and transportation of food stuffs including beverages, meat, poultry products, dairy products, fish, fruits, vegetables, fruit juices, etc.
- Preservation of photographic films, archeological documents.

3) Providing comfortable environment

- Industrial air-conditioning.
- Comfort air-conditioning of hospitals, residences, hotels, restaurants, theatres, offices, etc.

3. Rating of Refrigerating Machine

The rating of refrigerating machines is expressed in terms of refrigerating effect. The refrigerating effect is given in refrigerating ton and denoted by TR or simply by ton and denoted by T. Therefore, unit of refrigeration is ton.

"One ton is defined as the heat extracted from 1 US ton (2000 lbs) of water at 0^0C to freeze to ice at the same temperature in 24 hours of fusion".

Latent heat of ice is 335 KJ/Kg and 1 lb = 0.4536 Kg

$$1\,T = \frac{2000 \times 0.4536}{24 \times 60} \times 335$$

$$1\,T \approx 210 \text{ kJ/min} = \frac{210}{60} = 3.5\,kW$$

$$1\,T = \frac{210}{4.187} = 50\ K\,Cal/\min$$

Because 1 Cal $= 4.187$ J

$$1\,T = \frac{2000 \times 144}{24 \times 60} = 200\ Btu/\min$$

Because latent heat of fusion of ice is 144 Btu/lb.

4. Simple Vapour Compression Refrigeration System

There are many types of refrigeration system being used these days. The commonly employed systems are air refrigerating system, vapour compression system, vapour absorption system; ejector compression system and cascade refrigerating system. Out of all systems, the vapour compression system is the most popular and commonly used in industries and domestic applications.The four important component of a vapour compression refrigeration system are the compressor, evaporator, condenser and an expansion device.

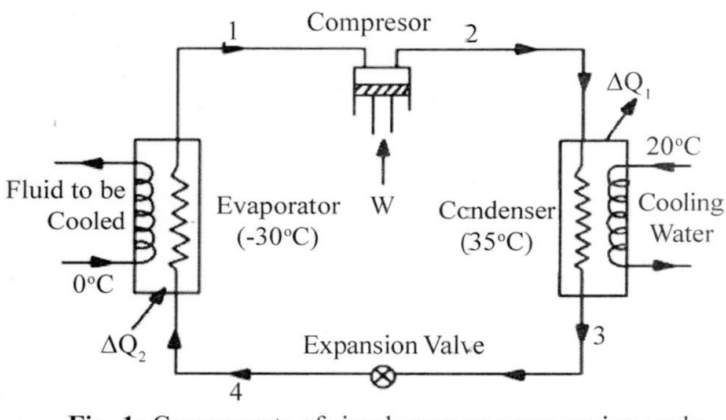

Fig. 1: Components of simple vapour compression cycle

The details of the various processes of the vapour compression refrigeration system are:

Process 1-2: Isentropic compression of saturated vapour in compressor

Process 2-3: Isobaric heat rejection in condenser

Process 3-4: Isenthalpic expansion of saturated liquid in expansion device

Process 4-1: Isobaric heat extraction in the evaporator

Low temperature and pressure refrigerant in vapour at state 1 enters the compressor. The vapour is compressed at constant entropy in the compressor. The superheated vapour enters the condenser at state 2. In the condenser, the vapour is cooled to liquid state at a constant pressure. High pressure liquid at state 3 is throttled through a capillary tube or expansion valve to a lower pressure at state 4. The vapour at state 4 enters the evaporator and converted into dry vapour at state 1 at constant pressure.

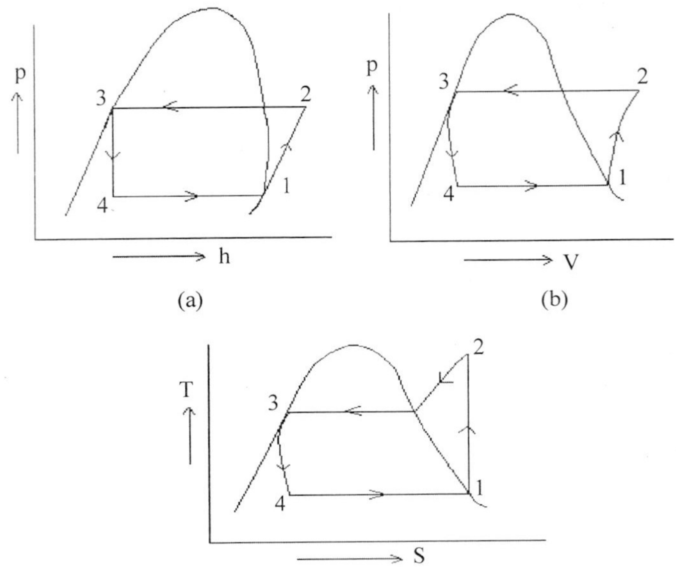

Fig. 2: a) P-h diagram **b)** P-V diagram **c)** T-S plot for Simple Vapour Compression Refrigeration System

4.1 Description of a Simple Vapour Compression Refrigeration System

In this system, the working fluid is a vapour called refrigerant. Liquid refrigerant evaporates in the evaporator, picks up heat from the surroundings and cools the surroundings. In the condenser, vapour form of refrigerant is converted into and loses heat to the atmospheric air. The process is repeated again and again. The commonly used refrigerants are R-22, R-502 and ecofriendly refrigerants. The figure 3 indicated the different conditions of the refrigerant before and after the compression but the idle compression is from 2'-3' represents net compression, which is highly dangerous for compressor. 2"-3" represent superheated vapour before and after compression.

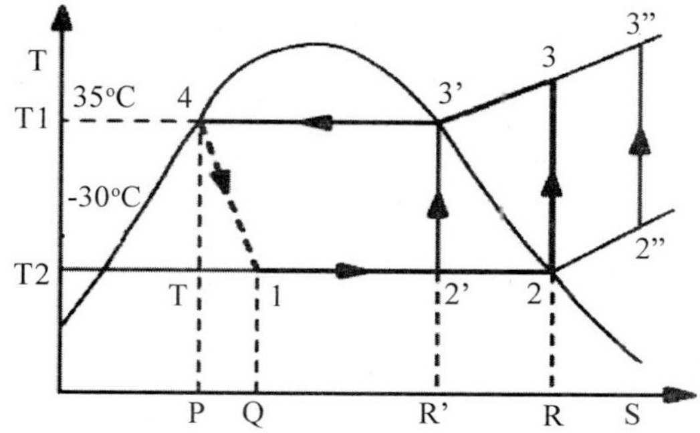

Fig. 3: Different condition refrigerant before and after the compression

Coefficient of performance of Vapour compression refrigeration system

In a constant pressure process, change in enthalpy is equal to heat supplied during the process. With reference to figure 2, the heat transfer in different components is

- Compressor

$Q_{1-2} = 0$, Process 1-2 being adiabatic also (Reversible Adiabatic compression)

- Condenser

$Q_{2-3} = - m_r(h_2 - h_3)$ (isobaric heat rejection)

Negative sign indicates that heat is leaving the condenser.

Where m_r= Mass flow rate of the refrigerant

Expansion valve

$Q_{3-4} = m_r (h3 - h_4) = 0$, as process is a constant enthalpy process

- Evaporator

$Q_{4-1} = m_r (h_1 - h_4)$ (isobaric heat addition)

Here positive sign has been taken because heat is entering the evaporator.

Now from first law of thermodynamics for a cycle, net work done

$$W = \oint \delta Q$$

$$= Q_{1-2} + Q_{2-3} + Q_{3-4} + Q_{4-1}$$

$$= 0 - (h_2 - h_3) + 0 + (h_1 - h_4)$$

$$= h_1 - h_2$$

$$= - (h_2 - h_1)$$

Minus sign shows that work is done on the system. Now

$$C.O.P = \frac{Heat\ extracted}{Work\ done}$$

$$C.O.P = \frac{h_1 - h_4}{h_2 - h_1}$$

5. Theoretical Vapour compression cycle with different case

1. Theoretical Vapour compression cycle with wet vapour after the end of compression

It is the case when the refrigerant after the compression will remain in the wet vapour form. When the liquid and vapour mixture is compressed in the compressor, it is called as wet compression. The phenomenon of wet compression damages the compressor valves as the liquid refrigerant present in the mixture is not compressible. The p-h plot for theoretical vapour compression cycle with wet vapour after the end of compression is depicted in figure 4.

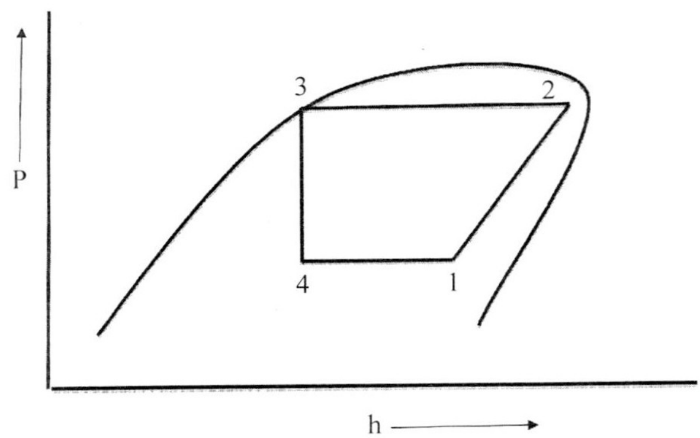

Fig. 4: Theoretical VCRS with wet vapour after compression

$$COP = \frac{Heat\ extracted\ in\ Evaporator}{Work\ net}$$

$$COP = \frac{h_1 - h_4}{h_2 - h_1}$$

$$h_1 = h_f + x_1 h_{fg} \text{ @ evaporator pressure, } p_1$$

$$h_2 = h_f + x_2 h_{fg} \text{ @ condensor pressure, } p_2, \text{if } x_2 \text{ is unknown than use } s_1 = s_2$$

$$h_3 = h_4 = h_{f3} \text{ @ condensor pressure, } p_2$$

2. Theoretical Vapour compression cycle with dry saturated vapour after the end of compression

It is the case when the refrigerant becomes dry saturated at the end of compression, i.e. at the end of compression the refrigerant reaches the saturated vapour line at condenser pressure. Theoretical vapour compression cycle with dry saturated vapour after the end of compression is depicted in the figure 5.

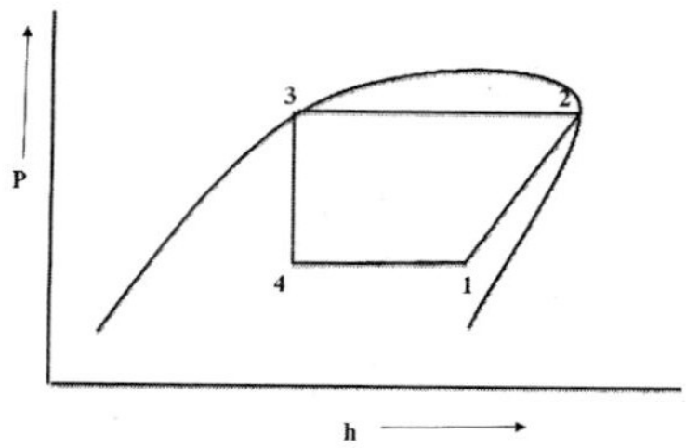

Fig. 5: Theoretical VCRS with dry saturated vapour after compression

$$\text{COP} = \frac{h_1 - h_4}{h_2 - h_1}$$

$h_1 = h_f + x_1\, h_{fg}$ @ evaporator pressure p_1

h_2 = hq @ condensor pressure, p_2

$h_3 = h_4 = h_{f3}$ @ condensor pressure, p_2

3. Theoretical Vapour compression cycle with superheated vapour after the end of compression

It is the case when the refrigerant becomes superheated at the end of compression, i.e. at the end of compression the refrigerant crosses the saturated vapour line at condenser pressure. The superheated refrigerant state means

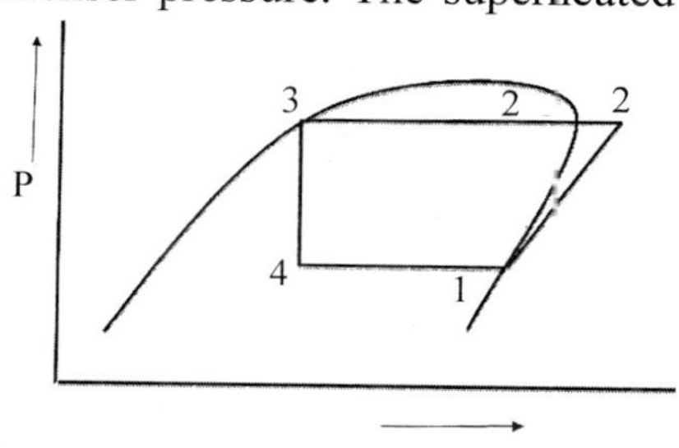

Fig. 6: Theoretical VCRS with superheated vapour after compression

that for the given condenser pressure the temperature of superheated refrigerant is higher than the saturation temperature of the refrigerant. Theoretical Vapour compression cycle with superheated vapour after the end of compression is depicted in the figure 6. In actual practice the refrigerant is in superheated state at the end of the compression.

$$COP = \frac{h_1 - h_4}{h_2 - h_1}$$

$h_1 = h_q$ + @ evaporator pressure p_1

$h_2 = h_{q2} + c_{ps} (T_2 - T_2)$ @ condensor pressure, p_2

$h_3 = h_4 = h_{f3}$ @ condensor pressure, p_2

4. Theoretical Vapour compression cycle with superheated vapour before and at the end of compression

It is the case when the refrigerant becomes superheated before and after at the end of compression, i.e. before and after at the end of compression the refrigerant crosses the saturated vapour line at condenser pressure. The superheated refrigerant state means that for the given condenser/evaporator pressure the temperature of superheated refrigerant is higher than the saturation temperature of the refrigerant. Theoretical vapour compression cycle with superheated vapour before and after the end of compression is depicted in the figure 7. In actual practice the refrigerant is in superheated state before and after the end of the compression. The superheating of vapour takes place in the evaporator before compression, it increases the refrigeration effect and hence the COP of the system.

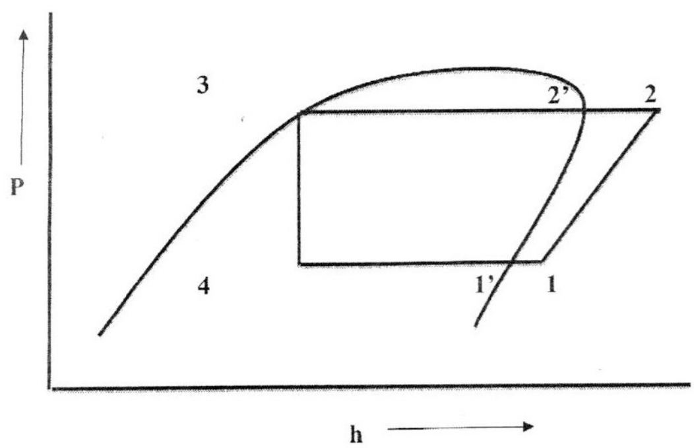

Fig. 7: Theoretical VCRS with superheated vapour before & after compression

$$COP = \frac{h_1 - h_4}{h_2 - h_1}$$

$$h_1 = h_g + C_{ps}\left(T_{\sup} - t_s\right) @ \ evaporator\ pressure, p_1$$

$$h_1 = h_g + C_{ps}\left(T_1' - T_1\right) @ \ evaporator\ pressure, p_1$$

$$h_2 = h_g + C_{ps}\left(T_{\sup} - t_s\right) @ \ condensor\ pressure, p_2$$

$$h_2 = h_g + C_{ps}\left(T_2 - T_2'\right) @ \ condensor\ pressure, p_2$$

$$h_3 = h_4 = h_{f3} @ \ condensor\ pressure, p_2$$

Cps - specific heat of the refrigerant

5. Theoretical Vapour compression cycle with sub-cooling after the end of compression

Sub-cooling is a process of cooling the refrigerant in the liquid state below its saturation temperature at a given condenser pressure. Sub-cooling is generally done in the condenser by the cooling water flowing in the condenser which takes away the heat and lowers the temperature of the liquid refrigerant below its saturation temperature. Sub-cooling increases the COP of the refrigeration system and hence also used in the actual vapour compression system. It is the case when the refrigerant is sub-cooled from 3' to 3 and hence the refrigeration effect increases and thereby the COP. Superheated before and after at the end of compression. Theoretical vapour compression cycle with sub-cooling after the end of compression is depicted in figure 8.

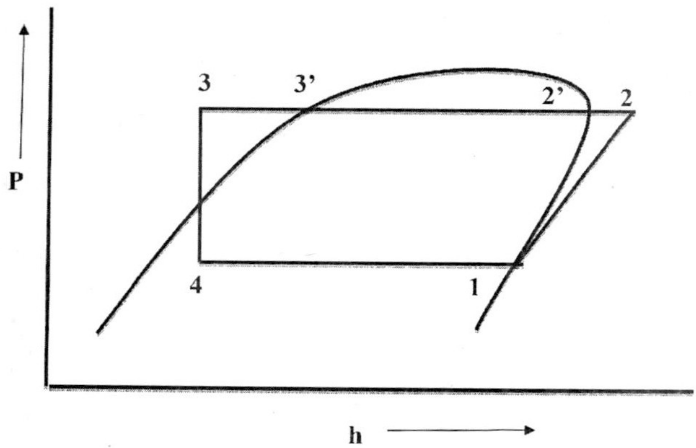

Fig. 8: Theoretical VCRS with sub-cooling after compression

$$COP = \frac{h_1 - h_4}{h_2 - h_1}$$

$$h_1 = h_g @ \ evaporator\ pressure, p_1$$

$$h_2 = h_g + C_{ps}\left(T_{\sup} - t_s\right) @\ condensor\ pressure, p_2$$

$$h_2 = h_g + C_{ps}\left(T_2 - T_2^{;}\right) @\ condensor\ pressure, p_2$$

$$h_3 = h_4 = h_{f3} - C_{ps}\left(T_3^{;} - T_3\right) @\ condensor\ pressure, p_2$$

Ex: A R12 vapour compression refrigeration system develops 15TR refrigeration while running at 40°C and 0°C. Determine the mass flow rate, Piston displacement, horse power per ton of refrigeration effect, heat rejected in the condenser, maximum COP and COP of the system. Assuming day saturated after and of compression.

Sol:From the thermodynamic properties table of R 12 it can be noted:

At the condenser temperature of 40°C

h_1= 187.5 kJ/kg, v_1=0.055 m^3/kg

h_2= 213.96 kJ/kg

h_3= 74.6 kJ/kg

Refrigerating effect = qo = $h_1 - h_4$ = (187.5-74.6) = 112.8 kJ/kg

i) Mass flow rate of refrigerant = m'= Qo'/qo = 15* 211/(112.8) = 28 kg/min

Theoretical piston stroke=V'=m'*v_1 = 28*.055=1.514 m^3/min

ii) Piston displacement per TR of refrigeration effect =1.514/15= 0.101m^3/min

Power consumption=W' =m'($h_2 - h_1$)=28(213.96-187.5)/(60)= 12.36 KW

Theoretical HP of the compressor = 12.36*103/746 =16.6

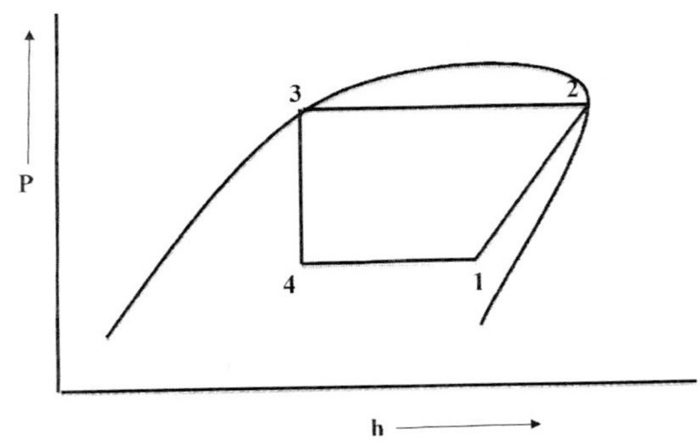

Fig. 9: VCRS on p-h plot with dry siturated vapour after compression

iii) HP/TR = 16.6/15 =1.1

iv) Heat rejected= Q2'= m'($h_2 - h_3$) =28(213.96-74.6)/(60)= 65.1 kW

v) Carnot COP max= 273/(40-0)=6.8

vi) COP of the system= $= \frac{h_1 - h_4}{h_2 - h_1} = \frac{112.8}{213.96 - 187.5} = 4.3$

6. Actual Vapour compression refrigeration cycle

The cycles considered so far are internally reversible and no change of refrigerant state takes place in the connecting pipelines. However, in actual VCRS several irreversibilities exist. These are due to:

1. Pressure drops in evaporator and condenser
2. Pressure drop across suction and discharge valves of the compressor
3. Heat transfer in compressor
4. Pressure drop and heat transfer in connecting pipe lines

Figures 7.10 and shows the actual VCRS cycle on P-h and T-s plots respectively indicating various irreversibilities. From performance point of view, the pressure drop in the evaporator, in the suction line and across the suction valve has a significant effect on system performance. This is due to the reason that as suction side pressure drop increases the specific volume at suction, compression ratio (hence volumetric efficiency) and discharge temperature increase. All these effects lead to reduction in system capacity, increase in power input and also affect the life of the compressor due to higher discharge temperature. Hence this pressure drop should be as small as possible for good performance. The pressure drop depends on the refrigerant velocity, length of refrigerant tubing and layout (bends, joints etc.). Pressure drop can be reduced by reducing refrigerant velocity (e.g. by increasing the inner diameter of the refrigerant tubes), however, this affects the heat transfer coefficient in evaporator. More importantly a certain minimum velocity is required to carry the lubricating oil back to the compressor for proper operation of the compressor.

Heat transfer in the suction line is detrimental as it reduces the density of refrigerant vapour and increases the discharge temperature of the compressor. Hence, the suction lines are normally insulated to minimize heat transfer.

In actual systems the compression process involves frictional effects and heat transfer. As a result, it cannot be reversible, adiabatic (eventhough it can be isentropic). In many cases cooling of the compressor is provided deliberately to maintain the maximum compressor temperature within safe limits. This is particularly true in case of refrigerants such as ammonia. Pressure drops across the valves of the compressor increase the work of compression and reduce the volumetric efficiency of the compressor. Hence they should be as small as possible.

Compared to the vapour lines, the system is less sensitive to pressure drop in the condenser and liquid lines. However, this also should be kept as low as possible. Heat transfer in the condenser connecting pipes is not detrimental in case of refrigeration systems. However, heat transfer in the subcooled liquid lines may affect the performance.

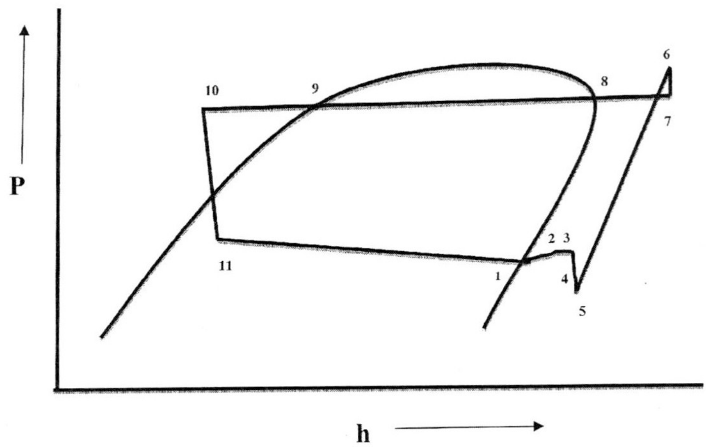

Fig. 10: Actual VCRS on T-s plot

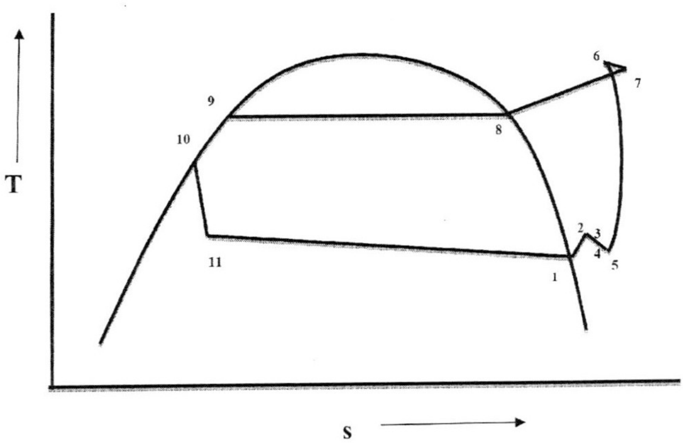

Fig. 11: Actual VCRS on T-s plot

1.	Pressure drop in evaporator	11-1
2.	Superheat of vapour in evaporator	1-2
3.	Useless superheat in suction line	2-3
4.	Suction line pressure drop	3-4
5.	Pressure drop across suction valve	4-5
6.	Non-isentropic compression	5-6
7.	Pressure drop in the delivery line	6-7
8.	De-superheating of vapour in delivery pipe	7-8
9.	Pressure drop in the condenser	8-9
10.	Sub-cooling of liquid refrigerant	9-10
11.	Throttling of the refrigerant	10-11

8

Cleaning and Grading in Food Processing

Harsh Prakash Sharma, Subir Kumar Chakraborty and H. Pandey

1. Introduction

Cleaning and grading are the first and vital post harvest operations undertaken to remove foreign and undesirable materials such as crop seed, chaff, mud, stones, straws, immature grains, dust, husk and even iron and steel particles from grains as well as fruits and vegetables and to separate the products into various fractions. Generally, the hand threshing and the traditional handling used in most developing countries cause a larger percentage of foreign matter with these crops. Thus, more cleaning is required. Cleaning is essentially a separation process, in which some difference in physical properties of the contaminants and the food units is exploited. There are a number of cleaning methods available, classified into dry and wet methods such as screening, aspiration, washing, soaking, spraying and hand picking etc. Grading of foods is the determination of the quality of a batch. Grading refers to the classification of cleaned products into various quality fractions depending upon the various commercial values and other usage. This can also be done by human graders who assess the quality of random samples of foods.

Foods by the nature of the way are grown or produced on farms in open environments often require cleaning before use. Cleaning can be accomplished with brushes, high- velocity air, steam, water, vacuum, magnetic attraction of metal contaminants, mechanical separation, and so on, depending on the product and the nature of the dirt. Increased mechanization in harvesting and subsequent handling has generally led to increased contamination with mineral, plant, and animal contaminants while there has been a general increase in the use of sprays, leading to increased chemical contamination. Microorganisms may be introduced preharvest from irrigation water, manure fertilizer, or contamination from feral or domestic animals, or postharvest from improperly cleaned equipment, wash waters, or cross-contamination from other raw materials.

Cleaning is essentially a separation process, in which some difference in physical properties of the contaminants and the food units is exploited. Therefore, some food materials require special cleaning. Many types of soil can be removed with mildly alkaline detergents, but strong alkali may be required for more tenacious deposits and heavy deposits of fats and oils. Alkaline films and hard water scales may require mildly acid detergents. Scalping is the first and most important grain cleaning operation after threshing. It removes most of the foreign matter and reduces drying cost, eliminates clogging or damage to conveying equipment, and prevents grain deterioration during storage. The second cleaning operation occurs after storage and is the first part of the grain milling process. At this point a cleaner removes any remaining foreign material that could damage the milling machinery and eliminates foreign material from the grains.

2. Types of Cleaning

There are a number of cleaning methods available, classified into dry and wet cleaning, but a combination would usually be used for any specific material. Selection of the appropriate cleaning regime depends on the material being cleaned, the level and type of contamination and the degree of decontamination required.

2.1. Dry cleaning

- Dry cleaning methods are used for products that are smaller, have greater mechanical strength and possess lower moisture content (for example grains and nuts).
- After cleaning, the surfaces are dry, to aid preservation or further drying.
- Dry methods are generally less expensive than wet methods and the effluent is cheaper to dispose off.

The main groups of equipments used for dry cleaning are:

1) Air classifiers: Aspiration
2) Magnetic separators: Magnetic cleaning
3) Separators based on screening of foods: Screening
4) Electrostatic Cleaning

2.1.1. Aspiration

The use of air to separate lower density materials from the grains in most cleaners and scalpers is referred to as aspiration. The separation is based on the difference in aerodynamic properties of various constitutes of the mixture.

The aerodynamic properties of the particle depend upon its shape, size, density, surface and orientation with respect to air current. It is widely used in the cleaning of cereals, but is also incorporated into equipment for cleaning peas and beans. The principle is to feed the raw material into a carefully controlled upward air stream. Denser material will fall, while lighter material will be blown away depending on the terminal velocity (Fig. 1). Terminal velocity in this case can be defined as the velocity of upward air stream in which a particle remains stationary, and depends on the density and projected area of the particles. By using different air velocities, it is possible to separate wheat from lighter chaff or denser small stones. Very accurate separations are possible, but large amounts of energy are required to generate the air streams. Obviously the system is limited by the size of raw material units, but is particularly suitable for cleaning legumes and cereals. Air streams may also be used simply to blow loose contaminants from larger items such as eggs or fruits.

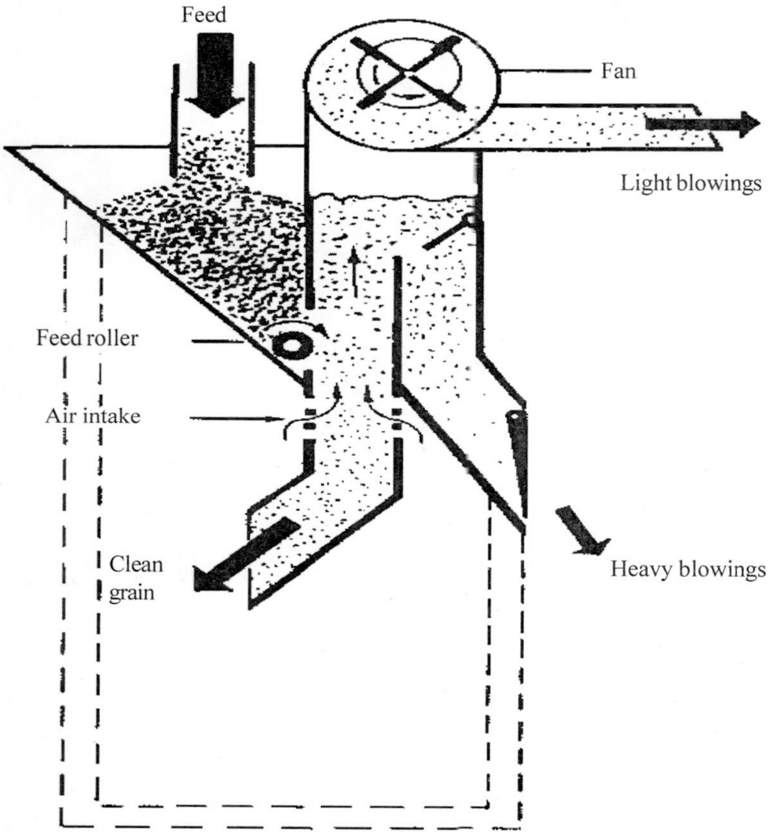

Fig. 1: Aspirator

2.1.2. Magnetic cleaning

The magnetic separator performs separation on the basis of surface texture and stickiness properties of the grain. Grains often contain scrap iron, nails, screws, bolts, pieces of wire, etc. These cause considerable damage to cleaners, sieves, elevator belting, rubber rollers, and polishing cones. If iron or steel particles are not removed by the normal sieving of grains, then they should be separated by magnetic separators. The geometry of magnetic cleaning systems can be quite variable as particulate foods may be passed over magnetized drums or magnetized conveyor belts, or powerful magnets may be located above conveyors. The grain mixture is fed onto the top of a horizontal revolving magnetic drum (Fig. 2). The materials with the iron powder and iron pieces attracted by the magnetic drum and stick to it and are removed by a flexible scraper. Electromagnets are easy to clean by turning off the power. Most of the magnetic separators have two or three revolving magnetic drums operating in a series. The grain mixture is passed over these magnetic drums to increase the efficiency of operation. Metal detectors are frequently employed prior to sensitive processing equipment as well as to protect consumers from this physical hazard at the end of processing lines.

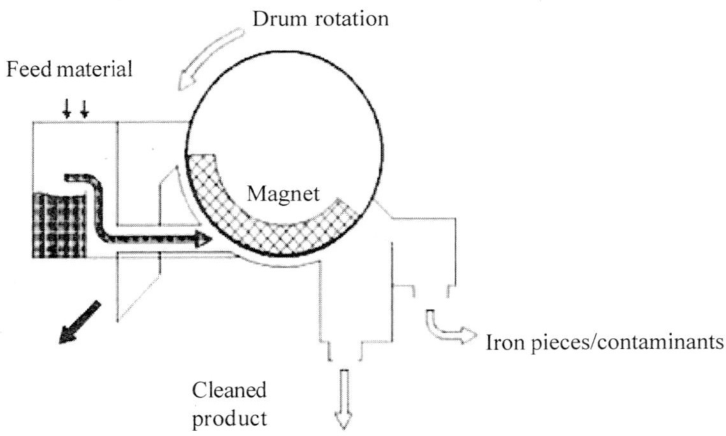

Fig. 2: Magnetic separator

2.1.3. Screening

Screening refers to the separation of solids on the basis of size by means of screens of known apertures. Screens are essentially size separators based on perforated beds or wire mesh by which larger contaminants are removed from smaller food items (e.g., straw from cereal grains, or pods and twigs from peas). Alternatively "de-dusting" is the removal of smaller particles (e.g., sand or dust) from larger food units. (Fig. 3)

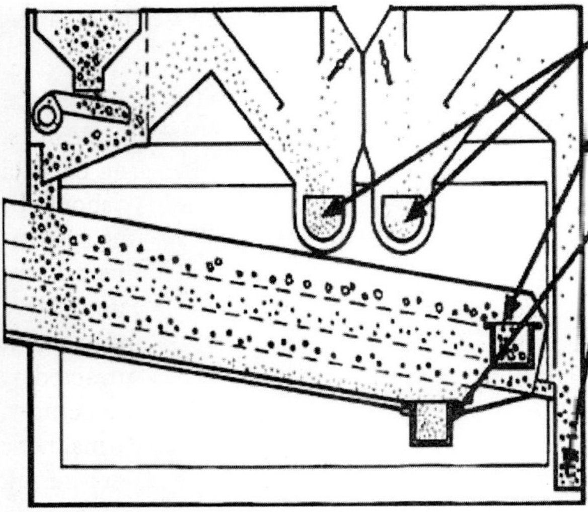

Fig. 3: Air lifting and screening system

Grizzlies and shaking screens are used for screening large lumps. Trommels (revolving screen) are used for fairly large particles. Vibrating and oscillating screens are used for fine sizing. Abrasion, either by impact during the operation of the machinery, or aided by abrasive disks or brushes, can improve the efficiency of dry screens. Screening gives incomplete separations and is usually a preliminary cleaning stage.

2.1.3.1. Cleaning Efficiency (η)

The cleaning efficiency or screen effectiveness may be defined as the ability of a screen in closely separating the feed into overflow and underflow according to its size. The cleaning efficiency for an air screen grain cleaner as suggested by the Bureau of Indian Standards (BIS) is given below,

$$\eta = \frac{E(F-G)(E-F)(1-G)}{F(E-G)^2(1-F)}$$

Where,

E fraction of clean seed at clean seed outlet

F fraction of clean seed in feed

G fraction of clean seed at foreign matter outlets

Example 1: During evaluation of an air screen grain cleaner with 2 screens 250 g samples were collected for analysis of clean seed fraction from different outlets. The data are presented in the following Table. Calculate the cleaning efficiency of the cleaner.

Sample fraction	Feed,g	Clean grain outlet, g	Blower outlet, g	Oversize outlet, g	Undersize outlet, g
Cleaned seed, g	231.25	246.5	1.25	4.5	2.0
Impurities, g	18.75	3.5	248.75	245.5	248.0

Solution

(1) Fraction of clean seed at clean seed outlet:

$$E = \frac{246.5}{250.0} = 0.986$$

(2) Fraction of clean seed in feed:

$$F = \frac{231.25}{250.0} = 0.925$$

(3) Fraction of clean seed in foreign matter outlets:

$$G = \frac{1.25}{250.0} + \frac{4.5}{250.0} + \frac{2.0}{250.0} = 0.031$$

Therefore,

$$Cleaning efficiency = \frac{0.986(0.925-0.031)(0.986-0.925)(1-0.031)}{0.925(0.986-0.031)^2(1-0.925)} = 82.34\%$$

2.1.4. Electrostatic cleaning

This can be used in a limited number of cases where the surface charge on raw materials differs from contaminating particles. The principle can be used to distinguish grains from other seeds of similar geometry but different surface charge, and has also been described for cleaning tea. The feed is conveyed on a charged belt and charged particles are attracted to an oppositely charged electrode according to their surface charge (Fig.4).

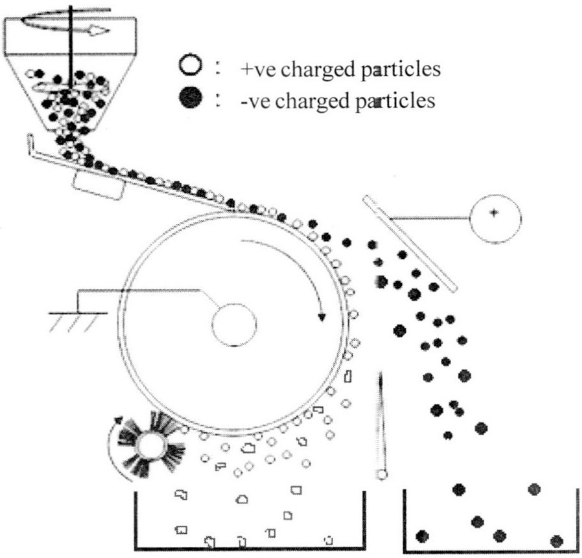

Fig. 4: Electrostatic cleaner

2.2. Wet cleaning

Wet methods are necessary if large quantities of soil are to be removed, and are essential if detergents are used. Wet cleaning is more effective than dry methods for removing soil from root crops or dust and pesticide residues from soft fruits or vegetables. It is also dustless and causes less damage to foods than dry methods. Different combinations of detergents at different temperatures allow flexibility in operation. Sanitizing chemicals such as chlorine, citric acid, and ozone are commonly used in wash waters, especially in association with peeling and size reduction, where reducing enzymatic browning may also be an aim.

The main methods of wet washing are:

1) Soaking
2) Spray washing
3) Flotation washing
4) Froth flotation
5) Hot water dipping
6) Peeling

2.2.1. Soaking

Soaking is a preliminary stage in cleaning heavily contaminated materials such as root crops, permitting softening of the soil, and partial removal of stones and other contaminants. Metallic or concrete tanks or drums are employed, and these may be fitted with devices for agitating the water, including stirrers, paddles, or mechanisms for rotating the entire drum. For delicate produce such as strawberries or asparagus, or products which trap dirt internally (e.g., celery), sparging air through the system may be helpful. The use of warm water or including detergents improve cleaning efficiency, especially where mineral oil is a possible contaminant, but adds to the expense and may damage the texture.

2.2.2. Spray washing

Spray washing is very widely used for many types of food raw material. Efficiency depends on the volume and temperature of the water and time of exposure. As a general rule, small volumes of high-pressure water give the most efficient dirt removal, but this is limited by product damage, especially to more delicate produce. With larger food pieces it may be necessary to rotate the unit so that the whole surface is presented to the spray. The two most common designs are drum washers and belt washers (Fig. 5). Abrasion may contribute to the cleaning effect, but again must be limited in delicate units. Other designs have included flexible rubber disks which gently brush the surface clean.

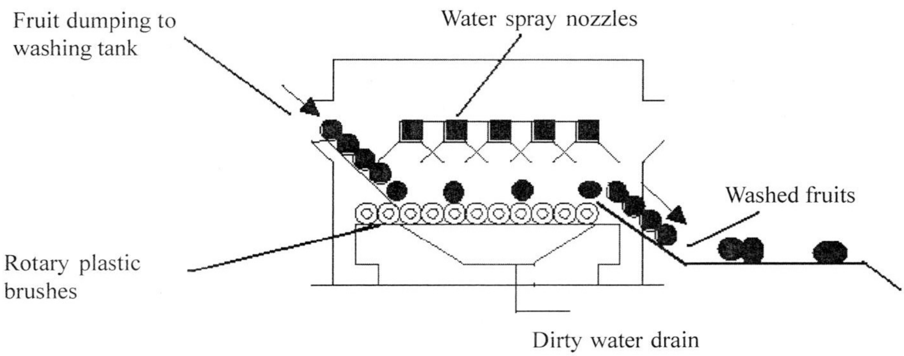

Fig. 5: Spray washing

2.2.3. Flotation washing

Flotation washing employs buoyancy differences between food units and contaminants. For instance, sound fruit generally floats, while contaminating soil, stones, or rotten fruits sink in water. Hence fluming fruit in water over a series of weirs gives very effective cleaning of fruit, peas, and beans. A disadvantage is high water use, thus recirculation of water should be incorporated.

2.2.4. Froth flotation

Froth flotation is carried out to separate peas from contaminating weed seeds, and exploits surfactant effects. The peas are dipped in oil/detergent emulsion and air is blown through the bed. This forms foam which washes away the contaminating material, and the cleaned peas can be spray washed.

2.2.5. Hot water dipping

Prestorage hot water dipping has been used as an alternative to chemical treatments for preserving the quality of horticultural products. One recent development is the simultaneous cleaning and disinfection of fresh produce by a short hot water rinse and brushing treatment. This involves placing the crops on rotating brushes and rinsing with hot water for 10–30 s. The effect is through a combination of direct cleaning action plus the lethal action of heat on surface pathogens. Fungicides may also be added to the hot water.

2.2.6. Peeling

The objective of peeling is to remove the outer layer. Peeling of fruits and vegetables is frequently carried out in association with cleaning. One of the following principles depending on the structure of the food and the level of peeling is used:

i) **Hand peeling:** It is done mostly in case of fruits of irregular shape, i.e. mango and papaya where mechanical peeling is not possible.

ii) **Abrasion peeling:** Abrasion peeling employs carborundum rollers or rotating the product in a carborundum-lined bowl followed by washing away the loosened skin. It is effective but here is a danger of high product loss by this method.

iii) **Steam peeling:** Free-stone and clingstone peaches are steam peeled in different ways. Steam is particularly suited to root crops. The units are exposed to high-pressure steam for a fixed time and then the pressure is released causing steam to form under the surface of the skin, hence loosening it such that it can be removed with a water spray. Potatoes and tomatoes are peeled by steam or boiling.

iv) **Mechanical peeling:** This is done in case of apples, peaches, pineapples and cherries and also for root vegetables like carrots, turnips and potatoes. Mechanical knives are suitable for peeling citrus fruits.

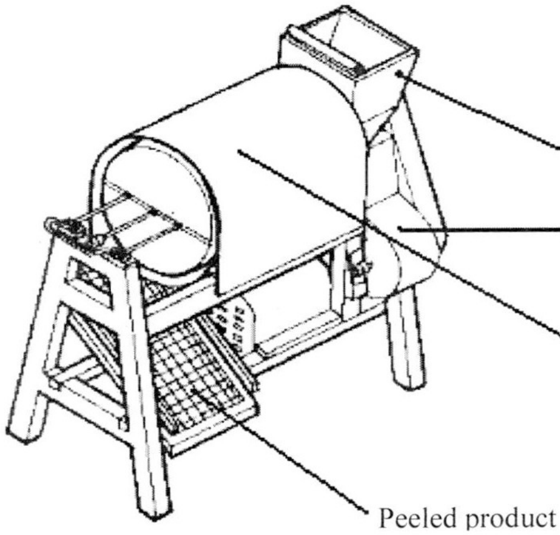

Fig. 6: Abrasion peeler

v) **Lye peeling:** Fruits like peaches, apricots, sweet oranges, mandarin oranges and vegetables like carrots and sweet potatoes are peeled by dipping them in 1 to 2 percent boiling caustic soda solution (lye) for 30 seconds to 2 minutes depending on their nature and maturity. Hot lye loosens the skin from the flesh by dissolving the pectin. Peel is then removed easily by hand. This is a quick method where by cost and wastage in peeling is reduced.

vi) **Flame peeling:** Flame peeling is useful for onions and garlic, which have a papery outer covering. In which the outer layers are burnt off and charred skin is removed by high-pressure hot water.

vii) **Brine peeling:** Brine solutions can give a peeling effect but are probably less effective than the above methods.

2.2.7. Sorting

Sorting is the separation of foods into categories on the basis of a measurable physical property. Like cleaning, sorting should be employed as early as possible to ensure a uniform product for subsequent processing. The four main physical properties used to sort foods are size, shape, weight and colour. The effectiveness of assorting procedure is calculated using

$$Effectiveness = PRX_p\left(1-X_r\right)/\,F^2X_f\left(1-X_f\right)$$

Where:

P Product flow rate (kg/s)

F Feed flow rate (kg/s)

R Reject flow rate (kg/s)

Xp Mass fraction of desired material in product

Xf Mass fraction of desired material in feed

Xr Mass fraction of desired material in rejected material

3. Grading

Grading is classification on the basis of quality (incorporating commercial value, end-use, and official standards, and hence requires that some judgment on the acceptability of the food is made, based on simultaneous assessment of several properties, followed by separation into quality categories. Grading is carried out by operators who are trained to simultaneously assess a number of variables. For example, eggs are visually inspected over tungsten lights (termed 'candling') to assess up to twenty factors and remove those that are for example, fertilized or malformed and those that contain blood spots or rot.

This term is often used interchangeably with sorting but strictly means 'the assessment of overall quality of a food using a number of attributes'. Sorting (that is separation on the basis of one characteristic) may therefore be used as part of a grading operation but not vice versa. In rice processing industries, after the whitening operation the unbroken rice is still mixed with different sized broken rice, bran, and dust. Separation of these particles after whitening is termed "grading." The degree of grading is determined by the rice market or consumer preference. Many rice markets do not require any grading; others require a sophisticated grading system that will produce a clean, bran-free rice with no brokens. Most rice markets will accept a small percentage of brokens but demand clean and bran-free rice.

Grading of foods is also the determination of the quality of a batch. This can be done by human graders who assess the quality of random samples of foods such as cheese or butter, or meat inspectors who examine the quality of individual carcasses for a number of criteria. Alternatively, batches of some foods may be graded on the basis of laboratory analysis.

There is increasing interest in the development of rapid, nondestructive methods of assessing the quality of foods, which could be applied to the grading and sorting of foods. Many scientists have described potential application of advanced optical techniques to give information on both surface and internal properties of fruits, including textural and chemical properties. This could permit classification of fruit in terms of maturity, firmness, or the presence of defects, or even more specifically, noninvasive detection of chlorophyll, sugar, and acid levels.

For International market three general grades are considered as:

a) Extra class b) Class 1st c) Class 2nd.

a) Extra class

The extra class is of superior quality posses the shapes and colour of the variety and without internal defect likely to affect the inherent texture and flavour. A 5% tolerance is allowed for errors. It must be carefully presented taking into accounts the uniformity of the produces in size colour, condition arrangement of the produce in the package quality and appearances of the packing or pre-packing material.

b) Class 1st

Almost having a same quality is like the 'extra class' except that a 10% tolerance is allowed. Individual fruit is allowed a slight defect in shape, colour and miner skin defect which do not affect the general appearance for keeping qualities. In packing the size range may be wider and product need not always be arranged in the package.

c) Class 2nd

This class product may exhibit some external or internal defects provided they are fit for consumption while fresh. This class is best fitted for local or short distance market. This category will satisfy the needs of customers who are not too demanding and for whom price is more important than quality.

3.1. Grading of Fruits and Vegetables

Grading of fruits and vegetables after harvesting is an essential step in post-harvest management. Grading of fruits and vegetables on the basis of physical characteristics like weight, size, colour, shape, specific gravity, and freedom from diseases depending upon agro-climatic conditions. The known methods of grading of fruits and vegetables are manual grading, size grading (Fig. 6). Grading of fruits and vegetables in the fresh form for quality is essential, as the peoples are becoming quality conscious day by day. Further, upon arrival of fruits and vegetables at the processing centres, they should be graded strictly for quality.

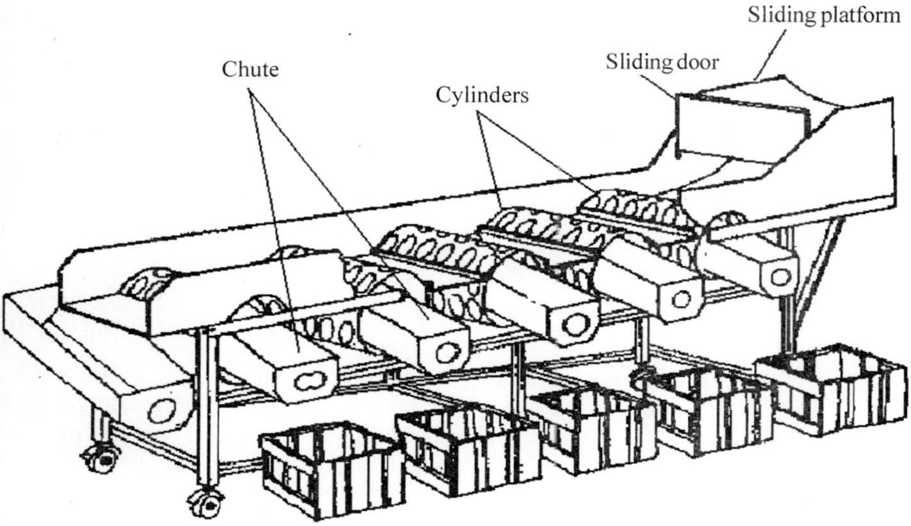

Fig. 6: Grader for fruits and vegetables
Source: Reyes, M. U. (Ed.) 1988 Design Concept and Operation of ASEAN Packinghouse Equipment for Fruits and Vegetables. Postharvest Horticulture Training and Research Center, University of Los Banos, College of Agriculture, Laguna, Philippines.

The immature, properly mature and over mature fruits and vegetable should be sorted out for the best attributes. Generally, the fruits are graded on the basis of size, weight, sp. gravity, colour, variety, etc. Size grading is predominantly followed in almost all types of fruits on the basis of size. The fruits are graded as a small, medium, large and extra-large. On the basis of maturity, the fruits are graded as immature, properly mature and over mature. Grading on the basis of maturity decides both quality and shelf life. The fruit vegetables such as bitter gourd, okra, bell pepper, brinjal, green chilli, etc. are also graded on the basis of size into three grades as small medium and large. The vegetables like tomato are graded on the basis of colour.

3.1.1. Advantages of Grading

i) Losses the selling price due to presence of substandard products or specimen can be easily avoided.

ii) It increased marketing efficiency by facilitating buying and selling a produce without personal selection.

iii) Grading enhanced to set good price for graded products.

iv) Heavy marketing cost in packing and transportation can be avoided by grading.

v) By grading there is fairness to both buyers and sellers.

vi) Properly graded vegetables and fruits are purchased by the consumer easily without inspection.

References

Alistair S. Grandison (2012). Postharvest handling and preparation of foods for processing.In: James G. Brennan and Alistair S. Grandison (Eds.), *Food Processing Handbook* (Second Edition, pp. 1-30. Published by Wiley-VCH Verlag GmbH & Co. KGaA.

Cubeddu, R., Pifferi, A., Taroni, P., and Torricelli, A. (2002). Measuring fruit and vegetable quality: advanced optical methods, in *Fruit and Vegetable Processing: Improving Quality* (ed. W. Jongen), pp. 150–169, Woodhead Publishing, Cambridge.

Nicolai, B.M., Beullens, K., Bobelyn, E., Peirs, A., Saeys, W., Theron, K.I., and Lammertyn, J. (2007). Nondestructive measurement of fruit and vegetable quality by means of NIR spectroscopy: a review. *Postharvest Biol. Technol.*, 46: 99–118.

Norman, N. Potter and Joseph, H. Hotchkiss (1996).Unit operations in food processing. *Food Science* (Fifth edition, pp.69-89). CBS Publishers, New Delhi.

Orea, J.M. and Gonzalez Urena, A. (2002). Measuring and improving the natural resistance of fruit. In: W. Jongen(Eds.), *Fruit and Vegetable Processing: Improving Quality* (ed.), pp. 233–266, Woodhead Publishing, Cambridge.

Pare, A. and Mandhyan, B.L. (2011).Introduction. *Food Process Engineering and Technology* (First edition, pp.1-36), New India Publishing Agency, New Delhi.

Sahay, K.M. and Singh, K.K. (1994).Cleaning and grading. In: Sahay, K.M. and Singh, K.K. (Eds.), *Unit operations of Agricultural Processing* (First edition, pp.66-102). Vikas Publishing House, New Delhi.

Saravacos, G.D. and Kostaropoulos,A.E. (2002). *Handbook of Food Processing Equipment*, Kluwer Academic, London.

9

Size Reduction

Subir Kumar Chakraborty and Harsh Prakash Sharma

1. Introduction

Size reduction is applied to all the ways in which particles of the solids are cut or broken into smaller pieces. Size reduction operation of solids is referred to as '*Comminution*', some of the common comminutive processes are - crushing, grinding, milling, mincing, and dicing. Breaking down of large sized particles of food substances into smaller particles is a common unit operation in many food processes, e.g. wheat flour, turmeric powder, spices, etc. are mostly used in powdered form. Some of the common objectives that are fulfilled by size reduction are:

i) to increase surface area to volume ratio.

ii) to extract desired component from the food matter.

iii) to obtain definite size range, to have desired functional properties.

iv) to make it suitable for intimate mixing.

v) to reduce bulk of the material for easy handling and transportation.

Size reduction is not a preservation action, for; enzymes will be exposed due to cell destruction which may lead to spoilage of the product, Perhaps this is also because, micro organisms find more area to act upon. Deformation of food pieces is carried out by a variety of forces till the food breaks or tears. The types of forces by which size reduction takes place and their corresponding equipments are summarized in Table 1.

Table 1: Forces in size reduction

Force	Features	Equipments
Compression	It is used for coarse size reduction of hard solids, to give relatively fewer fine products.	Roller crushers, e.g. refined wheat flour milling
Impact	It gives coarse, medium, or fine products depending on the residence time.	Hammer mills, e.g. spice grinding
Attrition or Shear	It yields very fine product from soft, non abrasive materials.	Attrition mills, e.g. wheat flour mills
Cutting	It gives definite particle size and sometimes definite shape, with a few or no fines.	Cutting blades, e.g. chaff cutter

1.1 Definitions

Grain shape: The grain shape can be expressed in terms of spherecity (φ).

$$\varphi = \frac{6V_p}{D_p S_p} \qquad (1)$$

Where,

V_p is the volume of a particle unit

S_p is the surface area of a particle

D_p is the equivalent diameter of particle

The equivalent diameter is defined as the diameter of a equivalent sphere having the ratio of surface to volume same as that of the actual particle.

***Reduction ratio (RR)**:* It is the ratio of the surface area of the ground product (output) and surface area of the feed (input). This ratio is used as a measure to classify the size reduction operation on the basis of fine ($RR>60$), intermediate ($8<RR<60$) and coarse ($RR<8$) grinding.

$$RR = \frac{Surface\ area\ of\ the\ ground\ product}{Surface\ area\ of\ the\ feed} \qquad (2)$$

***Size classification**:* The particles of the product formed after a size reduction process can be classified into, - Dimensional range, Sieve range and Microscopic range. Not all food types can fit into all the classifications, some details of the classification and the type of food product is given in Table 2.

Table 2: Classification of the range of particle size of some food items

Size classification	Size	Food items
Dimensional Range	> 0.125	dried food material, chopped forage, minced meat, etc.
Sieve Range	0.125 - 0.0029	ground feed, CTC tea, sugar grains, etc
Microscopic Range	< 0.0029	spray dried milk powder, chemical powders, etc.

Tyler series sieves: This series was given by Tyler in 1910. There are a set of sieves used for segregating small particles based on their linear dimension. Every sieve has square openings double in size from that of the next one, starting from the top. Ergo, the sides of openings of two consecutive sieves are in the ratio of $\sqrt{2}$. Every sieve, of a particular opening size has a Tyler Number based on the number of opening per inch.

Fineness Modulus (FM): The uniformity of distribution of size in a ground product is indicated by this modulus. It is defined as the sum of the weight fractions retained above each sieve divided by 100. The fineness modulus is determined by the following method:

$$FM = \frac{\sum_{i=1}^{n} w_i}{100} \tag{3}$$

Were,

w_i is the weight fraction of the material from the ith sieve,

n is the number of sieves.

The average size of the grain D (in) is indicated by FM and can be calculated using the following equation.

$$D = 0.0041(2)^{FM} \tag{4}$$

Table 3: Tyler standard series of sieves

Size of openings (μm)	Tyler Number (*meshes per in*)
3360	6
2380	8
1680	10
1191	14
841	20
594	28
420	35
297	48
212	65
150	100
103	150
73	200
53	270

1.2 Laws of Size Reduction

The energy supplied to during size reduction is used by the solid food particle for creation fissures along the planes of failures and also causing fractures over the existing fissures. Proper understanding of the energy consumption pattern of the various size reduction processes is important to work out the economics. During the size reduction process a huge amount of energy is utilized to counter the inter particle friction of the solid particles and to overcome the elastic limits of the particles, only a small amount of energy is utilized for actually breaking solids.

It is assumed, to calculate the energy requirement for any size reduction process, that the ground (output) particles and the unground (input) particles have symmetrical shapes and a common dimension is used to calculate the energy Fig. 1.

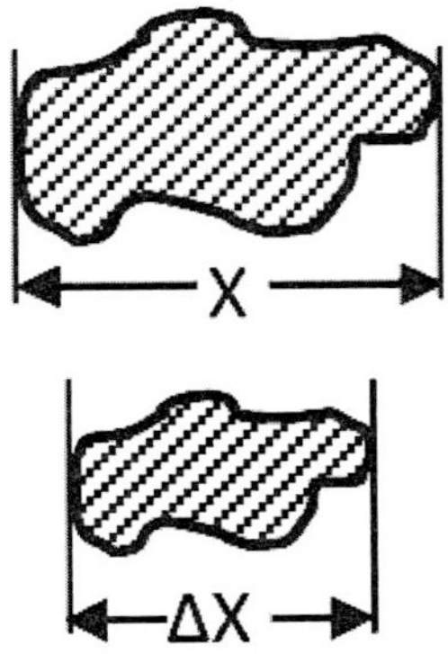

Fig. 1: Symmetrical shapes of the feed particle and the product particle.

The energy required for size reduction is proportional to the ratio of a common linear dimension of the feed particle and that of the product particle.

$$\Delta E \; \alpha \; \frac{\Delta X}{X} \tag{5}$$

$$\Delta E = -C\frac{\Delta X}{X^n} \tag{6}$$

This is the most basic equation for determination of energy requirement in a size reduction process. It has been used for estimating the energy requirement for more specific cases. The negative sign indicates that the energy requirement would increase if the size of the product decreases, or in other words energy requirement and size of the product are inversely proportional.

The total energy (E) required for any size reduction operation can be expressed as,-

$$E = -C\int_{X_1}^{X_2} \frac{\Delta X}{X^n} \tag{7}$$

Where,

X_1 and X_2 are the initial and final dimension of the food particle being ground, and

n takes the value 1, 3/2, 2 and for coarse, intermediate and fine grinding, respectively.

1.2.1 Rettinger's Law

It states that the work required in crushing is proportional to the new surface created. The energy requirement is calculated based on the assumption that the size reduction process is essentially a shearing process; therefore, energy requirement is proportional to the square of the common linear dimension. Here, n=2 in eq. 7.

$$E = -C\int_{X_1}^{X_2} \frac{\Delta X}{X^2} \tag{8}$$

$$E = C\left(\frac{1}{X_1} - \frac{1}{X_2}\right) \tag{9}$$

Energy requirement for fine (RR is greater than 60) grinding operations are effectively determined by Rettinger's law.

1.2.2 Bond's Law

It is based on the assumption that the work required to form particles of size X_2 from a very large feed is proportional to the square root of the surface – to – volume ratio of the product. This law is valid for size reduction process where RR is between 8 and 60 (intermediate grinding). Here, n=3/2 in eq. 7.

$$E = -C \int_{X_1}^{X_2} \frac{\Delta X}{X^{3/2}} \tag{10}$$

$$\frac{P}{f} = 0.3162\ w_i \left(\frac{1}{\sqrt{X_2}} - \frac{1}{\sqrt{X_1}} \right) \tag{11}$$

Where

P is power (kW),

f is feed rate (t/hr),

w_i is Bond's work index,

X_1 (mm) is diameter of mesh through which 80% of the feed pass through,

X_2 (mm) is diameter of mesh through which 80% of the product pass through.

Bond's work index (w_i) is defined as the gross energy required in kilowatt-hour per tonne of the feed required to reduce a very large feed to such a size that 80% of the product passes through a 100μm screen.

1.2.3 Kick's Law

This law is based on the theory of plastic deformation of the food particle. It is therefore assumed that the energy required for any size reduction process is a function of a common dimension of the particle. Here, n=1 in eq. 7.

$$E = -C \int_{X_1}^{X_2} \frac{\Delta X}{X} \tag{12}$$

$$E = C\ \ln \left(\frac{X_1}{X_2} \right) \tag{13}$$

It has been observed that Kick's Law is valid for coarse size reduction process or for those processes where RR is less than 8. A closer look at eq.13 yields that energy requirement estimation by Kick's law is same for a given reduction ratio, irrespective of the original size of the particles.

Example 1

In wheat milling, it was found that to grind 4.33 mm sized grains to IS sieve 35 (0.351 mm opening), the power requirement was 8 kilo watt. Calculate the power requirement for milling of wheat by the same mill to IS sieve 15 (0.157 mm opening) using (i) Retinger's Law (ii) Kick's Law.

(i) For Rettinger's Law we have, $E = C\left(\frac{1}{X_1} - \frac{1}{X_2}\right)$

For the first case,

Initial grain size is, $X_2 = 4.33$ mm

Final grain size is, $X_1 = 0.351$ mm

Power required is, E = 8 kilo watt

Therefore, to calculate C,

For the second case,

$$8 = C\left(\frac{1}{0.351} - \frac{1}{4.33}\right)$$

C = 3.056

Again, we can use calculated C to find the power required, $X_1 = 0.157$ mm

$$8 = C\left(\frac{1}{0.351} - \frac{1}{4.33}\right)$$

E = 18.76 kilo watt

(ii) For Kick's Law we have, $E = C \ln\left(\frac{X_1}{X_2}\right)$

For the first case,

Initial grain size is, $X_1 = 4.33$ mm

Final grain size is, $X_2 = 0.351$ mm

Power required is, E = 8 kilo watt

Therefore, to calculate C,

$$8 = C\ ln\frac{4.33}{0.351}$$

C = 3.184

For the second case,

Again, we can use calculated C to find the power required, $X_1 = 0.157$ mm

$$E = 3.184 \ln\frac{4.33}{0.157}$$

E = 10.56 kilo watt

Example 2

How much power is required to crush 2 t/hr of material in 80 % of the feed passes through IS sieve no. 48 (4.75 mm opening) and 80 % of the product passes through IS sieve no. 50 (0.5 mm opening)? Given that the work index of the material is 6.30.

For Bond's Law we have, $\frac{P}{f} = 0.3162\ w_i \left(\frac{1}{\sqrt{X_2}} - \frac{1}{\sqrt{X_1}}\right)$

Initial grain size is, $X_1 = 4.75$ mm

Final grain size is, $X_2 = 0.5$ mm

Work index is, $w_i = 6.30$ kilowatt-hour per tonne

Feed, f = 2 t/hr

Therefore, to calculate P,

$$\frac{P}{2} = 0.3162(6.30)\left(\frac{1}{\sqrt{0.5}} - \frac{1}{\sqrt{4.75}}\right)$$

P = 3.08 kilo watt

1.3 Machineries

Variety of equipments is used in food processing industries for size reduction purposes. Depending upon the output sought and product handled the choice of the machinery varies (Table 4). Apart from other things the choice of machinery is guided by the following criteria:

- A machine should have large capacity.
- Small power input per unit of product formed.
- At least 80 per cent of the product formed must conform to a single size.

Table 4: Suitability of equipments for different degree of grinding for different foods

Equipment	Type of grinding				Food material (*used for*)
	Coarse	Intermediate	Fine	Ultrafine	
Crushing rolls	√	√			Chocolate, pulses and cereal grains
Hammer mills		√	√	√	Dried fruits, dried vegetables, dried milk, spices, sugar, salt
Attrition mills		√	√		Cocoa, corn, cereal grains
Ball mills		√	√	√	Chocolate, cocoa

Measures that can be adopted for an improved performance of size reduction machines:

i) Size of the feed should be more or less of uniform.

ii) There should be a constant feeding rate.

iii) The feed should not contain hard or unbreakable material.

iv) The product should be removed from the mill as soon as desired size reduction is achieved.

v) Adequate cooling arrangement should be there for the machine.

Size reduction machineries can be classified into - Grinders and Crushers.

1.3.1 Grinders

The grinders find wide use in food industries. They are used generally for purposes where the final ground product is in the "intermediate" or in the "fine" range of grinding. The grinders can be classified into - Attrition mill (Single running, Double running and Buhr mill), Hammer mill and Ball mill.

Attrition Mill: In an attrition mill food materials are feed between the two stone discs. Control over the size reduction process is exercised by maintaining the adjustable gap between the stones and speed of rotation of the stone(s). The food material is broken down by intense shearing forces, leading to the formation of the desirable ground product. The most common example of this type of mill are the wheat flour mill. Now, these attrition mills can be classified into two types, - Single running (Fig. 2a), only one stone rotates and Double running (Fig. 2b), both the stones rotate in opposite direction resulting in the introduction of an increased shear force on the product.

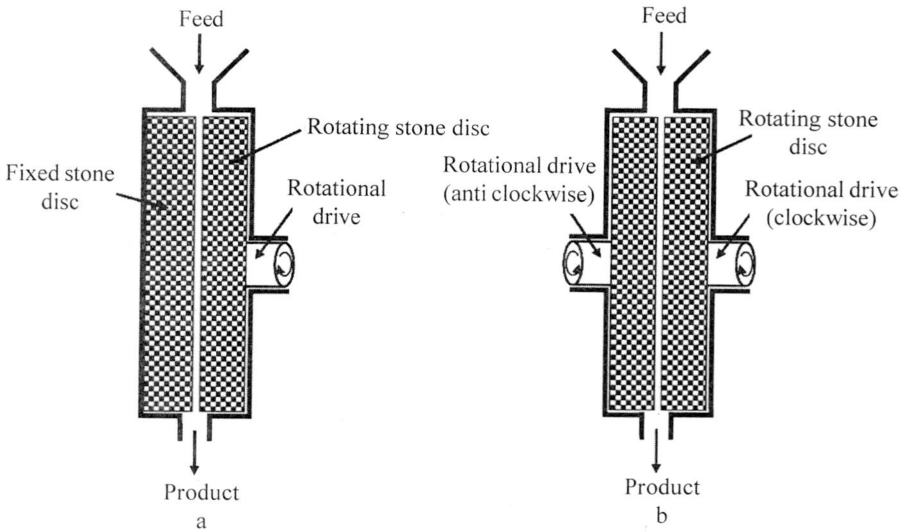

Fig. 2: Schematic diagram of (a) single (b) double running attrition mill

If the stones of the attrition mill are horizontally placed and material is feed at their common centre only to be thrown out from the periphery of the lower stone (the upper / top stone being fixed) as ground product, then such attrition mills are called, Buhr Mill (Fig. 3).

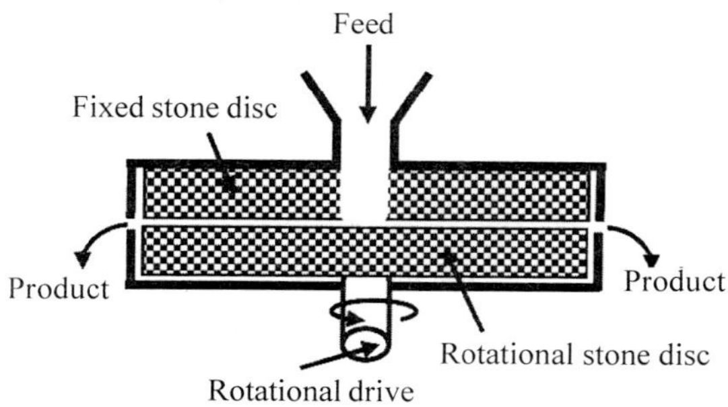

Fig. 3: Schematic diagram of Buhr mill

Hammer Mill: A hammer mill comprises of a set hammers (swinging or fixed) fastened on a rotating shaft and housed in a closed cylindrical case (Fig. 4) lined with a screen of openings of size as desired for the product. There is a hardened material coating in the inside of the cylindrical case; the feed is hit by the rotating hammers against the cylinder casing causing breakage of the particles into fines. The hammer mills are very versatile equipment can handle a wide range of materials and can achieve high *RR*. The size reduction is carried out mainly by the impact forces but if the machine is working under choking conditions then shear forces also come into picture.

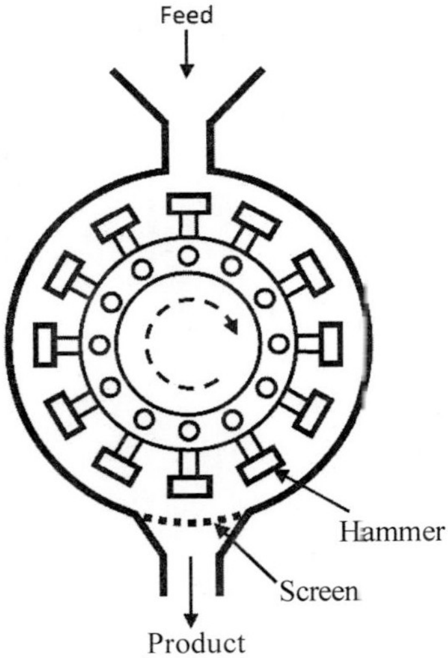

Fig. 4: Schematic diagram of hammer mill

Ball Mill: In a ball mill conical or cylindrical (radius, R) shell rotates about a horizontal axis; a percentage of volume of the shell is filled with a number of high carbon steel, porcelain or silica balls (radius, r) and feed (Fig. 5).

As the cylinder rotates the balls are lifted up because of centrifugal force, the feed being lighter does not move clinging to the surface of the cylinder and is not carried up. Size reduction is achieved when the balls repetitively (the cylinder is continuously rotating) fall down because of their weight on the feed material. The size reduction takes place because of the impact forces cause by falling balls and a bit of shearing force as well. The speed of rotation of ball mill is a critical factor in its operation. A high rpm (revolution per minute) would not let the balls fall down while a low rpm would not let the balls go up. The rpm is a very important consideration in ball mill operation.

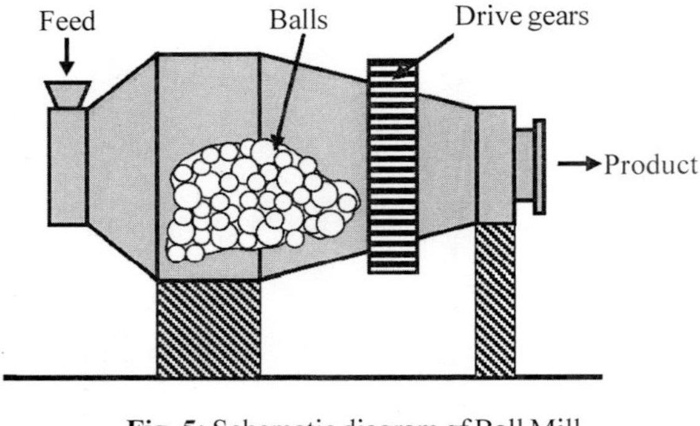

Fig. 5: Schematic diagram of Ball Mill

What if the balls do not fall at all? This can happen at a rotational speed when the centrifugal force (F_c) would balance the weight (w_b) of the balls (Fig.6).

$$F_c = \frac{mv^2}{R-r} \tag{14}$$

$$W_b = mg \tag{15}$$

For the ball to remain suspended,-

$$mg = \frac{mv^2}{R-r} \tag{16}$$

Where,

m is the mass of the ball, kg 9.81 m/s^2

g is the acceleration due to gravity, 9.81 m/s^2

v is the rotational speed of the cylinder

R is the radius of the cylindrical case

r is radius of the ball

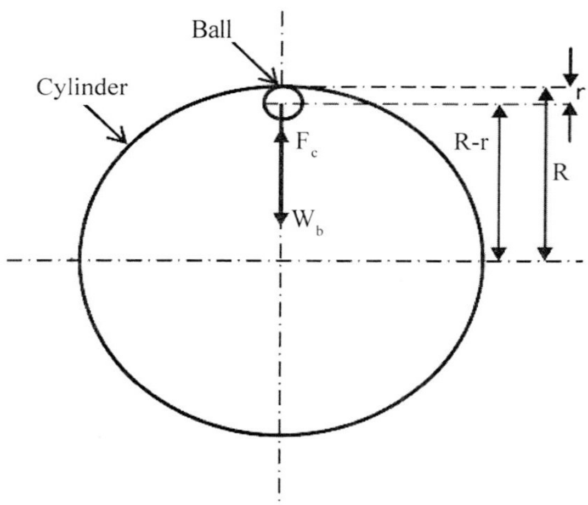

Fig. 6: Critical speed of rotation of Ball Mill

$$v = \omega(R-r) = 2\pi n(R-r) \tag{17}$$

Where, ε is the angular velocity, *n* is the revolution per minute of the rotating cylinder.

There is one critical speed (n_c) of rotation for ball mills, at this speed of rotation F_c is just equal to w_b. This can be obtained by combining eq. 16 and 17.

$$n_c = \frac{1}{2\pi}\sqrt{\frac{g}{R-r}} \tag{18}$$

For effective size reduction to take place the speed of rotation should be less than n_c. The rotational speeds of the ball mills should be 65 to 80 per cent of n_c, higher values for solid / dry grinding and lower values for wet / viscous grinding.

1.3.2 Crushers

The most common type of crushers used in food industries are the, - Crushing Rolls (Fig. 7). Other than this there are also, - Jaw Crushers and Gyratory Crushers, they find limited or no use in food industries. In the crushing rolls food material is nipped and pulled through two heavy rotating steel cylinders revolving towards each other. The food material is crushed by compressive force causing a reduction in size. Depending on the product characteristics, crushing rolls can have a serrated or smooth roll. A shearing action can be introduced while crushing is taking place by a speed differential between the rolls. The crushers are generally used for coarse grinding purposes; a common example is its use for the manufacture of CTC (cut-tear-crush) tea leaves.

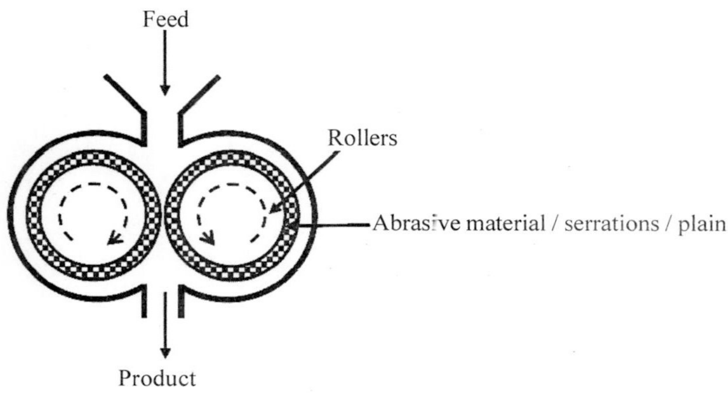

Fig. 7: Schematic diagram of roller crusher

10

Homogenization

Anit Kumar, Ashutosh Upadhayay, Rachna Sehrawat, Babar Onkar Avinash and Rupesh Chavan

1. Introduction

Raw milk is an emulsion. It is a mixture of milkfat globules, Solid Not Fat (SNF) and water. Over time, milkfat globule separate out from milk serum and comes to the surface of milk as a layer of cream due to the low density of milkfat or due to heating of milk. When milk is heated rapidly and continuously, some of the water in it evaporates from the surface. This exposes proteins and fat molecules, which bind and dry out and forms a skin like film on the surface. The skin is not harmful, but it is undesirable for many people. It can be prevented with constant stirring and through homogenization. First homogenizing machine was invented by Auguste Gaulin and patented in 1899. Homogenization is a process in which large molecules of fat globules (0.2-15 microns in diameter) convert into small fat globules ($\geq$ 2microns in diameter) and distributing it uniformly throughout in milk. In this process hot milk (approx 40^{0}C, where fat exit in liquid state) is forced through narrow opening with the help of piston pump under high pressure (10-20MPa) and speed, thus large fat globules are sub-divided into small fat globules of more uniform size due to the action of shearing, impact and distortion effect (Figure 1). Homogenization is carried out with the help of a mechanical device called a homogenizer.

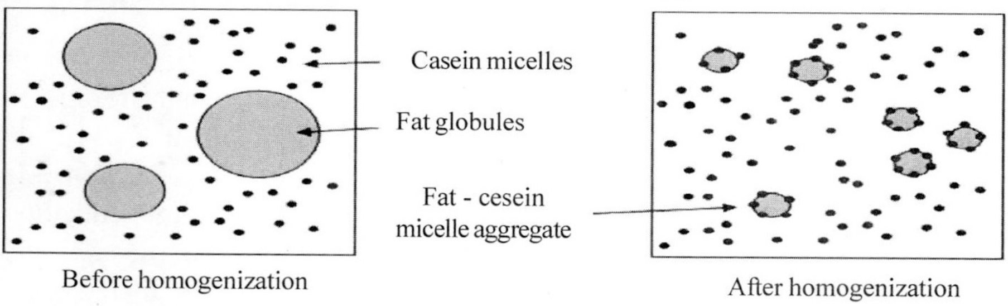

Fig. 1: Effect of homogenization on fat and casein protein of milk (Reproduced with the permission from Lalita Oraon, MT (PPD))

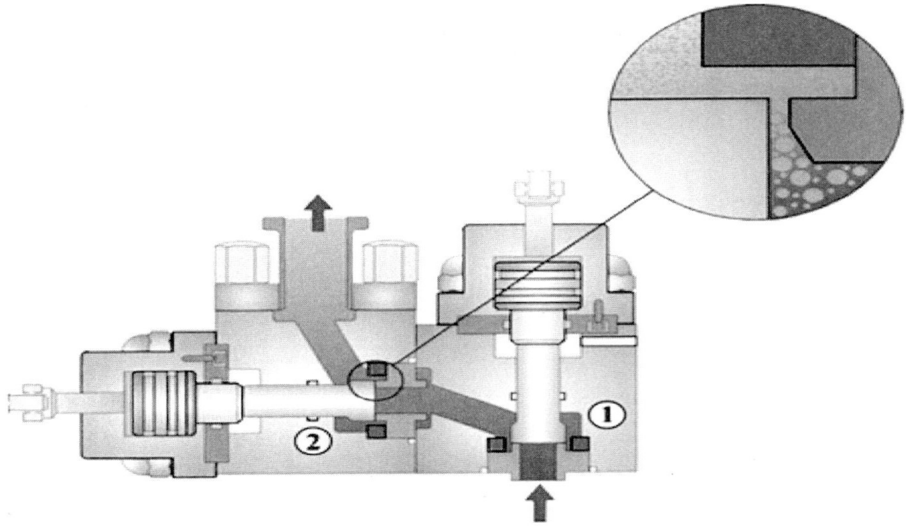

Fig.2. Operation principle of a two-stage homogenizer for liquid milk, where (1) representing the first satge and (2) representing the second stage. (Reproduced with the permission from Dairy Processing Handbook, Tetra Pak Processing system AB, Lund, Sweden, 1995)

Homogenization is a single-stage or two-stage processes, in single-stage process 2000-2500psi pressure are applied and in two-stage process, 2000-2500psi pressure is applied at the first stage and 500psi pressure is applied at the second stage (Figure 2). Generally two stage process follow when milk contain more than 6 percent fat. Small fat globules molecule has tendency to clump together due to the action of Laplace pressure, second-stage of the homogenization process at low pressure breaks up this unwanted assembly and gives uniformly homogenized milk. The Laplace pressure is the difference between the pressures at the concave and convex sides of a curved interface of two fluids. For a milk fat globule, the Laplace pressure (P_L), $P_L = 2\gamma/R$

Where, γ = Interfacial tension and, R = Radius of the droplet. The γ-value for native milk fat globule is very small, but it increases as globules are deformed (Walstra, 2003).

The size reduction and retard separation of fat globule on homogenization, results in a significant increase in the fat globule surface area and prevent the formation of skin layer on the surface of milk during storage. Small molecules of fat get stabilized in the milk due to the advancement of new surface of fat molecule with the combination of casein micelles and whey protein (If homogenization process occurs with heat treatment). Size reduction of fat globules and the adsorption of milk proteins on the surface of fat globules, increased their density and thereby decreased the rate of rising of fat globules towards the surface.

Homogenization process usually follows pasteurization otherwise the lipase enzyme (which is present in milk) may be activate and give bad flavor in milk due to the formation of hydrolytic rancidity. Now a day's microfluidization process are using by several dairy industries for better texture and viscosity of milk product (Chavan et al., 2014). Microfluidization is a high pressure homogenization (up to 300MPa) which is based on a double-acting intensifier pump or single acting intensifier pump and an interaction chamber.

2. Principle of Homogenization

The basic principle of homogenization is size reduction of milk fat globules by mechanical device. The reduction of milk fat globules is attributed to a several factors: (1) Shear stress between surface of milk fat globule and homogenizing device (2) Occurrences of pressure fluctuations under turbulent flow conditions (3) Vibrations of fat globules (>10 kHz) in milk may be generated as a result of the flow pattern or be a result of cavitations and (4) High velocity may provide additional impact such as shock force.

3. Design of Homogenizer

A heavy pressure spring is attached with the homogenizing valve which works against the seat of the valve. Generally, the homogenizing valve and the seats are made up of stelite and their contact faces are carefully ground, so that the homogenizing valve sits accurately on its seat. Homogenizer pump is an important part of the homogenization; generally it is in triplex design (sometime use five or seven pistons).

Single piston pump normally avoided due to it's generating a pulsating output with fluctuating pressure, which is difficult to damp. During processing, piston is lubricated by a fine water jet to reduce wear and a tubular sieve is frequently attached in the inlet manifold of homogenizer to stop the entry of foreign bodies in the pumping chamber. The homogenizer may be fitted with ball valves, which can apply more pressure. Recently, TetraPak has introduced a mushroom valve, which gives combine effect of the poppet and ball valves. Another type of valve such as Gaulin valve is still generally utilized and gives a decent model.

4. Location of Homogenizer in milk circuit

When the homogenizer is operated in conjunction with the HTST pasteurizer, it shall be installed so that it will not reduce the holding time below the legal minimum. When the homogenizer is used as a conventional timing pump, it must be interwired with the flow diversion valve (FDV) so that they homogenizer cannot operate at sub-legal pasteurization temperature unless the flow diversion valve is in its fully diverted position. Under these conditions, a time-delay relay may be installed in this circuit to permit the homogenizer to continue running during the normal time it takes for the flow diversion valve to move from forward to diverted flow. The time delay should not be more than one second. If the homogenizer motor stops it should not restart automatically. It shall not be possible to restart the homogenizer motor at sub-legal temperature unless the flow diversion valve is in its fully diverted position. The time-delay relay may be of the fixed time or the adjustable time type. If the time-delay relay is adjustable, means of sealing the unit must be provided.

If the homogenizer is of lower capacity than the timing pump at its maximum sealed flow, and the timing pump feeds product to the suction side of the homogenizer, it shall be installed upstream from the flow diversion valve. A sanitary relief line from the discharge side of the timing pump to the balance tank may be provided. This line is equipped with a relief valve capable of maintaining sufficient back pressure to assure a full supply of product to the homogenizer. Since the homogenizer an produce flow through the holding tube when the timing pump is stopped, an electrical interlock is required between the homogenizer and the timing pump, which causes the homogenizer to stop when the timing pump stops. A time-delay relay should be installed in the electrical circuit so that during the normal movement time of the flow diversion valve (one second or less from forward to diverted flow), the homogenizer motor will remain running.

When the "output" of the homogenizer exceeds the maximum sealed capacity of the timing pump, a recirculating line, connecting the suction line and the pressure line of the homogenizer, is installed to prevent "starving" the homogenizer. This line must be the same size as, or larger than, the inlet line to the homogenizer. When such a recirculation line is used, it is not necessary to interwire the homogenizer and the timing pump, since the homogenizer cannot flow through the holding tube.

When the homogenizer is installed upstream from the flow diversion valve, it normally is placed between the raw product outlet of the regenerator and the inlet of the heating section. When the homogenizer is installed downstream from the flow diversion valve, it normally is placed between the outlet of the flow diversion valve and the inlet of the pasteurized side of the regenerator. Installation shall be made in such a way that the homogenizer cannot apply negative pressure at the forward flow port of the FDV.

5. Efficiency of Homogenization

To determine the ideal operating pressure of the homogenizer on a particular product, the first step is to establish a method of evaluation. The method may be viscosity increase or decrease, average particle size (as determined microscopically or by instrument), rate of separation, grind gauge or change in physical appearance, just to mention a few. The next step is to collect samples from the homogenizer at different pressure settings; for example, every 1000 psi (or less, if a narrow pressure range is being evaluated). Finally, using the appropriate method of analysis, the evaluated parameter is compared or plotted versus homogenizing pressure and the lowest pressure that generates the required product can then be selected. For some products it may be desirable to determine the effect of multipassing at an appropriate pressure.

The efficiency of the homogenization may be useful in checking the continuous effectiveness of the equipment. In dairy industry, the simplest methods are carried out for the checking of efficiency of the homogenization. In which, store a 100 ml sample of milk in a measuring cylinder for 24 hour in a refrigerator and then quantify the fat content of the top and bottom portions of the milk by Gerber method. According to R.A. Wilbey (2011), homogenization efficiency is expressed as a homogenization index:

$$\text{Homogenization index} = \frac{\text{Difference in fat contents}}{\text{Fat value in the upper layer}} \text{ X } 100$$

6. Application of homogenization

Now a day's various industries are using homogenization for manufacturing of several products, such as:

(6.1) Application of homogenization in Dairy/Food Industry: In dairy industry homogenization is apply for the processing of milk, in which it helps in uniformly distribution of fat in whole milk by the mechanical breaking of fat globules and prevents the natural separation of cream from the milk. It provides stability of fat molecule in the milk serum. Cream and other dairy products, like peanut butter, may be homogenized to produce a stable emulsion. Homogenization also helps in the improvement of the texture and viscosity of various dairy products such as yoghurt. In beverage industry homogenization, stabilize the fruit juice by the reduction of particle size of fruit juices. During storage period, phase separation problems of juice may be reduced with the help of homogenization.

(6.2) Application of homogenization in Pharmaceutical industry: A similar process is used in the manufacture of some cosmetics and pharmaceutical products called as micronization. It reduces the particle sizes of pharmaceutical products to make them more stable and clinically effective.

(6.3) Application of homogenization in Biotechnology: It is mostly used in cell rupture operations and increases the yield from valuable source material. Sometime it is used for killing of bacteria by destruction of its cell wall.

(6.4) Application of homogenization in chemical industry: In chemical industry, homogenization is used in color extraction processes and also it is used for several processes such as to control the viscosity in the petrochemical industry.

7. Advantages and disadvantages of homogenization

Advantages: (1) Slight increase the smoothness or creaminess of the milk, so that it gives better mouthfeel (2) No formation of skin or cream layer for a long time (3) Stabilize the emulsion hence no separation take pace (4) It increase the surface area of fat globules and helps in easily digestion of fat (5) Provide more brighter appearance than unhomogenized milk (6) It produce soft curd, hence it may recommended for infant feeding (7) Increased the viscosity of milk.

Disadvantages: (1) Increased the production cost (2) It may gives bad flavor by increasing the chance of lipolysis of milk with the help of lipase enzyme (Lipoprotein lipase is a heat labile enzyme and its effect may be reduce by pasteurization before or instantly after homogenization) (3) Recovery of fat from homogenized milk is a big problem (4) More susceptible to produce sunshine flavor due to the conversion of methionine to methional (5) Curdling in cookery (6) Increased the area for microbial contamination.

References

1. Walstra, P. 2003. Physical Chemistry of Foods. Marcel Dekker, Inc., New York.
2. Principles and Mechanism of Homogenization, Effects and Assessment of Efficiency: Valve Homogenizers. R A Wilbey (2011), The University of Reading, Reading, UK, 2011 Elsevier Ltd.
3. Rupesh Chavan, Anit Kumar, Vijendra Mishra and Prabhat K. Nema. (2014), Effect of microfluidization on mango flavoured yoghurt: rheological properties and pH parameter, International journal of food and nutritional sciences. Volume 3, No 4, 84-90.

11

Mixing

Smriti Pillai, Sunil C.K., and Akash Pare

1. Introduction

Mixing is a fundamental and widely practiced operation in many fields, such as chemical, pharmaceutical, and food processing industries. Themixture's primary aim is to increase the homogeneity of the material in bulk (Uhl and Gray, 1986; Patterson *et al.,* 2004). Mixing means the random distribution of two or more ingredients throughout a system, which differs from the term' agitation,' which is defined as establishing a particular flow pattern with the liquid, especially in a circulatory motion within a container. Mixing is a necessary step in the context of food processing operations. The ingredients used in mixing can be present in any physical state of matter, like solid, liquid, and even gas. It can be mixed or dissolved to form a single uniform phase or dispersed homogeneously to create a multi-phase mixture, e.g., processed ready-to-eat products such as emulsions (soups), solid dispersed in liquids (baked beans), bubbly dispersions (ice-cream, sponge cake, etc.). Mixing includes a broad spectrum or range of materials from free-flowing powders to thin, viscous liquids to viscous pastes like dough and batter.

The food processing industry relies significantly on mixing to ensure the delivery of a product with constant properties. A mixing process is believed to be satisfactory when the process produces a uniform mixture in minimum time and minimum cost for overhead, power, and labor. Mixing can be done either in batch or continuous flow process. Yet, presently, almost all food mixing processes are batch, due to their higher rate of ûexibility and lower installation costs than constant methods. The actual knowledge in mixing is a combination of "know-how- to" specifics and science, in which "know-how-to" is predominant, and only a few individuals and companies around the world have thorough information on this matter.

1.1. Objective of mixing

Generally, the main two objectives of using mixing in the food industry are:

i) to reduce the gradients in properties such as concentration, color, texture, or taste between different systems (Uhl and Gray, 1986).

ii) to obtain a nearly uniform distribution of particles, i.e., attaining homogeneity in the product. Typically, the homogenization consists of particle size reduction, which is achieved by the action of shearing forces.

Mixing requires the movement of ingredients to achieve the desired homogeneity or a uniform distribution for a particular food commodity. It is due to food particles' motion between different parts of a whole particulate mass (Lindley, 1991).

1.2. Difference between solid and liquid mixing systems

Studies conducted have shown the following differences between mixing particulate systems and mixing liquid systems.

i) There is no particulate motion in the pattern that follows the molecular diffusion of gases and liquids. Thus, a randomized state in motion does not occur without external energy input.

ii) Granular or solid particulate components may not usually have constant properties and vary vastly regarding their physical characteristics. A mixer may cause a grading or segregation of particles based on their characteristics as size, density, resiliency, etc.

iii) Samples withdrawn from a randomized particulate mixture have a coarser texture or lower mixture quality than liquid mixtures.

2. Types of mixing

In food processing industries mixing operations occurs differently for different systems; the most frequently used types of mixing systems are-

i) Solid-solid mixing: all ingredients (dry, flake/powder/granule form) are mixed uniformly to provide consistent nutrition according to label specifications, for textural effects, and to give a specific taste.

ii) Solid-liquid mixing: the addition of solid to liquid produces the reconstitution of fluids, whereas the vice versa creates critical products to many food batters, pastes, and dough.

iii) Liquid-liquid mixing: where the addition of liquid in the liquid is the basis to the production of the emulsion, e.g., margarine and spreads.

iv) Gas-liquid mixing: when an appropriate amount of air is mixed into a chamber containing liquid such as a fermenter, there is no oxygen supply for microbial growth.

As mentioned earlier, the components being mixed can exist in any of the three forms - liquid, gas, or solid, hence such several mixing possibilities can arise, like ûuid, e.g., liquid-liquid and solid-liquid, which are most frequently encountered; therefore, they have been studied in detail. Despite the supreme importance of mixing particulate materials in many processing areas, fundamental contributions of real value to designers or users of solids mixing equipment are still relatively sparse. Significant advances in understanding the mixing of solids and pastes have been made in very speciûc ûelds such as powder technology and multi-phase ûow. Mixing is more difûcult to deûne and evaluates with powders and particulates than it is with ûuids. However, advanced studies in quantitative measures of dry solids mixing are helping to control mixer performance. Still, practically, a mixer is tested by the properties of the mixed material which it produces. Much significant proportions of research efforts in the food sector are initiated towards developing new and novel mixing devices for food materials. These devices may be useful for many applications as they deliver mixed products with the required blending characteristics. Powdered foods are complex systems, and properties in food particles can themselves change during mixing. Research and development in mathematical modeling of food mixing processes are scarce. The already available procedures for process design and scale-up models lack and result in the virtual impossibility to devise any relationships between mixing and quality of food products (Niranjan, 1995), especially when blending food powders. Table 1. presents some examples of different types of mixing used in the food industries.

Table 1: Examples of various types of mixing used in the food industry

S.No.	Types of Mixing	Mixing Mechanism	Examples
1.	Solid-solid mixing	Uses convective mixing and intensive mixing	Breakfast Cereals
2.	Solid-liquid mixing	Uses associated eddy diffusion which increases the rate of mass transfer between the solid and liquid	Batters, pastes, and dough
3.	Solid-gas mixing	Carried out by mix gaseous reactants with solid catalyst particles	Wood smoke used in the preservation of meat
4.	Solid-liquid-gas mixing	Occurs due to multi-phase mixing	Whipping cream
5.	Liquid-liquid mixing	Occurs due to single-phase mixing	Emulsions;e.g., margarines and spreads
6.	Liquid-gas mixing	Uses mass transfer between liquid and gas particles	Carbonated beverages
7.	Gas-gas mixing	Uses a mass fraction of the gases to be mixed	Regulation of gas mixture in MAP

3. Concept of mixing

As established, mixing is a unit operation used in many lines of work from medicine to construction. Whereas, while dealing with foods, mixing is an essential unit process to ensure food of homogeneous quality attributes is being supplied to consumers. However, mixing subsequently differs depending upon the system's physical phase, i.e., solid, liquid, and particulate foods.

3.1.1. Mixing of solid and particulate foods

In contrast to miscible liquids, the mixing of solid and particulate foods is often readily reversible. In particle materials, mixing occurs due to three main mechanisms (Fig.1), namely

i) dispersion,

ii) convection, and

iii) shear mixing

In any particular mixing process, one or more of these three basic mechanisms may be responsible for the course of the processing, mechanisms, such as segregation, can also be involved during particle mixing. Mixing mechanisms can also receive other classifications depending on the type of equipment used.

During convective mixing, bulks, or masses or groups of particles combined transfer from one point or location to another. In contrast to diffusion mixing, where individual particles are randomly distributed over a surface created within the mixture. In shear mixing, groups of particles are mixed through thedevelopment of slipping planes developed within the bulk or mass of the mixture. Shear mixing is, at times, considered as part of the convective mechanism.

Dispersion, which is also known as diffusion or micro-mixing, is known as the particle's overall random location change. The frequency with which each particle of ingredient A changes place with another ingredient is said to be the number of particles of the opposite ingredients within the direct and nearest reach of the particle of ingredient A. Hence; dispersion is defined as a neighborhood effect, i.e., micro-mixing, that occurs within the case of premix systems where a variety of particles of various ingredients which is in proximity, results in a ûne mix, that are localized to very small areas (Harnby et al., 1992). If the ingredients are separated spatially from the start of the method, a longer time would be required to mix them through dispersion alone, since they would be a higher low number of varied and valid neighbors. Generally, dispersion corresponds to diffusion in liquid mixtures. However, in variation to diffusion, mixing within the case of dispersion isn't caused by any concentration gradient as in liquids. The particles must be in motion for getting dispersed via vibration.

While on the other hand, convection, also known as macromixing, causes a movement of vast groups of particles relative to each other (macro-mixing). The complete volume of material is repeatedly segregated and remixed after the portions have changed (Fig. 1). Rotating elements will achieve this process of forced convection. The groups, mainly composed of only one pure ingredient, are set by the continuously reduced splitting action of the rotating paddles present in the mixer. This mechanism increases the number of assorted neighbors. It thereby increases the exchange processes of dispersive mixing as material mass is split up or convectively mixed through the rearrangement of a solid's layers by rotating devices within the mixer or by the falling stream of materials a static gravity mixer.

While shear mixing is induced by the momentum inter-change of powder particles having different velocities, i.e., differential velocity distribution, it is developed by the development of slipping planes in the bulk material; the original coherent particle groups are gradually broken down along these planes. Due to compression and extension of bulk powders, the velocity distribution forms around the agitating impeller and its vessel walls. The same is also seen in the powder layer in the rotary vessel mixers and at blowing ports in gas-flow mixers. Shear mixing can increase semi-microscopic mixing and be advantageous for both batch and continuous operations. In free-flowing powders, both diffusive mixing and shear mixing giverise to size segregation or de-mixing; therefore, for such powders, convective mixing is the most suitable mechanism for attaining uniform-sized mixed particles.

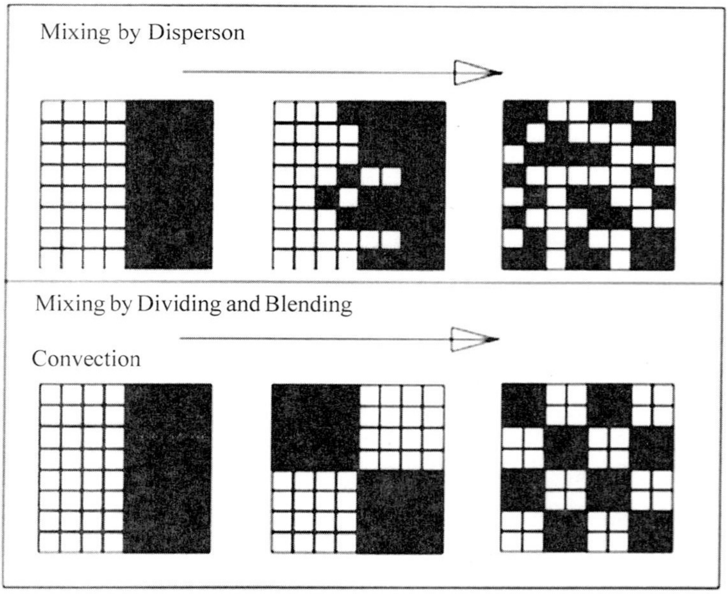

Fig.1: The diagrammatic representation of overlap of dispersion and convection. The mixture consists of two components A and B, where A symbolises the white blocks, and B symbolizes the hatched blocks. Dispersion results in the random arrangement of the particles, while convection results in a regular pattern. (Weinekotter and Gericke, 2013)

3.1.2. Mixing in liquids

When material or substance are dissolved or dispersed in liquids, the spatial distribution of solute concentration at any given time and instant can be determined by using statistical parameters. With the attainment of uniformity, the variation in spatial concentration should reduce and/or diminish progressively. In food processing industries, solid dispersion particles, insoluble liquids, and gas bubbles are largely practiced. On the one hand, the rate of dissolution has been studied in depth. On the contrary, limited studies have been conducted in terms of the relationship between the dispersed phase's spatial homogeneity and the movement of substances caused inside the mixer. The time of mixing indicates the variation of concentration with time at a given space. This is determined as the time measured from that instant, via a pulse tracer, which indicates that the level has reached a particular degree of homogeneity. Whereas, for impeller- agitated systems, dimensionless mixing time is known as the impeller rotational frequency product, which depends on the Reynolds number and the mixing time. Fig.2 shows the typical plot of dimensionless mixing time v/s

Reynolds number. These are used to determine the time required to mix smaller quantities of soluble matter into the bulk liquid, such as dispersing concentrates of flavors into solutions (Barbosa and Cánovas, 2009).

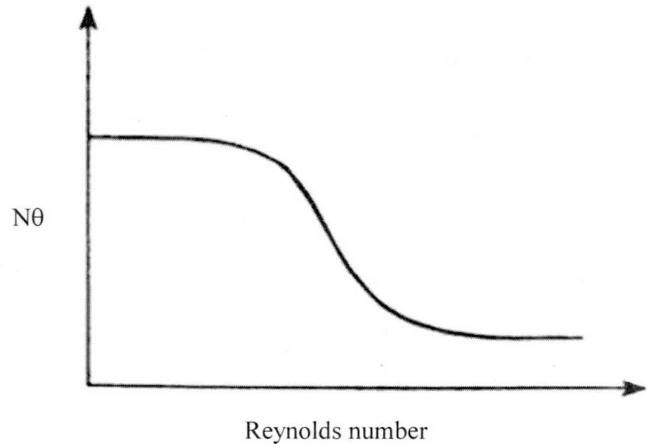

Fig. 2: Mixing time versus Reynolds number defined based on impeller diameter and fluid properties (Barbosa and Canovas, 2009)

The suspension of solid, dense particles in a liquid is a regularly conducted process in the food industry, which is necessary to produce uniform suspensions. Though, many studies have been conducted on the least or minimum speed requirement for suspensions of solid, dense particles in the liquid. Conversely, less is known on the power requirement for the subsequent homogenization of particles.

While dealing with the mixing of fluids, the evaluation of gas- containing mixes are also crucial in the production of aerated or carbonated beverages, along with the character assessment of bubble- providing dispersions, e.g., ice-creams, meringues, sponge cakes, etc., which are highly demanded in the food sector.

3.2. Mixing index

The mixing index is an indicator of the quality of the mixing process. It is defined as the degree to which fluid flow will promote the dispersion of dissolved solutes or suspended materials in the fluid, promoting to homogeneity.

A mixing index can help determine whether mixing is taking place and what degree it is taking place. A simple mixing index is a Peclet number (Pe), which is defined as the ratio of convection to diffusion:

$$Pe = \frac{UI}{D} \quad (1)$$

where U is the (absolute) fluid velocity, D is the solute diffusivity, and l is a relevant length scale, where a possible length scale is the Prandtl length (Plawsky, 2001), which is the length of the path a mass of fluid would travel before losing its characteristics by mixing with its neighboring particles. A large Peclet number dictates that convection dominates, and more mixing is occurring. Research of mixing strategies often determines the extent of mixing enhancement, which is produced in addition to the mixing intrinsic to the original flow field of the material. Due to this, it is obvious to use the perturbation velocity vector u, rather than the absolute velocity U, to quantify mixing. The perturbation's kinetic energy is referred to as the "turbulent kinetic energy" (k), which is a simple measure of mixing:

$$k = \frac{1}{2}\int_{\Omega} \| u \|_F^2 \, dv \tag{2}$$

Where Ω is the volume of fluid under analysis and $\|u\|_F$ is the Frobenius norm of the perturbation velocity vector. However, kinetic energy itself is not sufficient for adequate mixing. Mixing caused by material stretching and folding, which can be quantified through the stretching length based on the gradient of perturbation velocity (Ottino, 1989), which is given by,

$$S = \int_{\Omega} \| \Phi \|_F^2 \, dv \tag{3}$$

where $\phi = [\nabla u + (\nabla u)^T]/2$ is the stretching tensor. A mixing estimation parameter that uses the vorticity to describe mixing induced by vortices (Alexiadis *et al.* 2007). Defining $\psi = [\nabla u - (\nabla u)^T]/2$ as the folding tensor, mixing related to folding can be quantified by

$$F = \int_{\Omega} \| \Psi \|_F^2 \, dv \tag{4}$$

The sum of perturbation-induced stretching and folding can also be used as a mixing index (Balogh *et al.* 2005; Ouyang *et al.* 2013) and is shown

$$\Gamma = \int_{\Omega} \{(U.V)u + V)U\}.udv \tag{5}$$

Therefore, mixing indices defined by equation(iii), (iv), and (v) prove mixing enhancement due to perturbations but do not reflect mixing inherent in the steady-state system and related to the original velocity gradients and momentum diffusion. The interaction between steady-state flow and the distress can also lead to stretching and is described by

$$\Gamma = \int_{\Omega} \{(U.V)u + \overline{V})U\}.udv \tag{6}$$

3.3 Pearson's square method for mixing

Standardization refers to the adjustment, i.e., raising or lowering of fat and solids-not-fat percentage of milk to the desired value, to conform to some legal or other requirements. It is generally done by mixing of the addition of milk or cream with a higher and lower fat percentage than that of milk material to be standardized. A simple scheme, Pearson's square method, can calculate the relative quantities of the materials involved in a standardization problem. It should be remembered that all measurements based on these calculations are by weight and not by volume.

3.3.1. Example 1

How many parts by weight of 40% cream and milk of 3% fat mix make standard milk of 5% fat? If the processor needs 500kg milk of 5% fat, then how many kgs of 40% cream and 3% fats of milk will be required to mix?

Solution: The problem is solved by using Pearson's square method as

1. Draw a square and place the required fat percentage (i.e., 5% fat) in the centre.
2. On left-hand corners, place the given fat percent of cream and milk to be mixed (i.e., 40% cream and 3% fat of milk)
3. Find the difference of number on the diagonals and put them on the right-hand corner, as shown in Fig.3 (i.e., 40-5 = 35 and 5-3 = 2)
4. Now the right-hand corner (i.e., 2 and 35) represents the parts of left-hand corner values (i.e., 40% cream and 3% fat of milk), respectively.
5. The sum of right-hand corner values (i.e., 2+35 =37) represents the parts of standard milk of 5% fat.

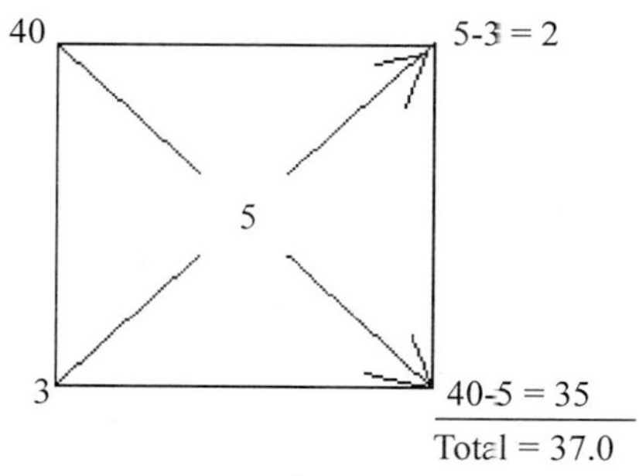

Fig. 3: Pearson's square method

Therefore, two parts of 40% cream mixed with 35 parts of 3% milk will give 37 parts of 5% milk.

To make standard milk of 500 kg with 5% fat, one will need 500 x 2/37 = 27.03 kg of cream of 40% fat and 500 x 35/37 = 472.97 kg of milk of 3% fat.

Problem 2. Determine the respective masses of 10% and 50% (by weight) aqueous sucrose solution, which must be mixed to prepare 100 Kg of 22% sucrose solution.

Solution: Let the masses of 10% and 50% solution of A and B, respectively. Then, the overall material balance is

$A + B = 100$

Where each term has a unit of Kg. The component mass balance (for sucrose) is

$0.10\,A + 0.50B = 0.22 \times 100$

Which can be solved to give

A = 70Kg and B= 30Kg

Alternately: The Pearson square method can also be used to solve such problems involved in the standardization process. It is of greatest value when only two ingredients are to be mixed. The method is as follows (refer Fig. 4a and 4b):

1. Draw the square and place the required sucrose percentage (i.e., 22%) in centre point E.

2. On the left-hand corner place, the given sucrose percentage to be mixed as A and B. (10% and 50%)

3. Find the difference of the number on diagonals as C (the difference between B and E) and D (the difference between A and E), as shown in Figure 2.3. Disregard any negative numbers that are generated on the right side of the square.

   Therefore, in order to prepare 100Kg of 22% sucrose solution, the quantity of two solutions will be

   Solution A = 100 x (28/40) = 70Kg

   Solution B = 100 x (12/40) = 30Kg

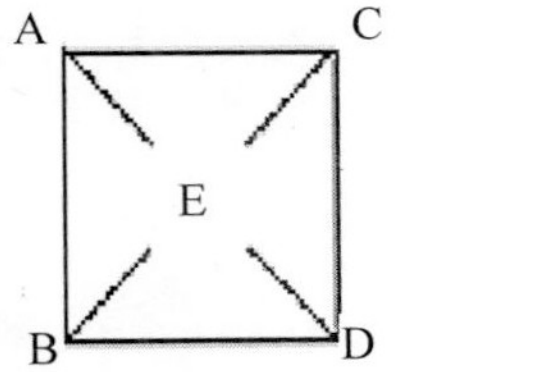

Total = C+D

Fig. 4a

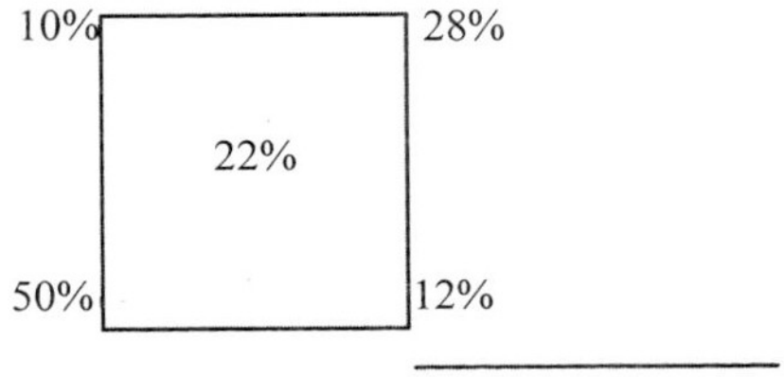

Total = 40%

Fig. 4a

4. Types of equipment used in mixing

4.1. Equipment for Mixing Particulate Solids

Mixers for particulate and solid materials fall into two categories:

i) Diffusive (passive) mixers: These mixers achieve homogeneity due to the random movement of the particles. When the particle is in motion under the effect of gravity or vibration, without external mechanical force or agitation. The mechanism is called "diffusion" (Hwang and Hogg, 1980), concerning molecules' random motion. There is nearly no shear. Consequently, these mixers are mainly used for particulates that require subtle mixing, like fragile agglomerates, but they do not function well with cohesive powders.

ii) Convective (active) mixers: These mixers carry out mixing with the help of mechanical agitation. Sections of the bulk material are transferred to each other by the action of impellers or turbulent gas flow, where shearing occurs considerably.

Along with mixing, powder mixers may take up other powder processing technology (Harnby, 1997). Some of these other functions are:

i) particle coating

ii) agglomeration

iii) an admixture of liquids (such as fat into a dry soup mixture)

iv) drying

v) size reduction, change of particle shape.

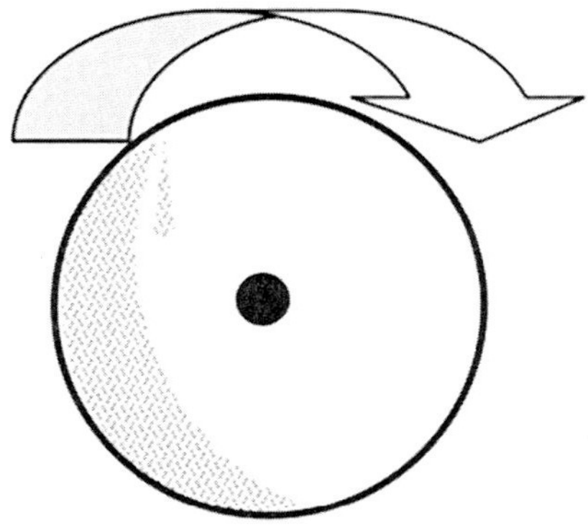

Fig. 5: Cross-section of drum blender (Berk, 2018)

4.1.1. Types of diffusive mixers

Drum blender is a type of diffusive mixing equipment used in food processing industries to mix solid particulate materials. Fig.5 shows a horizontal rotating cylinder on its axis. The mixing action is essentially diffusive. The powder to be mixed is placed in the drum. As the drum rotates, the powder travels up the drum until the angle of repose is increased. The powder drops back on the rest of the bulk and enters a new cycle of rising and dropping. Here diffusive mixing takes place during the residence of the powder in the air while dropping. Continuous operation can be made possible by tilting the drum.

i) Tumbler mixers are mainly used for mixing powders, granules, lubricant additive, binders, and oils. In food processing industries, two variations of tumbler mixers are used, as shown below.

Firstly, in the double-cone tumbler (Fig.6a), the powder undergoes repetition of expansion and compaction as the vessel rotates, whereas, in the V-shaped tumbler (Fig.6b), the powder is subjected to cycles of division and reassembly. Convectional factors like turning or stationary flow distortion bars (intensifier bars) are sometimes installed in both mixers.

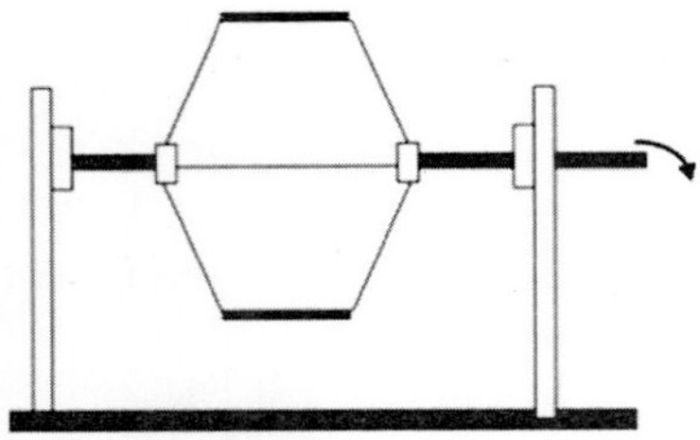

Fig. 6: Cross-section of double-cone tumbler mixer(Berk, 2018)

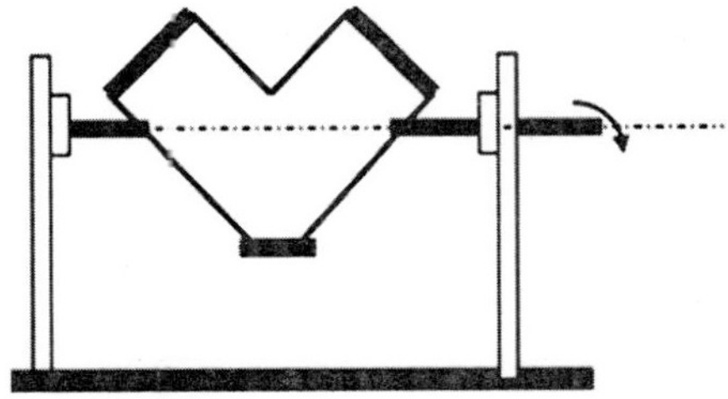

Fig. 7: Cross-section of V-shaped tumbler mixer(Berk, 2018)

4.1.2. Types of convective mixers

i) Paddle mixer (Fig.8) is a powerful mixer for mixing solids, powders, and liquids. These multi-functional mixers consist of rotating parts mixes the powder both by moving the bed and by fluidizing action. A liquid component may be sprayed on the powder while mixing. Mixers of these types can be used both in batch and continuous operations. Recent advancements in this type of machine allow the entire mixing chamber can be rotated upside down for faster discharge of the product and then reverted for charging a new batch of feed. Further explanation on paddle mixers in 4.2 (iii)

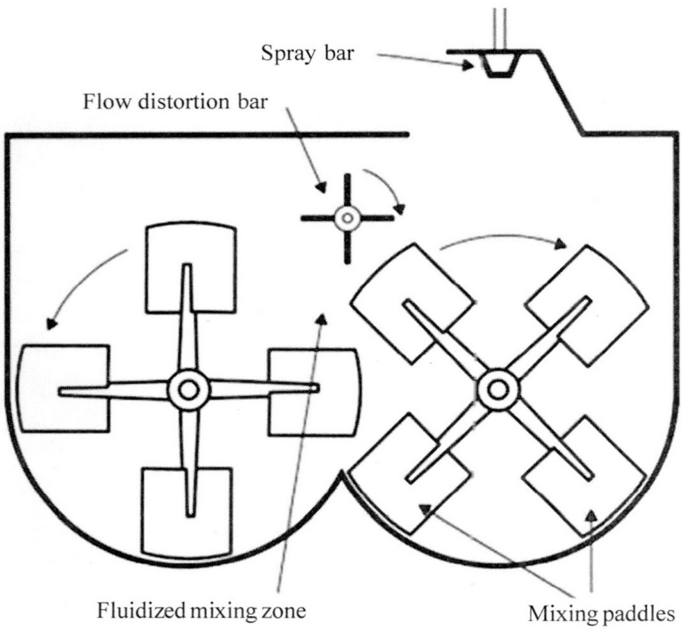

Fig. 8: Cross-section of paddle mixer (Berk, 2018)

ii) Trough mixers, also known as the ribbon mixer, are used for mixing powders, paste, and heating and cooling of materials, with the help of shear and gentle mixing. It consists of conduct with a U-shaped cross-section and a longitudinal rotating shaft (or shafts), carrying various mixing elements. The mixing parts may be a series of paddles or a screw, such as in screw conveyor. (Fig. 9).

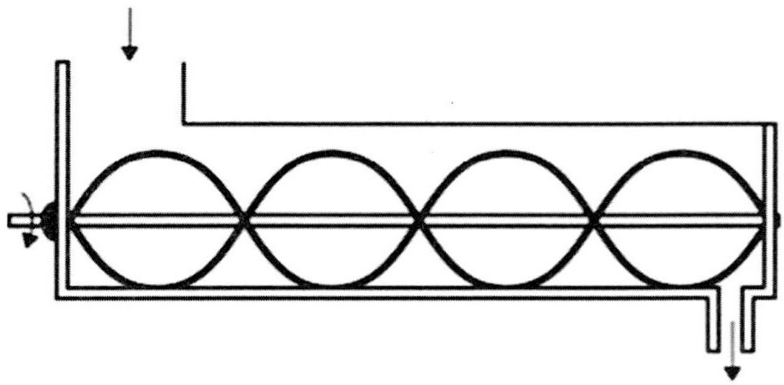

Fig. 9: Cross-section of trough mixer (Berk, 2018)

4.2. Types of equipment used for liquid/fluid mixing-

i) Propeller mixers (Fig.10) are used for mixing liquids of a wide range of viscosities. It is used due to its more streamlined shape, the power requirements of a propeller mixer less than other types of mixers at the same Reynolds number. They have a high-speed operation without using a gearbox and are more cost-effective as there are no mechanical losses in the gear system and transmission. Propeller mixers produce an axial flow, which has a higher pumping effect and provides shorter mixing times. Its disadvantages in comparison to paddle and turbine mixers include higher initial cost, the sensitivity of operation to the container (vessel or tank), and its location within the tank/vessel.

Draft tubes are installed to enhance the mixing of large quantities of liquids by directing the fluid's movement. The mixer's high pumping capacity is mainly used in such applications, mainly where the mixture of low viscosity is used. The draft tube directs the flow to the vessel (or tank), which or else would not be agitated by the liquid stream. The absence draft tubes and at high rotational velocities of the propeller, baffles are placed at

Fig. 10: Propeller mixer (Berk, 2018)

various vessel or tank points. Baffles minimize vortex formation and divide it into several local eddies, increasing the container's total turbulence.

ii) Turbine mixers (Fig.11) is an essential class of liquid mixers which is similar to a centrifugal pump impeller; it consists of a constant blade angle concerning to a vertical plane, over its entire length or specific parts, having blades either vertical or set at an angle less than 90Ú with the vertical. Edges may be curved or flat. The mixing action is acquired by the turbine blades, which entrain and discharge the liquid. The predominantly radial flow from the impeller impinges onto the vessel walls, where it splits into different streams. This energy causes the mixing of the liquids. When turbine mixers are operated at high rotational speeds, both radial and tangential flows occur, along with vortex formation in the center. This situation is controlled by installing baffles to ensure a more uniform flow distribution of liquid throughout the mixing container.

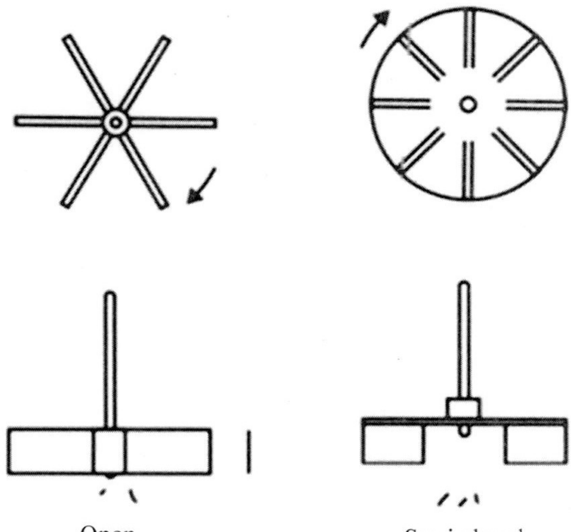

Fig. 11: Turbine mixer (Berk, 2018)

iii) Paddle mixers (Fig.12) are mixers containing two or more blades mounted on a vertical or inclined shaft. The main advantages of paddle mixers are their simple and low cost. Their disadvantage is they have small pumping capacity, which means a slow axial flow, due to which thorough mixing does not occur along the volume of the tank or vessel. The perfect mixture is obtained only in relatively thin strands of liquid, which reach the blades' immediate reach. The turbulence spread occurs outward, slowly and imperfectly into the container's entire contents, resulting in the slow circulation of the liquid. Due to the concentration gradient, which is regularly created in the liquid when these mixers are used, they are usually not suitable for continuous operation. This can be corrected by tilting the paddle blades 30Ú to 45Ú to the shaft axis, resulting in higher axial flow and lower concentration gradients. Such a mixer can maintain particle suspension, provided the settling velocities are not high. Tilted blade mixers are used for slow reactions, which are not contained by diffusion. Paddle mixers have rotational velocity in the range of 15 to 60 rpm. However, some particular types might be operated outside of this speed range, such as the leaf-shaped (broad-blade) paddle mixers, which operate at speeds up to 120 rpm and incorporate baffles. There is a reduction in vortex formation also.

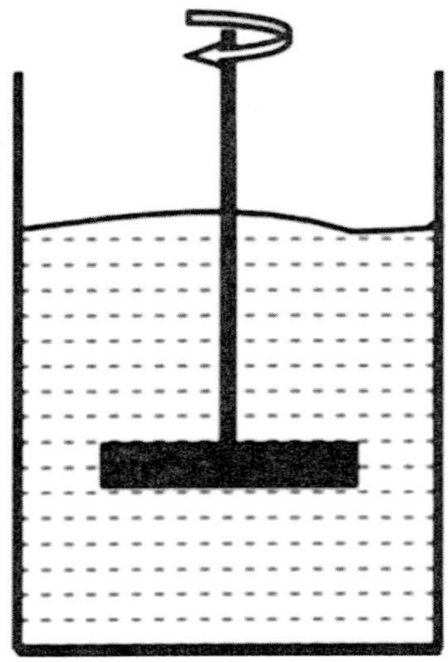

Fig.12: Paddle mixer (Berk, 2018)

iv) Anchor mixers, which are similar to paddle mixers, are used for mixing liquids in hemispherical kettles and vessels with a bowl-shaped bottom (Fig.13), that imitates to the bottom of the vessel walls which are used. These mixers are usually operated at low speed, often tens of revolutions per minute. Anchor mixers are commonly used in jacketed cooking kettles and are typically equipped with scrappers or wipers that scrape the product off the heated surface to prevent scorching.

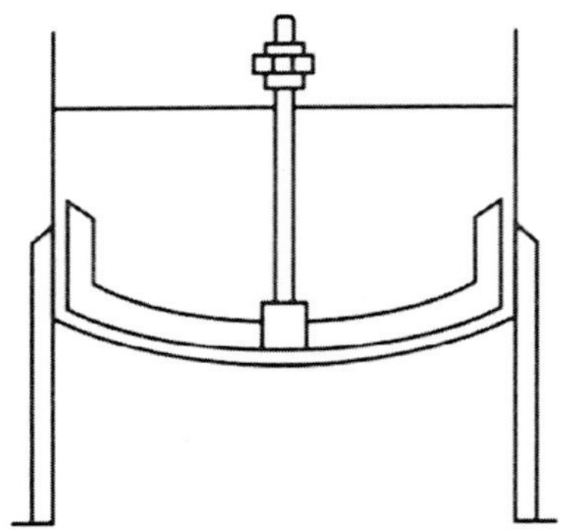

Fig. 13: Anchor mixer (Berk, 2018)

The above discussed different types of mixing types of equipment used in food processing industries are shown in Fig.14.

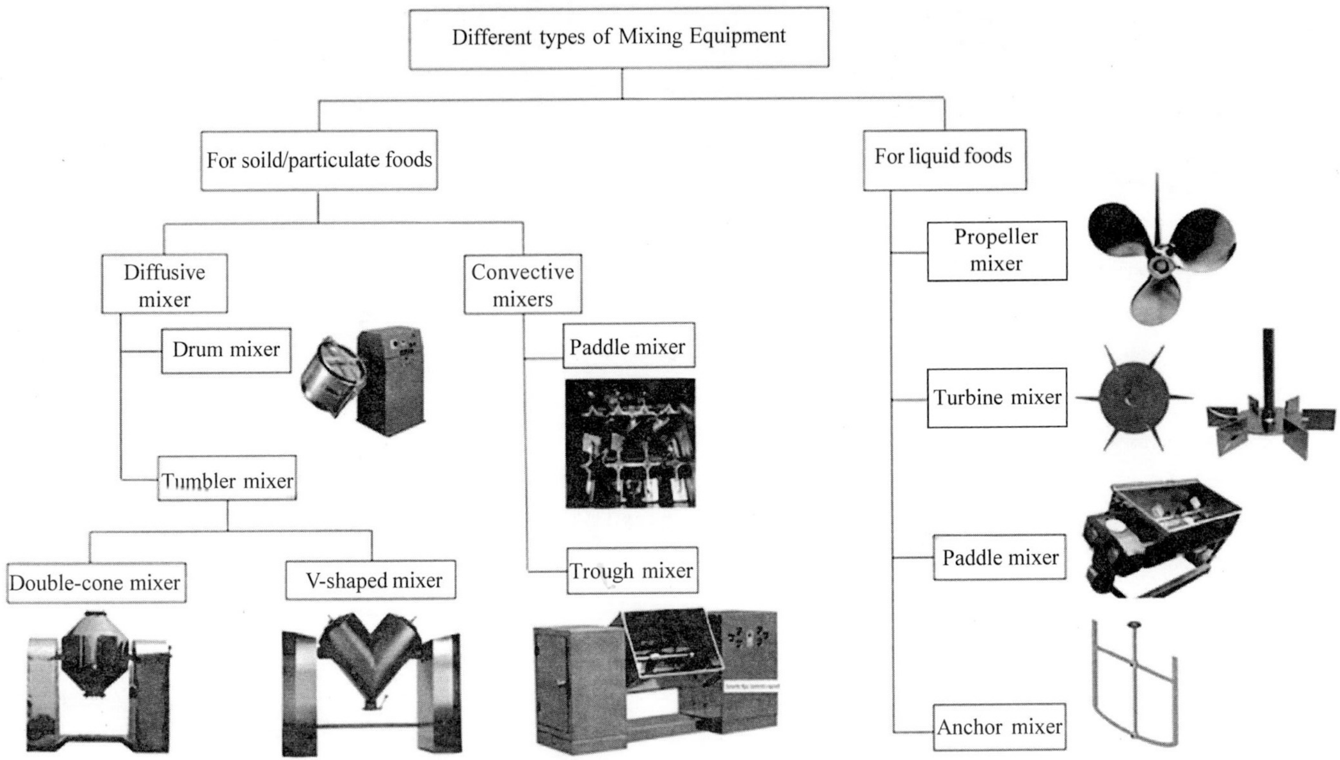

Fig. 14: Classification of mixers

5. Problems during mixing

Although mixing processes are not always designed with the appropriate care, due to this, it causes a significant financial loss which arises in two ways:

i) The quality of the mix is low: when the mixing procedure produces the end product, the main focus would depend on the end quality of the product. This will be noticed immediately during the product's quality inspection and evaluation. Frequently, however, mixing is only one in a series of further processing stages. In such cases, the effects of unsatisfactory blending are less apparent, and it might be overlooked to the detriment for ûnal product quality.

ii) The homogeneity is satisfactory, but the effort employed is too high: Overmixing in batch blending system is induced by an overlong mixing time or too long residence time in a continuous blending system. This leads to a higher strain on the mixture of material, which can have an adverse and irreversible effect on sensitive products' quality. Furthermore, larger or more numerous pieces of equipment must be used than would be necessary for an optimally configured mixing process.

iii) Material flow factor: a wide variation in particle properties such as particle-size distribution, density, shape, and surface characteristics, i.e., electrostatic charge, might cause significant difficulties in blending. Certain additional properties which might affect the mixing process are, mainly, flow characteristics, i.e., angle of repose and flowability; friability, which is when the materials tend to break into smaller pieces; agglomeration-energy which is essential to cause the breakdown of particles; moisture content, and temperature limitations of the ingredients or changes that might occur because of temperature changes (Perry *et al.* 1984). The relative absence of cohesive or adhesive forces in particulate materials will be dependent on the size and moisture content of the material. Segregation, which is when the particles tend to separate depending on the size and/or density, is a problem that generally occurs in solids mixing only (Nienow *et al.* 1985). If mixtures of solid particles have the same size and shape but different densities, or vice versa, difficulty in mixing may occur. More massive particles tend to accumulate and stay near the bottom of the container during a mixing operation. A study conducted stated that "round or small particles tend to travel toward the top during mixing" (Henderson and Perry, 1976). However, an alternate study reported that for particles of the same density but different sizes, the smaller particles would travel to the bottom (Perry,1950). Also, the round, smooth particles are found near the bottom, while the jagged or polyhedral ones seek the top of the container. The mixing process must overcome these natural tendencies of particles; this is obtained by lifting the materials, mostly in terms of their mass, from the bottom to the top of the mixing vessel while allowing it to fill the voids with the help of gravity falling (Henderson and Perry, 1976; Perry, 1950). Flow properties can be modified by the addition of anti-caking agents to prevent unwanted agglomeration and flow promoters.

6. Conclusion

Although mixing of materials is a common and ancient practice in many fields, considering the world-wide application of mixing, the amount of knowledge available appears to be quite limited. Though mixing of solids and melts has always been problematic, leading to diverse models leading to its the melting process, and the changes mixing causes in the viscosities of fluids remain a matter of concern. Hence, we can say there is still an overall lack of general understanding of what happens in this process, as the details and mechanisms have not been thoroughly studied, and developments in its research have still been carried out into creating appropriate design procedures. Although mixing might not always be the only process in determining the product's end quality, it still has an unnoticeable effect on the food material in terms of its texture, nutrient, and overall mouthfeel. Therefore, it can be confirmed that mixing ensures the delivery of a product with constant properties. If mixing fails to achieve the needed product yield, quality, organoleptic, or functional properties, the production costs may rise significantly.

References

Alexiadis A, Wiley DE, Fletcher DF, Bao J (2007) Laminar flow transitions in a 2D channel with circular spacers. Ind Eng Chem Res 46(16):5387–5396.

Balogh A, Aamo OM, Krstic' M (2005) Optimal mixing enhancement in 3-D pipe flow. IEEE Trans Control Syst Technol 13(1):27–41.

Barbosa-Cánovas, G. V. (Ed.). (2009). *Food Engineering-Volume I*. EOLSS Publications.

Berk, Z. (2018). *Food process engineering and technology*. Academic Press.

Harnby, N. (1997). Mixing in the process industries. Harnby N, Edwards MF, Nienow AW, editors.

Harnby, N., Edwards, M.F., Nienow, A.W. (eds.): Mixing in the Process Industries, 2nd edn. Butterworth- Heinemann, ISBN 0 7506 1110 3 (1992).

Henderson, S. M., & Perry, R. L. (1976). Agricultural Process Engineering. The AVI Publ. Co. *Inc, Wesport, Connecticut, USA*.

Hwang, C. L., & Hogg, R. (1980). Diffusive mixing in flowing powders. *Powder Technology*, *26*(1), 93-101.

Lindley, J.A. (1991). Mixing processes for agricultural and food materials: 1. Fundamentals of mixing. J. Agric. Eng. Res. 48: 153–170.

Nienow, A. W.; Edwards, M. F.; Harnby, N. Introduction to mixing problems, Chapter 1. In: Mixing in the Process Industries (Harnby, N.; Edwards, M. F.; Nienow, A. W., eds). London: Butterworths, 1985, pp l-22' Perry, J. H. (ed.) Chemical

Niranjan, K. (1995). An appraisal of the characteristics of food mixing. In Food Process Design and Evaluation, Singh, R.K. (ed.). Technomics, Lancaster, PA, USA.

Ottino J (1989) The kinematics of mixing: stretching, chaos, and transport. Cambridge University Press, New York.

Ouyang H, Bao J, Fimbres Weihs GA, Wiley DE (2013) Control study on mixing enhancement in boundary layers of membrane systems. J Process Control 23(8):1197–1204

Patterson, G. K., Paul, E. L., Kresta, S. M., & Etchells, A. W. (2004). Mixing and chemical reactions. *Paul EL, Atiemo-Obeng VA, Kresta SM. Handbook of Industrial Mixing-Science and Practice. Hoboken: John Wiley & Sons, Inc*, 755-868.

Perry, J. H. (1950). Chemical engineers' handbook.

Perry, R. H.; Green, D. W.; Maloney, J. 0. (eds) Perry's Chemical Engineers' Handbook, 6th Ed. New York; McGraw-Hill, 1984,

Plawsky J (2001) Transport phenomena fundamentals. CRC Press, New York

Uhl, V. W., & Gray, J. B. (1986). Mixing, Vols. 1–3.

Weinekötter, R., & Gericke, H. (2013). *Mixing of solids* (Vol. 12). Springer Science & Business Media.

12

Freezing

Kshitiz Kumar and Vijay Singh Sharanagat

1. Introduction

Freezing is a method of food preservation that involves lowering the temperature below the freezing point of food generally -18 °C or below. During freezing of food the water present in them changes its state to form ice crystals. Human beings in the colder regions have used freezing as a means of preservation since they learnt hunting and gathering foods. The foundation of the modern frozen industry was laid in mid 1920 when Clarence Frank Birdseye, an American inventor used cold brine solution to chill packaged fish and extended it further for quick freezing of meat poultry fruit and vegetables. The concept of convenient food, ready to eat frozen meals became the reflection of modern life style and increasing women workforce shedding their housewife roles in last fifty years. Freezing is a convenient, flexible and one of easiest method of food preservation requiring very less time for processing. The retention of sensory and nutritional characteristics of fruit and vegetable through freezing is superior to drying and other methods, which involves addition of heat for preservation. The preservative effect of freezing is due to reduced water activity, reduction of microbial growth, enzymatic activity and chemical reactions at sub zero temperature. As opposed to refrigeration, which involves short-term storage, freezing is used as long-term preservation process retaining the quality of the food up to several months or a year.

2. Freezing Process

Food freezing can be understood more effectively by simultaneously studying the freezing of water. Hence, in this chapter we will study the freezing of food and compare it with freezing of water. Food is a complex system having water and various solute fractions and can be considered as an aqueous solution. The freezing process is divided into five major events (a) Prefreezing (b) Supercooling (c) Nucleation (d) Crystal growth (e) Temperature reduction to storage temperature

2.1 Prefreezing

During prefreezing the temperature of food material is brought down from the ambient temperature to the freezing temperature. It involves removal of sensible heat of food material. For example if any food material is kept at ambient temperature of 20 °C, it first needs to be brought at freezing temperature of water i.e. 0 °C. The energy required to remove this heat depend on the specific heat of the food material, which is a function of its constituents. Water is a major constituent of most food material having specific heat of 4200 J kg^{-1} K^{-1}. In the freezing curve this phase is represented by A-B portion of the curve (Fig. 1). This portion of curve is same in both the freezing curve of water and aqueous solution.

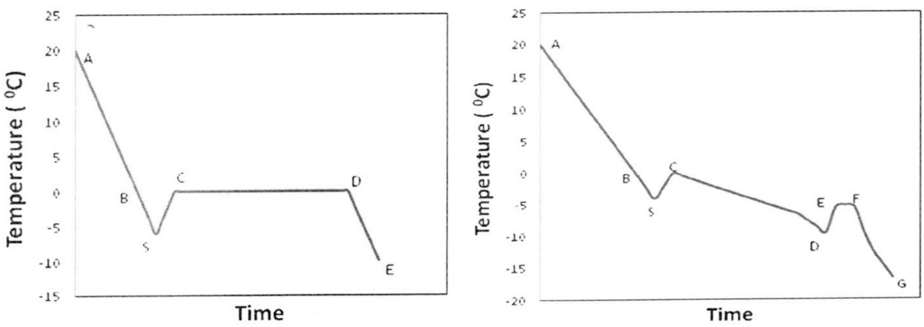

(a) (b)

Fig. 1 (a): Freezing curve of pure water (b) Freezing curve of aqueous solution (food)

2.2 Supercooling

Supercooling is also mentioned as subcooling or undercooling in literatures. If water or any food material is cooled down to 0 °C after removal of sensible heat it will not freeze instantaneously. Formation of ice crystal requires bringing down the temperature (T) below the freezing point (T_f, for water 0 °C), a process known as supercooling. Hence supercooling can be defined as lowering of temperature below the freezing point for the formation of first ice crystal (nucleation) and expressed as $\Delta T_s = T_f - T$. For example if the formation of first ice crystal takes place at -8 °C, we say that extent of supercooling (ΔT_s) is 8 °C [ΔT_s = 0 °C – (-8 °C)]. Supercooling is necessary for nucleation and the extent of supercooling depends on the type of nucleation whether homogeneous of heterogeneous (discussed in next section). B-S portion of the freezing curve denotes the supercooling (Fig 1). In aqueous solution this portion of curve is short as compared to pure water (reason discussed in next section).

2.3 Nucleation

Nucleation refers to the birth of minimum crystal with a critical radius which can expand and grow. Molecules, atoms and ions in a aqueous solution are always in random motion due to which they can loosely aggregate and associate to form cluster. Most of such cluster disappears quickly, but, if such clusters have enough particles so that they can arrange themselves in a lattice arrangement and do not redissolve, they are known as nucleus which on gaining more unit and stability becomes crystal. Thermodynamic analysis of nucleation process gives the value of critical radius as:

$$r = \frac{2\sigma T_{Kf}}{L_f \Delta T_s} = \frac{2\sigma T_{Kf} v}{\lambda_f \Delta T_s} \tag{1}$$

where T_{Kf} is the absolute freezing temperature of pure water (273 K), λ_A the molar latent heat of fusion for pure water, and v the molar volume of water, σ is surface tension, ΔT_s is supercooling.

It clear from the equation (1) that critical radius is inversely proportional to the supercooling (ΔT_s). So at 0 °C i.e. without supercooling (ΔT_s=0) the critical radius required for nucleation would be infinity, this explains why crystals formation doesn't start at 0 °C and certain supercooling is must for first ice crystal formation.

There are two types of nucleation homogeneous and heterogeneous. In homogeneous nucleation formation of new crystal is uninfluenced by solids of any sort, which may be any suspended particle in the aqueous solution or the walls of container. It occurs in water free of all impurities and requires larger supercooling as compared to heterogeneous nucleation. For pure water the homogeneous nucleation occurs at -45 °C. In heterogeneous nucleation, the nucleation rate is catalysed by the active surfaces (such as suspended solid, wall of container). Heterogeneous nucleation predominates in food, aqueous systems and water containing some impurities. The rate of nucleation can be expressed as:

$$J = CT \exp\left(\frac{-BT_f^2}{\Delta T_s^2 T}\right) \tag{2}$$

where B and C are product constant, T the system temperature, ΔT_s the supercooling, and T_f the initial freezing temperature of the system. Equation (2) states that rate of nucleation increases with extent of supercooling.

Point S indicates the nucleation point in the freezing curve (Fig 1). In pure water owing to homogeneous nucleation the point S is deeper than aqueous solution where heterogeneous solution prevails. As the first crystal is formed at point S, the release of heat of crystallization being faster than heat removed from the freezing system raises the temperature from point S to point C in the freezing curve. The temperature reaches point C which is the freezing point. In aqueous solution point C is below 0 °C due to freezing point depression, a colligative property dependent upon number of dissolved solutes. The depression in freezing point ($\Delta T_f = T_f - T_{fs}$; T_f and T_{fs} being freezing point of pure water and food system respectively) is given by the equation:

$$\Delta T_f = \frac{RT_{Kf}^2 M_A m}{1000\lambda_A} = 1.86m \quad (3)$$

where ΔT_f is freezing point depression, R is universal gas constant, λ_A is the latent heat of fusion for pure water (6003 kJ mol^{-1}); M_A the molecular weight of water (18 g mol^{-1}); m the molality of the solution (number of moles of solute/ 1000 g of solvent); T_{Kf} the freezing point of pure water (273 K).

The concept that nucleation and subsequently high number of small sized crystal formation when freezing rate is high and a few large sized crystal when freezing rate is low throughout the food sample is valid only for small specimen. For large specimen nucleation and crystal formation does not takes place throughout the surface rather it is restricted to solid-liquid interphase region. The supercooling in other parts of large sample is inhibited by the release of heat of crystallization and sensible heat of food sample.

2.4 Crystal growth

If cooling is continued after formation of nucleus of minimum crystal radius, the growth of the crystal takes place. Crystal growth is a mass diffusion process which can be thought as combination of two event. First the water molecules diffuse from surrounding bulk solution to liquid-ice interphase and second acceptance of this water molecules by the crystal and organization into space lattice. The latent heat of water is removed during this process.

The C-D portion of freezing curve which is like a plateau demonstrate the crystal growth. In pure water, this is almost a straight-line parallel to abscissa. In aqueous system as a portion of water is converted to ice, the concentration of remaining solution increases resulting in depression of the freezing point. As the crystal growth occurs, the solute concentration reaches to supersaturation point which is maximum solute concentration indicated by point D (Fig 1a & Fig 1b)

If cooling is continued further, in case of pure water, this leads to removal of sensible heat and further lowering of temperature indicated by D-E portion of curve (Fig 1a). In aqueous further cooling point D (Fig 1b) leads to the crystallization of solute (non-aqueous component of food). The crystallization of solute also goes through the same sequence of events as occurred in water crystallization earlier i.e. supercooling, nucleation, release of heat of solute crystallization and crystal growth. D-E portion of curve (Fig 1b) represents release of heat of solute crystallization. Point E (Fig 1 b) is termed as eutectic temperature where solute is freeze concentrated to its supersaturation concentration and this temperature remains constant as solute crystal growth is takes place indicated by E-F portion of the freezing curve. In food where various solutes may be present more than one eutectic point can be observed but the plateau of each would be small owing to low quantity of solute. Cooling the aqueous system beyond point F only lowers the temperature without causing any significant change as solution is already completely solidified.

Looking at the freezing curve of aqueous system it can be observed that such system do have a sharp freezing point like water and the temperature do not remain constant during removal of latent heat of crystallization and crystal growth. This explains the existence of straight plateau C-D in water while a negative slope in aqueous system. The point C can only be termed as 'initial freezing point' in aqueous system because in real sense there is a range of freezing point for such system. The initial freezing point for fruits and vegetables are in the range of -0.8 to -2.8 °C for vegetables and -2.7 °C for fruits.

2.5 Glass Transition During Freezing

It should be remembered that practically the crystallization of solute does not take place in food. At very low temperature when the freeze concentration is maximum i.e at eutectic temperature, the viscosity of the system becomes so high (in the range 10^{10} - 10^{14} Pa.s) that instead of crystallization the supecooled melt undergoes state transition from viscoelastic liquid to noncrystalline glass state, a process known as glass transition (or vitrification). The temperature at which this occurs called as glass transition temperature (T_g) . Glass is a non-equilibrium, metastable, amorphous, disordered solid state where all diffusion related activity is inhibited due to low molecular mobility. Frozen food only remain safe when they are stored below T_g, where ice and glassy state exists. If the food is stored at a temperature between freezing point of water (T_f) and T_g it will still have diffusion governed activities and hence get deteriorated on storage.

3. Methods of Freezing

The freezing methods can be broadly be classified as: (a) Natural convection freezing (b) Forced convection freezing (c) Liquid immersion freezing (d) Contact freezing (e) Cryogenic freezing

3.1 Natural convection freezing

It is based on natural convection where movement of fluid takes place owing to density difference. When a fluid is exposed to heat source, it gets warmed and becomes less denser than the fluid away from the heat source. The hot less denser fluid rise up forcing cooler fluid to sink and thus a natural fluid flow is created. In natural convection freezing food is the heat source while air is the fluid which circulate. The circulating air takes up the heat from food material, becomes lighter and then disperses its heat to the refrigerant flowing inside the evaporator coil of mechanical refrigeration system. This is a slow method, the rate of cooling is uncontrollable and temperature may be non-uniform across the different corners of the freezer.

Chest Freezer is an example of natural convection freezing where air between -20 °C to -30 °C is passed over food materials. Freezing time is too long (3-72 h) to make it commercially viable.

3.2 Forced convection freezing

In this method also air as fluid is circulated across the food material to be frozen and evaporator coil of the mechanical refrigeration system. The food acts as heat source while the refrigerant in evaporator acts as heat sink as in natural convection freezing. The movement of the air is augmented by a fan. The movement of air is due to both density difference and effect of fan. The forced circulation makes this method fast and controllable. The temperature is uniform across the different corners of the freezer. Following are the examples of forced convection freezing:

3.2.1 Air blast Freezer

High velocity air (1.5-6 ms^{-}) having temperature in the range of -30 °C to -40 °C is passed over food materials. In batch form it consist of an insulated room or cabinet such as cold stores where food is stacked in trays or hung. In continuous form the food is placed on a moving belt or in trolleys that passes through insulated tunnel containing air moving at high velocity. Heat transfer is increased by providing a vertical airflow to force air through the product layer. It is economical, flexible and suitable for most of the food products.

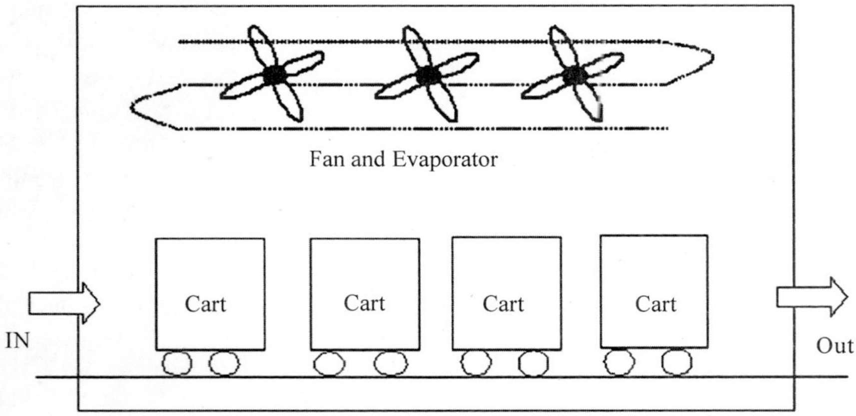

Fig. 2: Air blast Freezer

3.2.2 Fluidized bed freezers

It consists of a perforated bed through which cold air is blown in vertically upward direction so that product kept on bed gets lifted from the surface and remains suspended. The air velocity is adjusted in such a way that it is greater than the fluidization velocity of the product. The fluidization helps in increased heat transfer between the food product and moving air. Typical air velocity, air temperature and bed depth used in this process lies in range of 2-6 ms^{-1}, -25 °C to -35 °C and 2-13 cm respectively. This is a quick method of freezing having freezing time less than 10 minutes. The product suitable for this method are those which can easily be fluidized such as small particulate food of uniform size such as peas, diced carrots, corn kernel and berries.

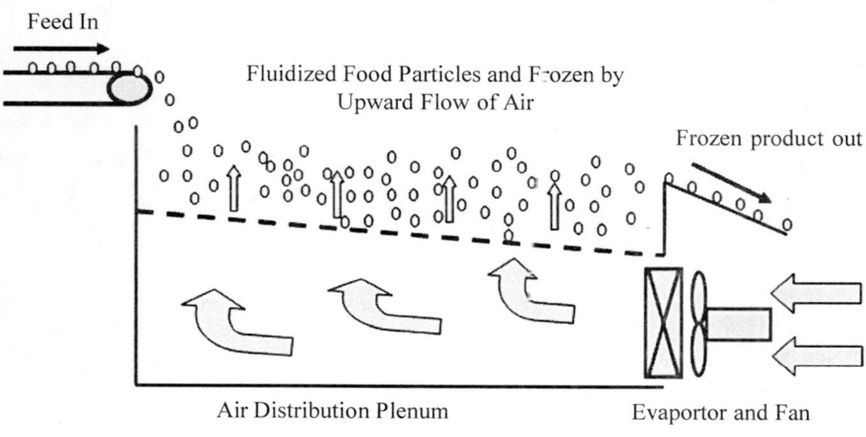

Fig. 3: Fluidized bed freezer

3.2.3 Individual quick freezing (IQF) is a term used for quick freezing of individual food products using the concept of air blast freezer and fluidized bed freezer. It involves sending the individual food pieces on a bedplate into a tunnel freezer that freezes the food items very fast. Since the individual food items are separate when they go in, they stay separate even after they have been frozen. The cold airflow keeps passing through the pieces of product in circular motions while the product is also advancing through the freezer towards the exit end.

3.2.4 Impingement Freezers

In this method air at high velocity is directed towards food materials through jet nozzles. The food are kept on conveyor belt and jets are positioned above and below the belt. It is suitable for only thin products.

3.3 Liquid immersion freezing

In this method the food to be frozen is immersed in a solution of cold liquid such as brine, glycol or sugar. The cooling of solution is achieved by passing the evaporator coil of the mechanical refrigeration system through it. It is quicker than air freezing method due to high heat transfer coefficient. The cooling liquid does not changes its phase unlike cryogenic immersion freezer where liquid nitrogen gets vaporized. It can operate in batch and continuous mode. In batch system the food products are kept in a basket which is then immersed in cold liquid bath. The basket is removed from the cooling medium after freezing is complete. In continuous form the food products are kept on conveyor belt which passes through cold liquid solution. Low density product where buoyancy would be large are unsuitable for immersion freezing. The food products to be frozen must be denser than the cooling fluid. The freezing time is shorter than air blast or fluidized bed freezer.

3.4 Contact freezing

In this method the foods are sandwitched between two metal plates which is itself cooled by flow of refrigerant. This process eliminates the two stage heat transfer (first from food to fluid and second from fluid to refrigerant) occurring in convection freezing and immersion freezing. Here heat from food is directly transferred to the refrigerant flowing through metal plates. Application of pressure across the plates ensures good solid to solid contact between food and metal plates. This increases the heat transfer from food to the refrigerant due to which high freezing rates can be achieved. The cooling temperature obtained for a given evaporation temperature of refrigerant is lower as compared to convection and immersion freezing.

3.4.1 Plate freezers

Plate freezers are a type of contact freezing. They can operate in both batch and continuous system. Spacers are used between the plates to minimize the bulging and prevent crushing of the food products. They can be used either to completely freeze the products or 'crust freeze' the outer few millimeters of the products. They are suitable for regular shape materials only such as meat steaks, fish fillets and leafy vegetables.

3.5 Cryogenic freezing

In this method the food materials are exposed to cryogens such as liquid nitrogen or liquid carbon dioxide either through spraying or immersion. The temperature of cryogen used for this purpose is very low such as -195.8 °C for liquid nitrogen and -78.5 °C for liquid carbon dioxide. When the food materials comes in contact with cryogens at such a low temperature the heat transfer is very rapid due to large initial temperature difference and large surface heat transfer during phase change of cryogens. Cryogenic freezing systems are characterized by a moderate capital investment, minimal preventive maintenance cost, and a smaller and more flexible commitment of plant space. It is used for freezing of high value products. Liquid nitrogen freezer and carbon dioxide freezer are two types of cryogenic freezer. The spraying arrangement can be made on flat belt tunnel, multi-pass tunnel, flighted belt tunnel or spiral freezer for both liquid nitrogen and CO_2 while the immersion system is only for liquid nitrogen. The cryogenic immersion system are also example of individual quick freezing (IQF).

3.5.1 Liquid nitrogen freezer

Here liquid nitrogen at -196 °C is fed to the product placed on a continuous belt toward the end of the tunnel. On absorbing the heat from the product liquid nitrogen changes its phase to vapour which is then forced back towards the entrance point of the conveyor belt thereby cooling the incoming food product. Liquid nitrogen freezer in batch cabinet and immersion bath are also available. For freezing of 1 kg of food product the liquid nitrogen required varies from 0.3-1.5 kg depending on initial temperature and composition.

3.5.2 Carbon dioxide freezer

It also operates in a tunnel system as in liquid nitrogen freezer. Liquid carbon dioxide under high pressure is piped through the tunnel, and as it exit the injection orifice, a mixture of cold CO_2 vapour and tiny dry ice solid referred as 'snow' is obtained at -78.5 °C. The dry ice and cold CO_2 vapour removes the heat from the food product when they come in contact with them. Dry ice at -78.5

°C sublimate to gas on extracting heat from food material at atmospheric pressure. Cold CO_2 vapour does not have the same cooling effect as vapour of liquid nitrogen because CO_2 vapour are at much higher temperature than nitrogen vapour (-78.5 °C against -196 °C). The major cooling effect is due to the high enthalpy of sublimation of dry ice which is 571 kJ/kg (25.2 kJ/mol). Unlike liquid nitrogen freezer, the liquid CO2 in carbon dioxide freezer is sprayed on the product from the entry point of tunnel and continued till 70% length of tunnel.

The commercial classification of freezer on the basis of freezing rate is given in Table 1.

Table 1: Types of freezer on the basis of freezing rate.

Freezer Type	Freezing rate(cm h^{-1})	Example
Slow freezer	0.2-0.5	Cold store, chest freezer/air still freezer
Quick freezer	0.5-3	Air blast and plate freezers
Rapid freezer	5-10	Fluidized bed freezer, IQF
Ultra rapid freezer	10-100	Cryogenic freezer

4. Freezing Time Calculation

Calculation of accurate freezing time for food product is complicated due to the following reasons:

i) Variation in the initial temperature, size, shape of individual piece of food,

ii) Different regions of food have different rate of nucleation and crystal growth

iii) Thermophysical properties such as density, thermal conductivity, specific heat and thermal diffusivity changes with lowering in temperature

Food do not have sharp freezing point but rather freezing takes place over a range of temperature and hence there is no clear definition of freezing time for food materials. Some of the accepted definition are given here. One definition says freezing time is the time required to lower the temperature of a food at its thermal centre to a desired temperature below the initial freezing point. Another definition of freezing time is the time between the surface of the food reaching 0°C and the thermal centre reaching 10°C below the initial freezing point. The freezing process can be considered complete when all the freezable water at the thermal center of the food gets converted to ice.

Freezing rate of a food is defined as the ratio between the minimum distance from the surface to the thermal center, and the time elapsed between the surface reaching 0 °C and the thermal center 10 °C colder than the temperature of initial ice formation. The freezing rate is also defined as the difference between the initial and the final temperature divided by the freezing time.

However, some equations have been developed to calculate the freezing time which are accurate for most practical purpose. This includes Plank's equation, Nagaoka et al. equation, Levy equation, Cleland and Earle equation and Pham method.

4.1 Plank's Equation

This is the most popular equation based on energy balance principle. The following assumptions are to be made before using this equation:

- Freezing starts from the freezing point of water and there is no prefreezing step and removal of sensible heat from food product is ignored
- The gradual removal of latent heat over a range of temperatures during the freezing process is to be ignored.
- Steady state heat transfer takes place throughout the freezing process.
- The shape of freezing front is similar to food.
- There is no depression in freezing point and there exists a single freezing point in food.
- Density of food remain constant
- There is a single step change in the thermal conductivity and specific heat of food material during conversion from unfrozen to frozen state.

The equation is:

$$t_F = \frac{\rho_f L_f}{T_F - T_a}\left[\frac{Pa}{h} + \frac{Ra^2}{k_f}\right] \tag{4}$$

where t_F is the freezing time (kJ/W), ρ_f is the density of the frozen material (kg/m^3), L_f is the change in the latent heat of the food (kJ/kg), T_F is the freezing temperature (°C), T_a is the freezing air temperature (°C), h is the convective heat transfer coefficient at the surface of the material (W/[m^2°C]), a is the thickness/ diameter of the object (m), k_f is the thermal conductivity of the frozen material (W/[m°C]), and the constants P and R are used to account for the influence of product shape with P = 1/2 and R = 1/8 for infinite plate, P = 1/4 and R = 1/16 for infinite cylinder, P = 1/6 and R = 1/24 for sphere.

5. Quality of Frozen Food

5.1 Freeze Concentration

It refers to the concentration of the nonaqueous component during freezing. As the water gets converted to ice, the solute in the remaining of the unfrozen water becomes more and more concentrated. This increased solute concentration may change the pH, ionic strength, viscosity, surface tension, redox potential, protein aggregation and water-solute interaction. The rate enzymatic reactions goes through a maximum at a temperature below the initial freezing point due to enzymes and substrate coming together at increased solute concentration.

5.2 Effect of freezing rate

Fast freezing produces better quality product than slow freezing. In slow freezing (lower than 1 °C /min) ice crystals are exclusively formed in extracellular space. The solute concentration in extracellular space rises, a phenomena known as freeze concentration. The increased solute concentration decreases the chemical potential of extracellular space in comparison to intracellular space. The osmotic pressure difference causes water to diffuse out from the cell to extracellular space causing shrinkage. In rapid freezing at low temperature, large number of small ice crystals are formed uniformly in both the intracellular and extracellular space and hence no chemical potential difference or osmotic pressure difference is generated. There is no severe damage to the tissue and better quality product is obtained.

5.3 Drip Loss

During freezing water diffuses from bulk solution and deposit on solid-liquid interphase during crystal growth. This dislocated water do not get reabsorbed at original position during thawing, leading to release of drip known as drip loss.

5.4 Freeze-Cracking

When some food products are subjected to high freezing rate or low temperature such as liquid nitrogen , a crust is formed on the outer surface. This crust prevents volume expansion of product when interior water freeze to ice developing internal stress. If the internal stress exceeds the material strength crack appears on food product which is known as freeze cracking. It can be prevented by precooling the product prior to freezing.

5.5 Freezer Burn

Freezer burn is defect that occurs due to loss of water vapour from unpackaged frozen food product due to temperature fluctuation during storage. As temperature inside storage room increases, the surface ice crystal gets sublimated to the surrounding air where water vapour pressure is less as compared to water vapour pressure over food surface. This water loss make the surface appear as opaque dehydrated and prone to oxidation changing the colour, flavor and texture of food product. It can be prevented by packaging the food in moisture barrier material and glazing a thin layer of ice on surface of food material.

5.6 Recrystallization

Recrystallization refers to any change in size, shape or orientation of ice crystal during frozen storage of food products. Small ice crystals are thermodynamically unstable and have natural tendency to fuse together to form large ice crystal. Thus small ice crystals that are formed during rapid freezing disappear while the large ice crystal grows in size minimizing the advantage of rapid freezing and jeopardizing the quality of food products. The tendency of large ice crystal to grow at the expense of smaller one due to difference in their surface energy is referred to as Ostwald ripening.

5.7 Biohemical activity

Freezing only decrease the activity of enzyme but do not inactivate them. The hydrolytic enzymes such as lipases, phospholipases, proteases, chlorophylases and anthocyanases may cause quality deterioration if they are not inactivated through blanching prior to freezing. The ice crystals formed may damage the membranes in which these enzymes are present and facilitates their disruptive actions such as oxidative rancidity, colour loss and development of off flavours. Protein denaturation in frozen product takes place mainly due to ice formation, recrystallization, dehydration and exposure to increased salt concentration

6. Thawing

Thawing is longer process than freezing for the same temperature difference. During thawing melting of outer layers of ice crystal into a layer of water creates a insulating effect on heat transfer. This is because at 0 °C thermal conductivity of water is four times less and thermal diffusivity of water is nine times less than ice. While during freezing the formation of outer ice layer expedites the process due to high thermal conductivity and thermal diffusivity of ice. The effect of cell damage by ice crystal and recrystallization gets manifestedin form of drip loss of cellular component during which may lead to nutrient loss. These drip loss also acts as site for microbial growth and enzymatic activity. Thawing of food products in microwave, dielectric heaters and vacuum chambers are more safe than open air thawing at ambient temperature.

Example 1. Calculate the average air velocity over the surface of the product kept in a tunnel air blast freezer with following specifications: Total cross sectional area of the tunnel air blast freezer is 1.2 m x 1m. The sum of cross sectional area of the trays and the product placed inside tunnel air blast freezer is 0.7 m^2. The air flow as obtained from fan rating is 2 m^3 s^{-1}.

Solution: In air blast freezer for all practical purpose air immediately adjacent to the surface of the product is stagnant due to the friction between the air and the surface of the product. The air speeds quoted for air blast freezers are therefore only average speeds for the spaces between the packages of product being frozen.

Space between packages = (1.2 m x 1 m) – 0.7 m^2= 0.5 m^2

Hence, the average air velocity over product surface = $\dfrac{2\left(m^3s^{-1}\right)}{0.5\left(m^2\right)} = 4 \text{ m s}^{-1}$

Example 2: Calculate the freezing time using Plank's equation of a cylindrical food product which is to be frozen in air blast freezer. The initial product temperature is 20 °C and the cold air to be circulated is at -35 °C. The specification of product to be frozen are: Diameter = 8 cm; Density = 1050 kg/m^3; Initial freezing temperature = -1.5 °C; Thermal conductivity = 1.3 W/(m K); Latent heat of fusion = 260 kJ/kg; Convective heat-transfer coefficient h_c = 50 W/(m^2K)

Solution

Given

Initial product temperature T_i = 20°C

Air temperature, T_a = -35 °C

Initial freezing temperature T_F = -1.5 °C

Product diameter, a = 8 cm = 0.08 m

Product density, ρ_f = 1050 kg/m^3

Convective heat-transfer coefficient h_c = 50 W/(m^2K)

Thermal conductivity of frozen product, k_f = 1.3 W/(m K)

Latent heat $L_f = 260$ kJ/kg

Shape constants for cylinder: $P' = \frac{1}{4}, R' = \frac{1}{16}$

Using Equation (4)

$$t_F = \frac{\rho_f L_f}{T_F - T_a}\left[\frac{P'a}{h} + \frac{R'a^2}{k_f}\right]$$

$$t_F = \frac{(1050\,kg\,m^{-3})(260\,kJ\,kg^{-1})}{[-1.5\,^0C - (-35\,^0C)}\left[\frac{0.08}{4(50\,W\,m^{-2}K^{-1})} + \frac{0.08^2}{16(1.3\,Wm^{-1}K^{-1}}\right]$$

$t_F = 5.76$ kJ/W

Since, 1000J = 1kJ and 1W= 1J/s

$t_F = 5.76 \times ?10^3$ s

$t_F = 1.6$ h

Example 3: 100 kg of food product is to be frozen, and heat content of every kg of product (from + 20 °C to -30 °C) is 100 kcal/kg. If the freezing time is 2.5 h and fan circulation rate is 2.5 m³/s, then calculate the total mass of air required and the average increase in temperature of air. The specific heat and density of air is 0.24 kcal/kg °C and 1.45 kg/m³ respectively.

Solution:

Total amount of heat to be extracted, $Q = 100\,(kg) x 100 \left(\frac{kcal}{kg}\right) = 10,000$ kcal

Mass of air circulated = Freezing time × Fan circulation rate × Density of air

= 2.5×3600 (s) ×2.5 ($m^3\,s^{-1}$) × 1.45 ($kg\,m^{-3}$)

= 32625 kg

Heat extracted from food product is gained by air, hence,

$Q = mass_{air} \times c_{p,air} \times \Delta T_{air}$

$$\Delta T_{air} = \frac{10000\ (kcal)}{32625(kg) \times 0.24\ (kcal\,kg^{-1}\,^0C^{-1})}$$

$\Delta T_{air} = 1.27\,^0C$

Example 4: Calculate the total heat removed from 1000 kg of fresh vegetable while it is transported in refrigerated van for 24 h. Inside temperature of the van is 2 °C and initial temperature of vegetable while loading is 6 °C. The vegetable has specific heat capacity of 4.0 kJ/kg °C and attain van temperature in 1.5 h. The area of the truck is 90 m^2 and overall heat transfer coefficient is 0.35 W/m^2 °C. The outside air temperature is 20 °C and heat of respiration of vegetable in the given temperature range is 32 x 10^{-3} W/kg. Also calculate the peak refrigeration load of the refrigerated van.

Solution

Cooling load

I. Sensible heat to be removed to bring vegetable temperature from 6 °C to 2 °C:

$q_1 = mc_p\,(T_i - T_f)$

$q_1 = (1000\text{ kg})\,(4.0\text{ kJ kg}^{-1}\,{}^0\text{C}^{-1})\,(6\text{-}2)\,{}^0\text{C}=16000\text{ kJ}$

II. Heat of respiration; $q_2 = mq_R$

$$q_2 = (1000\,kg)(32\times10^{-3}\,Js^{-1}kg^{-1})\left(\frac{1kJ}{1000J}\right)(24\,h)\left(\frac{3600\,s}{h}\right) = 2764.8\,kJ$$

III. Heat lost:

$q_3 = A\,U\,(T_a - T_f)\,t$

$$= (90\,m^2)\left(0.35\frac{W}{m^2\,{}^0C}\right)\left(\frac{1\,kJ}{1000\,J}\right)(20-2\,{}^0C)(24\,h)\left(\frac{3600\,s}{h}\right) = 48988.8\,kJ$$

Hence, the total amount of heat removed:

$$q = q_1 + q_2 + q_3 = 16000\text{ kJ} + 2764.8\text{ kJ} + 48988.8\text{ kJ} = 67753.6\text{ kJ}$$

Calculation of peak refrigeration load

I. Sensible heat:

$$q_{1P} = \frac{q_1}{t_F} = \frac{16000\,kJ}{1.5\times3600\,s} = 2.96\,kW$$

II. Heat of respiration:

$$q_{2P} = (1000\,kg)\left(32\times10^{-3}\frac{W}{kg}\right) = 32\,W = 0.032\,kW$$

III. Heat losses:

$$q_{3P} = AU\left(T_a - T_f\right) = (90m^2)\left(0.35\frac{W}{m^2\,{}^0C}\right)(20-2\,{}^0C) = 567\,W = 0.567\,kW$$

Therefore, peak refrigeration load:

$$q_P = q_{1P} + q_{2P} + q_{3P} = 2.96 \text{ kW} + 0.032 \text{ kW} + 0.567 \text{ kW} = 3.56 \text{ kW}$$

References

Fellows, P. (2000). Freezing. In: *Food Processing Technology Principles and Practice Second Edition* (pp. 418-439). CRC Press, Boca Raton, Florida, USA.

Fennema, O. (1996). Water and ice. In: Fennema, O. (Ed.), *Food Chemistry, 3rd ed* (pp. 17–94). Marcel Dekker, New York.

Fennema, O.R., Powrie, W.D, Marth, E.H. (1973). *Low Temperature Preservation of Foods and Living Matter*, (pp. 3–207). New York: Marcel Dekker.

Franks, F. (1985). *Biophysics and Biochemistry at Low Temperatures*, (pp. 39–52) Cambridge University Press, Cambridge.

Meste, M.Le, Champion, D., Roudaut, G., Blond, G., Simatos, D. (2002). Glass transition and food technology. A critical appraisal. Journal of Food Science 67(7):2444–2458.

North, M.F. & Lovatt, S.J. (2011). Freezing Methods and Equipment. In: Da-Wen Sun (Ed.), *Handbook of Frozen Food Processing and Packaging* (pp. 199-210). CRC Press, Boca Raton, Florida, USA.

Rahman, M.S. & Velez-Ruiz, J.F. (2007). Food Preservation by Freezing. In: Rahman, M. S. (Ed.), *Handbook of Food Preservation* (pp.635-666). CRC Press, Boca Raton, Florida, USA.

Singh, R.P. & Heldman, D.R. (2009). Food Freezing. In: *Introduction to Food Engineering Fourth Edition* (pp. 501-539). Academic Press, USA

Slade, L, Levine, H. (1995). Glass transitions and water-food structure interactions. In: Kinsella J.E., Taylor, S.L. (Eds.), *Advances in Food Nutrition Research* (Vol. 38, pp. 103–269). Academic Press, San Diego.

Yanniotis, S. (2008). Cooling and Freezing. In: *Solving Problems in Food Engineering* (pp 193-214). Springer, New York, USA.

Zaritzky, N. (2011). Physical–Chemical Principles in Freezing. In: Da-Wen Sun (Ed.), *Handbook of Frozen Food Processing and Packaging* (pp. 3-31). CRC Press, Boca Raton, Florida, USA.

13

Pasteurization

Ajit Kumar Singh, Anit Kumar, Rupesh Chavan and Kumar Sandeep

1. Introduction

Milk is a source of abundant nutrition has been a vital part of human civilization since prehistoric times. However, because of its perishable nature, it requires immediate processing to improve the shelf life and safe consumption. With the advancement of processing techniques in dairy and food processing, there is a widespread option being utilized to improve the milk shelf life by either thermal technologies (pasteurization, sterilization, and ultra-high temperature) or non-thermal technologies (high pressure processing, high pressure homogenization, pulse electric field and ultra sonication). Still, pasteurization and sterilization are the two foremost process is used for milk industries. Pasteurization is the thermal inactivation of microorganisms at temperatures below 100°C that enhance the safety and quality of milk and milk products and extends the shelf life to an extent. The fundamental purposes of pasteurization are reduction of both pathogenic and spoilage microbial load as well as inactivation of enzymes to slow down the physicochemical changes in the milk. Nowadays, pasteurized milk is common around the developed nations and have been also practicing extensively in the remaining world because of positive influence in terms of extending shelf life and enhancing sensorial attributes. Further, the selection of pasteurization methods and efficient equipment for attempting the precise balance of heating and agitation is vital for the processing of milk and milk products. The understanding of pasteurization method and equipment does not only induce changes to the product's shelf life but also enable one for judicial use of utilities leading to economic operation. Moreover, exploring the advance techniques of the pasteurization system is equally significant to reduce the cost of operation as well as to save the energy required in processing. In addition, development and innovation in pasteurizing equipment like minimizing fouling, undesired reactions, cross-contamination, and in maintaining temperature regulation during processing is essential for dairy processing plant. This chapter covers the fundamentals of pasteurization, followed by methods and equipment used as pasteurizer and then designing of the pasteurization process. In the end, different pasteurization equipment has been discussed to comprehend the practice of pasteurization as thermal processing in the dairy and food processing industry.

2. History Background and Origin

The pioneering investigations related to pasteurization were initiated by Spallanzani in 1765 to slow down the spoilage of meat extract through heat treatment method (Silva and Gibbs, 2009). However, the concept of Pasteurization has been termed so after its renowned French scientist, Louis Pasteur during the nineteenth century. This invention took place in the middle of the nineteenth century, when a group of commercial winemakers encountered severe problem of atypical fermentation and subsequent spoilage of wines and approached Pasteur to follow a solution. During 1862–1864, Pasteur studied the problem and found that the main cause of atypical fermentation and leading to early spoilage of wines are some microorganisms and it can be prevented by heating the wine around 50-60°C, which restricts these microorganisms. Considering this momentous innovation of shelf life extension and food safety, Pasteur, patented his heating and cooling process in 1865 and named the process as Pasteurization. Later, in 1876, Pasteur also investigated beer spoilage and adopted pasteurization as a solution. Further, milk processors also applied this process to remove foodborne illnesses causing microorganisms (Von,1886; Davis, 1995; West off, 1978).Milk is predominantly known as a source of highly nutritious growth medium for microorganisms, also comprise of health enhancing lactic acid forming bacteria along with pathogenic and spoilage causing microorganisms such as Salmonella, Listeria and Gram-negative psychrotrophs. Following this, pasteurization of milk on a commercial scale has been practicing since 1890, in all around the world.

In modern dairy processing industries, pasteurization stands as an essential milk processing operation in each dairy plant for commercial milk handling to meet legally defined hygienic requirements.

3. Classical Definition of Pasteurization

The term "pasteurization" has been introduced by the French researcher Louis Pasteur. It referred as, pasteurization is a mild heat treatment process at 50-90°C temperature for a specific time followed by rapid chilling used for the inactivation of pathogenic and spoilage microorganisms. In case of, "milk pasteurization", every particle of milk is heated at 60-90°C and kept holding at same temperature for a given time (1 sec-30 min) followed by fast cooling to less than 4°C.Consequently,several time-temperature combinations of heat treatment are applied for killing the pathogenic microorganisms for enhancing the keeping quality of milk. The schematic diagram of pasteurization has been represented in Fig. 1. Whereas milk processing at high temperature at less holding time gives a severe reduction in the microbial load and losses the some beneficial ingredients in milk, milder temperature with longer holding time inhibits microorganism in longer operation time nevertheless improves the keeping quality of milk (Simpson, 2009).

Thus, it is important to standardize a time-temperature combination that would reduce maximum microbial load along with minimal damage of nutrition. Theorization is the another mild heat treatment process, where milk is heated at 57-68°C for 15 min and targets the pathogenic bacteria while leaving the good bacteria in the milk. The low-temperature selection does not modify the behaviour and quality of milk constituents. Unlike termination, milk is heated at 100°C during sterilization process that much more severe than pasteurization process in which complete killing of microorganisms occurred including spore. Because of the balanced time-temperature combination, Pasteurization is considered as an optimized heat treatment method in milk processing that only kills all the pathogenic microorganisms without affecting milk nutritional and sensorial value.

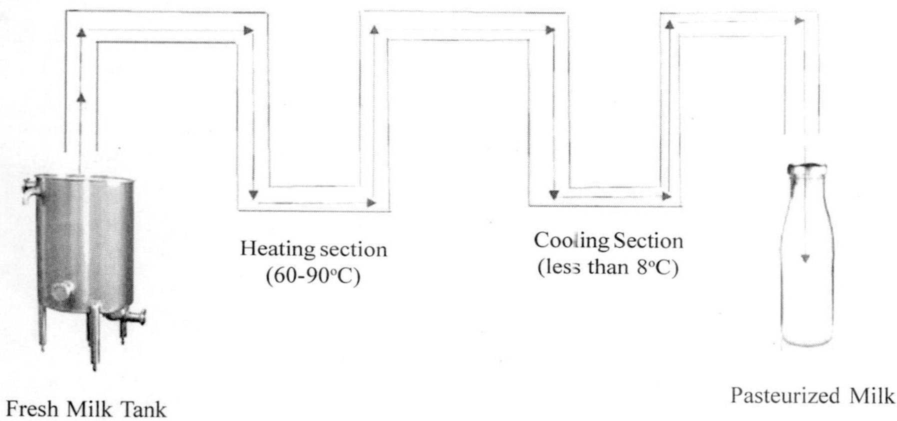

Fig. 1: Pasteurization process of milk

FSSAI elaborates the term "Pasteurization", "Pasteurised" and similar terms as the low temperature long time heating and holding treatment of every particle of milk at 63°C for 30 min or high temperature short time at 72°C for 15 sec followed to instant cooling to a temperature of 4°C, or less.

According to International Dairy Federation (IDF), pasteurization is a process, which is applied to milk with the objective of reducing possible microbial load (*Mycobacterium tuberculosis* and *Coxiellaburnetti*) with minimal physicochemical changes in the product.

During pasteurization process, some microorganisms are survived which are known as *thermoduric* and spore forming bacteria also survive even a severer treatment around 80-100°C for 30 min. The properties of the liquid milk that need the most consideration is safety to the consumer, shelf life, sensory characteristics and nutritional value and thus, by optimizing Pasteurization with an optimal time-temperature combination is a key aspect in the dairy and food processing industry.

4. Public Health Siginificance of Pasteurized Milk

The significance of pasteurization in safeguarding milk supplies for health benefits has been recognized over a long period of years.Milk is a perishable food that caters as an exceptional growth medium for an extensive range of microorganisms. These organisms either produce spoilage of the milk and milk products or causes disease in the consumers (Pal, 2012). Although, the purpose of Pasteurization is to prevent the diseases transmitted by raw milk, however, the shortage of insufficient infrastructure, lack of knowledge about hygiene and food safety results to the incidence of milk-borne diseases.

Table 1: Transmission of Milk Borne Diseases through the Contaminated Handling

S. No.	Disease	Pathogen	Incubation Period
1	Diphtheria	Corynebacterium diphtheriae	Variable
2	Tuberculosis	Mycobacterium tuberculosis	Several weeks
3	Salmonellosis	Salmonella typhimurium. S. enteritidis	12-14 hours
4	Typhoid fever	Salmonella typhi	2-3 weeks
5	Infectious hepatitis	Hepatitis A virus	3-6 weeks
6	Cholera	Vibrio cholerae	3-4 days
7	Colibacillosis	Escherichia coli	8-24 hours
8	Campylobacteriosis	Campylobacter jejuni	2-5 days
9	Staphylococcal infection	Staphylococcus aureus	1-6 hours
10	Streptococcal infection	Streptococcuspyogenes	3 days
11	Amoebiasis	Entamoeba histolytica	2-4 weeks
12	Rotaviral gastroenteritis	Rota virus	2 days
13	Giardiasis	Giardia lamblia	7-13 days
14	Yersiniosis	Yersinia enterocolitica	2-7 days
15	shigellosis	Shigelladysenteriae	2-3 days

Further, there are some of the additional aspects like insufficient heat treatment, contaminated pasteurizing equipment, poor operating hygiene as well as mishandling keeping temperature during refrigeration causing an outbreak of milk-borne diseases (Pal, 2012). Consequently, consumers in specific, infants, elderly people and pregnant women may become susceptible to the high incidence of milk-borne diseases. The brief summary of diseases transmitted by contamination of pasteurized milk through infected personnel (Pal, 2007), in a dairy plant has been tabulatedin Table 1. Since, Pasteurization is a basic preventive measure of public health significance intended to promote safe milk for consumption, it is vital to adopt a standard that measures the safety of milk and milk products with respect to food borne diseases around the world. The microbial standards of milk are one such frequently used method to measure the hygienic quality of milk. The microbial standards of pasteurized milk as recommended by as prescribed by the Bureau of Indian Standards (BIS, 1988) and now practiced in FSSAI have been tabulated in Table 2.

Table 2: Microbial Standards of Pasteurized Milk

S. No.	Test Parameter	Standard
1	Standard plate count (SPC)	Maximum 30,000 cfu/ml
2	Coliform count	Absent in 1:10 dilution
3	Methylene blue reduction test (MBRT)	More than 1 hour
4	Alkaline phosphatase	Test negative

5. Methods of Pasteurization

Over the course of more than 150 years of the invention of pasteurization, processing of milk has covered a long journey. There are several methods developed for effective pasteurization, one or other of which may have a distinct advantage in certain circumstances over the others but the emphasis of pasteurization is singular on the purpose, namely, destruction of microorganisms with minimum alter in the constituents of milk and control operation of time and temperature of heat treatment.

The lower the exposed temperature the greater the time needed, the higher the temperature, the less the time. The pasteurization methods are categorized on the basis of different parameters like nature of the material (viscous or non-viscous), holding method, time-temperature combination and heat exposure. With the advancement of techniques, pasteurization methods are classified in four categories.

Table 3: Process and time-temperature parameter of Pasteurization

Process	Temperature (°C)	Time
Vat Pasteurization	63°C	30 min
HTST Pasteurization	72°C	15 sec
Flash Pasteurization	85-90°C	1-4 sec
UHT Pasteurization	135-150°C	1-2 sec

5.1 Batch Pasteurization

Batch pasteurization also known as Vat pasteurization, contain a jacketed vat enclosed by either heating coils containing hot water or direct steam that helps in heating of every particle of milk. It involves heating and holding of milk at63°C for a minimum of30 min followed by rapidly cooling to below 10°C.It is also known as Low Temperature Long Time (LTLT) pasteurization, and has been widely used since the inception of commercial milk pasteurization till 1940 before the arrival of HTST pasteurization method. This treatment of 63°C and 30 min was the first time-temperature combination that causes the inactivation of microbial pathogens especially *Mycobacterium tuberculosis*, which is heat resistant pathogen of major public health issue in milk.

Table 4: Merits and Limitations of Vat Pasteurization

Merits	Limitations
• Lower capital cost of Pasteurizer	• Capacity is poor and limited for low production volume
• Low Temperature Longer Time (LTLT) processing is effective in cheese and ice cream production	• Slow heating and cooling rates for a long time may alter the physiochemical changes
• Suitable for small scale processing plant	• Time consuming operation

Vat pasteurizer is still used by small scale milk processors because of its low capital cost. The two main types of Vat-pasteurizer are commercially used at a small scale and they are (a) in-tank pasteurization and (b) in-container pasteurization, which is further used in the form of unsealed and sealed containers. The Vat pasteurizer is used extensively in the cheese and ice cream industry as it helps in dissolution and smooth blending of constituents during the heating stage.The Vat pasteurization technique has imparted a major role since the beginning of the pasteurization process, and it has its own merits and limitations as described in Table 4.

5.2 HTST Pasteurization

High Temperature Short Time (HTST) pasteurization has been considered as the most accepted technique of pasteurization by dairy industry that accurately and adequately kills microorganisms in milk and milk products. It is also known as Continuous Holding Pasteurization. HTST Pasteurization processes were developed between1920 and 1927 and by the early 1950s,it was adopted by most of the dairies because of shorter operation time. It involves the heating up of milk using the time-temperature combination as 72°C for 15 sec, and then immediately cooled to below 4°C.In HTST pasteurization, the process consists of five stages i.e. regeneration of heating, heating, holding, regeneration of cooling and chilling in sequence.

HTST pasteurization is operated with the help of a heat exchanger to maintain the temperature and continuous flow of milk. Principally, heat exchangers for HTST pasteurization associate three different types of heat exchangers such as tubular heat exchanger, scraped-surface heat exchanger and plate heat exchanger. The tubular heat exchanger comprises of two parallel tubes (tube-in-tube). Milk is passed within the inner tube while the heating or cooling medium is passed through annular space between the tubes. Dairy products have a wide range of viscosity and thus this type of heat exchangers are mostly handling for processing high concentrated liquid food products, whereas low concentrated liquid food products is processed through plate heat exchanger.

Scraped-surface heat exchangers (SSHEs) are widely handled for the processing of highly concentrated, sticky and lumpy food products. During the pasteurization process using SSHEs, the product is carried in direct contact with the surface of heat transfer that is rapidly and repeatedly scraped, thereby exhibiting the surface to transmit of unprocessed products. Pasteurization of sticky and concentrated products such as butter and ice cream, are effective using SSHEs.

Plate heat exchanger (PHE) is the most commonly used method for the continuous pasteurization process due to several aspects such as a compact model, a wide surface area over a small volume in contemporary heat exchanger applications. It consists of several thin plates marginally separated from each other with the provision of passage for fluid flow for heat transfer. The HTST pasteurization has been extensively used by the dairy plants for pasteurization of milk. The important influence of HTST pasteurization is the regeneration system of heating and cooling of milk that lower the adequacy demand by almost 80-90% alternative to batch pasteurization. In addition to this, HTST system renders the capacity to heat treat milk continuously and appropriately with built-in safeguards that restrict inappropriate pasteurization due to passing of heated milk.

5.3 Flash Pasteurization

Flash pasteurization is a pattern of HTST pasteurization and heating the milk products up to 85-90°C and holding for a very short time of span in the range of 1 to 4 seconds. Because of such a short residence of food products at higher temperature, this method is termed as flash pasteurization and commonly accomplished by direct heating of the product through steam injection or infusion. The method of flash pasteurization is specially adopted for milk products like butter and cream instead of milk to control the colour and flavour of the final product compared to HTST process.

5.4 Ultra High Temperature (UHT) Pasteurization

Ultrahigh temperature (UHT) pasteurization, also referred to as ultra-pasteurization, is a process in which milk or cream is heated above 135°C for one or two sec. UHT does minimal damage to the flavour and extends the shelf life of products such as butter and cream. The milk processed under UHT can be stored without refrigeration and has a shelf life of six months to a year. Products processed under UHT do not taste as fresh but are very useful for remote areas especially when refrigeration is not available. Although the UHT pasteurization is severe, it is different from the milk sterilization and shows negative turbidity test. It indicates less nutritional damage to UHT milk than sterilized milk.

6. Designing the Pasteurization

Like other heat treatment, pasteurization is a mild heat treatment process applied to foods to inactivate in specific disease-producing organisms, but less severe than sterilization. Pasteurization causes inactivation of most viable vegetative forms of microorganisms but not heat-resistant spores. Effective pasteurization of each particle of milk shall able to safeguard the shelf life of the milk & milk products as well as to enable compliance to the microbiological specifications both of hygienic indicators and pathogens, consistently. Further, heat is lethal to microorganisms, but each species has its specific heat tolerance. During a thermal destruction process, such as pasteurization, the rate of destruction is logarithmic, as is their rate of growth. Thus, bacteria subjected to heat are killed at a rate that is proportional to the number of organisms present. The pasteurization treatment is depend on time-temperature parameters. The target microorganism for pasteurization is *Mycobacterium tuberculosis* and *Coxiellaburnetti*.

6.1 Thermal Death Rate Kinetics of Microorganisms

Heat treatment like pasteurization is extensively used process for removal or destruction of spoilage and pathogenicbacteria. The death rate at constant temperature can be expressed as a first-order kinetic equation. The rate of destruction is directly proportional to the number of microorganisms,

$$\frac{dN}{dt} = -kN \tag{1}$$

Or,

$$\frac{dN}{N} = -kdt \tag{2}$$

Where,

N = No. of viable organisms at a given time

T = time in min

K = reaction velocity constant in min^{-1}

The reaction velocity constant is a function of temperature and the type of microorganism.

Integrating the above equation as follow gives,

$$\int_{N_0}^{N} \frac{dN}{N} = -\int_{t=0}^{t} k\,dt \tag{3}$$

$$ln\frac{N}{N_0} = -kt \tag{1.4}$$

Or,

$$ln\frac{N_0}{N} = kt \tag{5}$$

Above equation can also be written as,

$$N = N_0 e^{-kt} \tag{6}$$

Where N_0 is the original number of microorganisms at t = 0 and N is the number of microorganisms at time t.The effect of temperature on the reactionrate constant *k* may be expressed by the Arrhenius-type equation:

$$k = a.e^{-E/RT} \tag{7}$$

Putting the value of *k* in the previous equation gives:

$$\int_{N_0}^{N} \frac{dN}{N} = -a \int_{t=0}^{t} e^{-E/RT}\, dt \tag{8}$$

Several parameters are used to do thermal calculations and explain the rate of thermal lethality. The Decimal reduction time or D-value is one of the measures of the heat resistance of a microorganism and defined as:

It is the time in min at a given temperature required to destroy 1 log cycle (90%) of the target microorganism. In other words, it is the time in min during which the original number of viable microbes is reduced by 1/10.

$$\frac{N}{N_0} = \frac{1}{10} = e^{-kD} \tag{9}$$

Taking logarithm on both sides, the equation can be written as:

$$ln\left(\frac{1}{10}\right) = ln\left(e^{-kD}\right) \tag{10}$$

Or,

$$-\ln(10) = -kD \tag{11}$$

$$2.303 = kD \tag{12}$$

Or,

$$D = \frac{2.303}{k} \tag{13}$$

Putting the value of *k* in basic equation,

$$ln\frac{N_0}{N} = kt \tag{14}$$

Or,

$$ln\frac{N_0}{N} = \frac{2.303}{D}.t \tag{15}$$

Or,

$$2.303\,log\frac{N_0}{N} = \frac{2.303}{D}.t \tag{16}$$

Or,

$$t = D\,log_{10}\left(\frac{N_0}{N}\right) \tag{17}$$

Now, if the $log_{10}\left(\frac{N_0}{N}\right)$ is plotted versus t, a straight line should result from the above equation.

Thermal Death Time (TDT)

Thermal death time is the time required that is necessary to kill a specific number of microbes at a specific temperature. This value is obtained by keeping temperature constant and measuring the time necessary to kill the number of cells specified. For example, in the process of pasteurization, a D value at 63°C of 1-min means that for each min of processing at 63°C the bacteria population

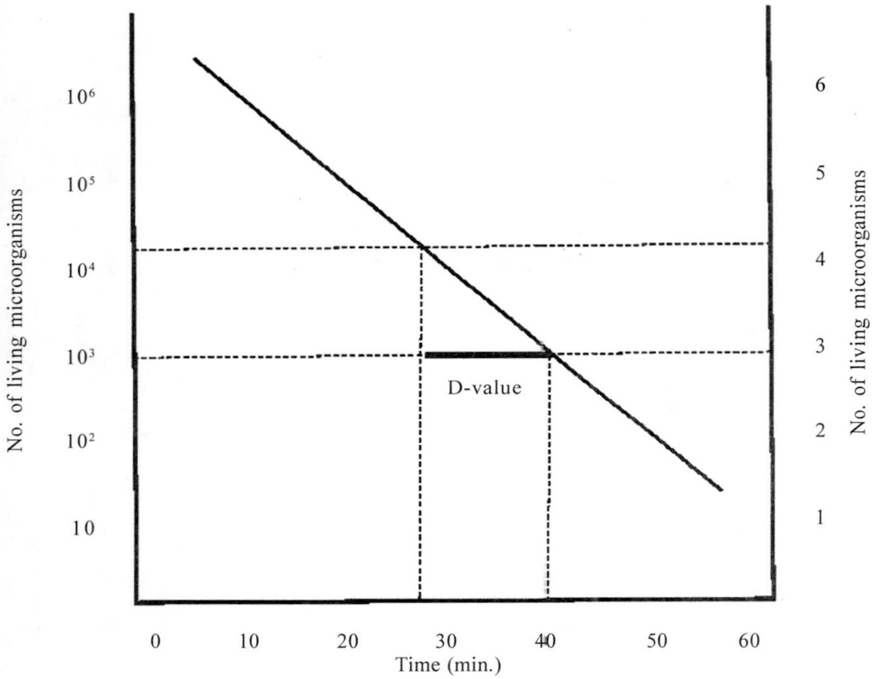

Fig. 2: Representation of D-value on the plot

of the target microorganism will be reduced by 90%. Thus, as represented in Fig. 2, the D value is 12 min (40-28) and would be representative of a process at 63°C.

The Z-value reflects the temperature dependence of the reaction. It is defined as the temperature change (Celsius or Fahrenheit)required in changing the D-value by a factor of 10. In the illustration below the Z-value is 10°C.It can be interpreted as the susceptibility of a spore population with changes in temperature. For example, if the Z-value of a population is 10 degrees, then increasing the sterilization temperature 10 degrees will result in a log reduction of the D-value as depicted in Fig. 3. The Z-value can be found by plotting D-values against temperatures on a semi logarithmic scale and adding a line that best fits the data. The absolute value of the reciprocal of the slope of this line will be the Z-value.

Further, Z-value is required to determine the F_0 value for the sterilization process. Biological indicators are tested to determine their Z value and thus, each microorganism has its unique Z value. Reactions that have small Z-values are

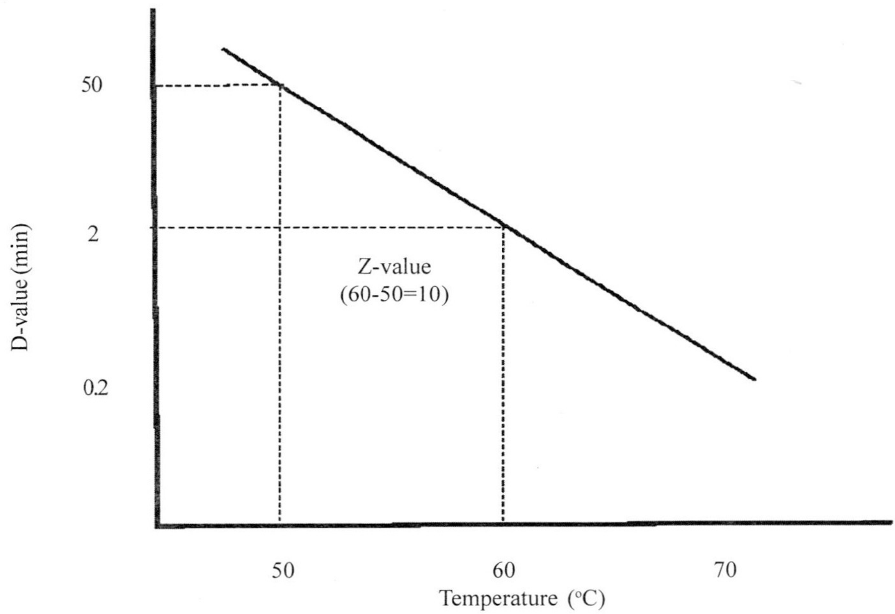

Fig. 3: Representation of Z-value on the plot of D-value *vs.* Temperature

highly temperature-dependent, whereas those with large Z-values require larger changes in temperature to reduce the time. Z-value at 10°C is typical for a spore-forming bacterium. Heat-induced chemical changes have much larger Z-values than microorganisms, as shown in Table 5.

Table 5: D-value and Z-value of some biochemical identities (Morris, 2011)

	Z-value (°C)	D-value (min)
Bacteria	5-10	1-5
Enzymes	30-40	1-5
Vitamins	20-25	150-200
Pigments	40-70	15-50

The higher Z-value of vitamins, enzymes and pigments is vital during the

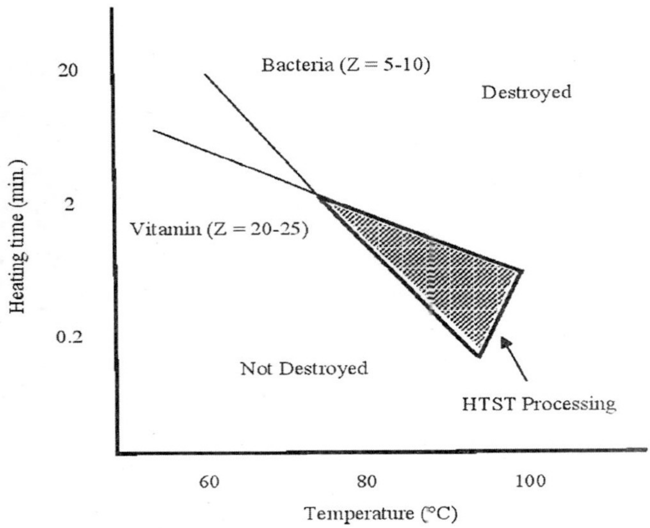

Fig. 4: Mapping heat treatment with quality factors

processing of milk or other food products. This is illustrated in Fig. 4 which highlight region exist (shaded area) where only microorganisms are destroyed due to considerable difference in Z-value whereas, vitamins present in milk is maintained.

It is evident from the plot that above and to the right of microorganisms and vitamin line would be destroyed in marked time-temperature zone whereas below and left of the line, vitamins are safe, but pasteurization is incomplete. The other quality factors like enzymes and pigments are also safe because of higher Z-value. Therefore, before having any heat treatment, the process must be analyzed as it is safe in terms of a microbial load as well as a quality factor. Further, it can be recognized that opting HTST Pasteurization process is beneficial in comparison to LTLT Pasteurization in terms of nutritional and sensorial value.

Pasteurization is generally used for the destruction of all disease-causing microorganisms (e.g. pasteurization of milk) and alkaline phosphatase is a naturally occurring enzyme in raw milk with a similar Z-value to heat-resistant pathogens, used as an indicator of adequate pasteurization. The assessment of pathogen numbers is not economic using microbial methods and thus, dairy processing plants often implement a simple test for phosphatase activity. The absence of phosphatase activity in pasteurized milk affirms the completion of the pasteurization process. The assessment of microbial load by using D and Z values in pasteurization has been illustrated in Example 1.

Example 1: During the processing of milk in a dairy plant, the bacterial population of raw milk is assessed as 6×10^5/mL. Milk is then pasteurized at73°C for 25seconds as referred by standard pasteurization protocol. Determine the number of organisms left after completion of pasteurization and find the time required at 63°C to accomplish the same degree of lethality. The average D value at 63°C for the mixed population is 8 min and the Z-value is 5°C.

Solution: At 73°C, the D value has been reduced by two log cycles from that at 63°C as the Z value is 5°C.Thus,D value at 73°C is 0.08 min or 4.8 sec. Further, the milk is processed for 25/60=0.42 min, so that it would accomplish 5 log cycle reductions to 6 organisms/mL. Therefore, the time required at 63°C, to accomplish a 5D reduction and the same degree of lethality would be 42 min.

The 12-D Process

As discussed in the previous section, the extent of the pasteurization treatment required is assessed by the heat resistance of the most heat-resistant microorganism in the milk. Thermal destruction of microorganisms takes place logarithmically, but it is hard to sterile any product completely no matter how long the process time. also, the probability of survival of a single microorganism can be predicted using details of the heat resistance of the micro-organism and the temperature and time of heating. This underlines to a concept, more commonly known as commercial sterility. As explained by FAO/WHO, "Commercial sterility refers to the absence of microorganisms capable of growing in the food at normal non-refrigerated conditions at which the food is likely to be held during manufacture, distribution and storage. To counter this, A 12D process is used especially when C. botulinum is likely to be present in foods, but in foods that contain more heat-resistant spoilage microorganisms, the application of a 12D process results in over-processing and quality deterioration. In such cases, a 2D to 8D process is used to provide the most economical level of food spoilage reliable with acceptable food quality and safety. Therefore, a process that reduces cell numbers by twelve decimal reductions (a 12D process) are applied for effective processing to reduce the probability of the survival by the most heat-resistant C. botulinum spores by 12 logs.

Applying the concept of 12D as the minimum heating process to reduce the number of the spores by a factor 10^{-12} in the equation (17), we have

i.e.

$$\frac{N}{N_0} = 10^{-12}$$

Or,

$$\frac{N_0}{N} = 10^{12}$$

Therefore,

$$t = D\log_{10}\left(\frac{N_0}{N}\right) = D\log_{10}\left(10^{12}\right) \tag{18}$$

$$t = 12D \tag{19}$$

6.2 Thermal Destruction of Microorganisms

During the thermal processing such as pasteurization, microorganisms are destroyed by heat at rates which depend primarily on the temperature. The destruction of microorganisms at higher temperatures causes killing spores more quickly. At any given temperature, the spores are killed at several times, some spores being more resistant to heat than other spores. When the graph ofthermal death rates of microorganisms is plotted as the decimal reduction time D_T at a given temperature versus the temperature T in °C on a semilog plot, give essentially straight lines over the range of temperatures used in pasteurization as represented in Fig.5.

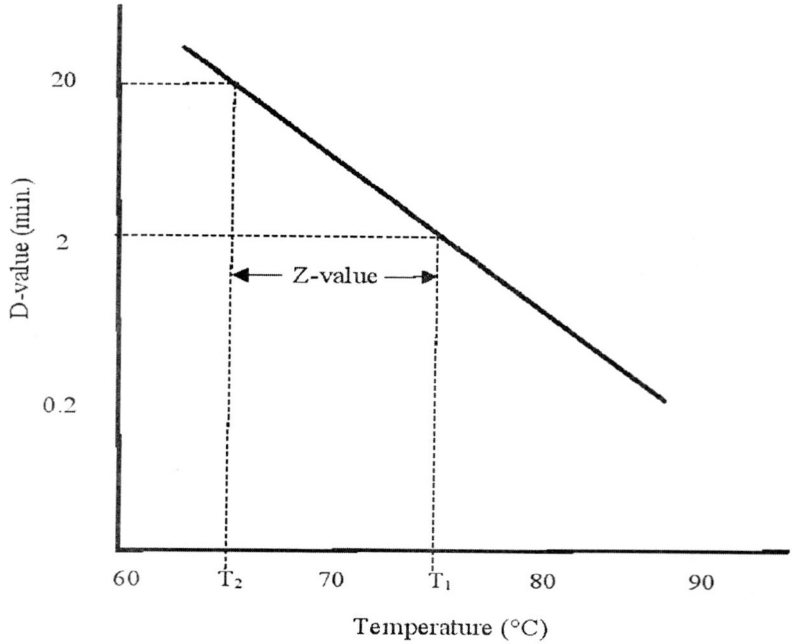

Fig. 5: Thermal death rates of microorganisms is plotted D_T *vs.* T

In Fig. 5, the term *z* represents the temperature range in °C for a 10:1 change in D_T. Since the plot is a straight line, the equation can be represented as:

$$t = D\log_{10}\left(\frac{N_0}{N}\right) = D\log_{10}\left(10^{12}\right) \tag{20}$$

Or,

$$\frac{D_{T_2}}{D_{T_1}} = 10^{\frac{(T_1 - T_2)}{z}} \tag{21}$$

Letting, $T_1 = 121.1°C$, which is the standard temperature against which thermal processes are compared, and calling $T_2 = T$, the above equation can be written as:

$$D_T = D_{121.1} 10^{\frac{(121.1 - T)}{z}} \tag{22}$$

For the microorganism, if $z = 10°C$, it means that each increase in temperature of 10°C will increase the death rate by a factor of 10.

Further, using

$$t = D_T log_{10}\left(\frac{N_0}{N}\right) \quad (23)$$

Substituting T = 121.1°C, as the standard temperature into the above equation and substituting t = F_0, Equation can be written as:

$$F_0 = D_{121.1} log_{10}\left(\frac{N_0}{N}\right) \quad (24)$$

Where F_0 Value is process time and is defined as:

F_0-Value at a specific temperature other than 121.1°C, is the time in min required to provide the lethality equivalent to that provided at 121.1°C for a stated time.

As,

$$t = D_T log_{10}\left(\frac{N_0}{N}\right) \quad (25)$$

Or,

$$D_T = \frac{t}{log_{10}\left(\frac{N_0}{N}\right)} \quad (26)$$

Again,

$$F_0 = D_{121.1} log_{10}\left(\frac{N_0}{N}\right) \quad (27)$$

Or,

$$log_{10}\left(\frac{N_0}{N}\right) = \frac{F_0}{D_{121.1}} \quad (28)$$

Putting the above value in the equation,

$$D_T = D_{121.1} 10^{\frac{(121.1-T)}{z}} \quad (29)$$

Or,

$$\frac{t}{log_{10}\left(\frac{N_0}{N}\right)} = D_{121.1}10^{\frac{(121.1-T)}{z}} \quad (30)$$

$$\frac{t}{\left(\frac{F_0}{D_{121.1}}\right)} = D_{121.1}10^{\frac{(121.1-T)}{z}} \quad (31)$$

$$\frac{t}{F_0} = 10^{\frac{(121.1-T)}{z}} \quad (32)$$

Or,

$$\frac{t}{10^{\frac{(121.1-T)}{z}}} = F_0 \quad (33)$$

Therefore, for z = 10°C

$$F_0 = t.10^{(T-121.1)/z} \quad (34)$$

This is the F_0-value in a min for the given thermal process at constant temperature T°C and a given time t in a min. Values for F_0 and *z* for adequate sterilization vary somewhat with the type of foods. Further, the effects of different but successive sterilization processes in a given material are additive. Therefore, for several different temperature stages T_1, T_2, T_3°C and so on, each having different times t_1, t_2, t_3...the F_0-values for each stage are added to give the final value of F_0.

$$F_0 = t_1.10^{(T_1-121.1)/z} + t_2.10^{(T_2-121.1)/z} + t_3.10^{(T_3-121.1)/z} + \ldots \quad (35)$$

In the general case, when foods are being sterilized, the temperature is not constant for a given time period but varies continuously with time. Therefore, the above equation can be modified and written for continuously varying temperature T by taking small time increments of *dt* min for each value of T and summing.

$$F_0 = \int_{t=0}^{t=t} 10^{(T°C-121.1)/z°C}\ .dt \quad (36)$$

6.3 Process Time for Pasteurization

The process value in the case for the pasteurization of foods such as milk and milk products, are determined using the same mathematical and graphical procedures as covered for the sterilization process. Thus, the times involved are much shorter and the temperatures used in pasteurization are much lower. For pasteurization, F_0-value is considered as at 150°F (65.6°C) or similar temperature rather than at 250°F (121.1°C) as in sterilization.

An F_0-value written as $F^5_{65.6}$ means the F-value at 65.6°C with a z-value of 5°C.

$$F^z_{T_1} = D_{T_1} log_{10}\left(\frac{N_0}{N}\right) \tag{37}$$

$$F^z_{T_1} = t.10^{(T-T_1)/z} \tag{38}$$

6.4 Q_{10} Value

The Q_{10} value has a distinct role in quality analysis to measure the rate of quality changes with temperature. It refers to changes occurred in the reaction rates, if the temperature of the system is increased by 10°C and expressed as the ratio of the increased rate of reaction to the earlier rate of a reaction. It can be written in equation form as:

$$Q_{10} = \frac{k_{(T+10)}}{k_T} \tag{39}$$

Where, T is in degree Celsius (°C).

The concept of Q_{10} value is commonly used to assess the quality parameters like flavor changes, vitamin losses, pigmentation and browning as well as accelerated shelf life during the thermal processing of food products. The Q_{10} value of most of such quality parameters are in the range of 2 to 3, and for the destruction of bacterial spores is around 8 to 30.

Example 2: For an initial spore load equal to 25 spores per container inoculated with *Clostridium botulinum* having D_{121}= 0.25 min, the spoilage probability of the container subjected to F_{121} = 1.5 min is:

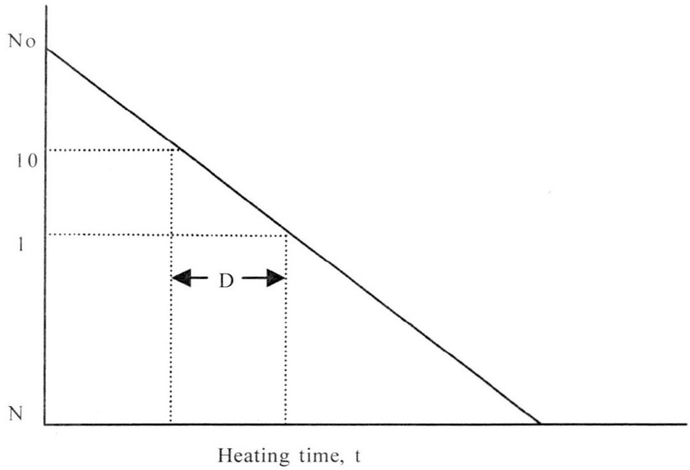

(A) 10^5 **(B) 10^6** **(C) 10^8** **(D) 10^9**

[GATE AG 2013]

Solution

Here,

N_0 = No. of spores per container at initial

N = No. of spores per container after time, t min

As we know that,

$$\frac{logN_0 - logN}{\log 10 - \log 1} = \frac{t}{D}$$

Or,

$$logN_0 - logN = \frac{t}{D}$$

If, time of heating at a temperature is F, we can write above equation:

$$logN_0 - logN = \frac{F}{D}$$

Or,

$$log\frac{N_0}{N} = \frac{F}{D}$$

$$\frac{N_0}{N} = 10^{F/D}$$

Therefore, the spoilage probability of the container can be calculated as:

$$\frac{N_0}{N} = 10^{F/D}$$

Thus,

$$\frac{N}{N_0} = \frac{1}{10^{1.5/0.25}}$$

$$\frac{N}{N_0} = \frac{1}{10^6}$$

Therefore, spoilage probability is 10^6.

Hence, Option [B] is right choice.

Example 3: A suspension contains 3.6×10^5 spores of *C. botulinum* having a D-value of 1.5 minute at 121.1 °C and 8.5×10^6 spores of *B. subtilis* having a D-value of 0.9 minute at the same temperature. The suspension is heated at a constant temperature of 121.1 °C. The heating time needed in minutes for the suspension to obtain a survival probability of 10^{-3} for the most heat resistant organism in it is______.

[GATE AG 2015]

Solution

Given,

for *C. botulinum*, $N_0 = 3.6\times10^5$ spores

and, $N = 10^{-3}$ spores

$D_{121.1} = 1.5$ min.

Therefore, using equation

$$logN_0 - logN = \frac{F}{D}$$

Or,

$$log\frac{N_0}{N} = \frac{F}{D} log\frac{N_0}{N} = \frac{F}{D}$$

$$F_{121.1} = D_{121.1} log\frac{N_0}{N}$$

$$F_{121.1} = 1.5\,x\,log\left(\frac{3.6\,x\,10^{5}}{10^{-3}}\right) = 12.83\,min.$$

Again,

for *B. subtilis*, $N_0 = 8.5 \times 10^6$ spores

and, $N = 10^{-3}$ spores

$D_{121.1} = 0.9$ min.

Therefore, using equation

$$logN_0 - logN = \frac{F}{D}$$

Or,

$$log\frac{N_0}{N} = \frac{F}{D}$$

$$F_{121.1} = D_{121.1} log\frac{N_0}{N}$$

$$F_{121.1} = 0.9\,x\,log\left(\frac{8.5\,x\,10^{6}}{10^{-3}}\right) = 8.93\,min.$$

Therefore, the heating time needed in min for the suspension to obtain a survival probability of 10^{-3} for the most heat resistant organism in it is **12.83 min**.

Example 4: Milk sterilization kinetics is based on inactivation of index microorganism, *Bacillus stearothermophilus*. The D-values at 121.1°C and 139.1°C are 1.2 min and 0.019 min, respectively. For 12 log-cycle reduction of this microorganism at 130°C, the processing time in second is ____________.

[GATE AG 2016]

Solution

Given,

$D_{121.1} = 1.2$ min.

$D_{139.1} = 0.019$ min.

As we know that,

$$\frac{D_{T_1}}{D_{T_2}} = 10^{\frac{T_2 - T_1}{z}}$$

$$\frac{D_{121.1}}{D_{139.1}} = \frac{1.2}{0.019} = 10^{\frac{139.1-121.1}{z}}$$

z = 10 °C

Now, for D_{130}

$$\frac{D_{121.1}}{D_{139.1}} = \frac{1.2}{0.019} = 10^{\frac{139.1-121.1}{z}}$$

$$\frac{D_{121.1}}{D_{139.1}} = \frac{1.2}{0.019} = 10^{\frac{139.1-121.1}{z}}$$

D_{130} = 0.155 min.

Therefore,

$$\frac{D_{121.1}}{D_{139.1}} = \frac{1.2}{0.019} = 10^{\frac{139.1-121.1}{z}}$$

$$F_{130} = D_{130} \log\left(10^{12}\right) = 12 D_{130}$$

F_{130} = 12 x 0.155 = 1.86 min. = 111.6 s

Example 5: The reaction rate for destruction of *Clostridium botulinum* increases 11 times for temperature rise of 10°C from 121.1°C. The decimal reduction time of this organism is 5.7 sec at 121.1°C. The minimum sterilization time in seconds at 135°C, for eight log cycle reduction of the organism is:

(A) 0.20 **(B) 0.31** **(C) 1.63** **(D) 2.49**

[GATE AG 2017]

Solution

As we know that,

$$\frac{D_{T_1}}{D_{T_2}} = \frac{k_{T_2}}{k_{T_1}} = 10^{\frac{T_2 - T_1}{z}}$$

If, $k_{T1} = k_{121.1} = k$, then $k_{T2} = k_{131.1} = 11k$

Therefore,

$$\frac{D_{T_1}}{D_{T_2}} = \frac{k_{T_2}}{k_{T_1}} = 10^{\frac{T_2 - T_1}{z}}$$

Taking log both sides, we have

$$\log 11 = \frac{10}{z}$$

$z = 9.60\ °C$

Now,

$$\frac{D_{121.1}}{D_{135}} = \frac{5.7}{D_{135}} = 10^{\frac{135 - 121.1}{9.60}}$$

$D_{135} = 0.203\ s$

Therefore,

$$F_{135} = D_{135} log \frac{N_0}{N}$$

$$F_{135} = D_{135} \log\left(10^8\right) = 8D_{135}$$

$$F_{135} = 8 x 0.203 = 1.63 s$$

Hence, Option [C] is right choice.

Example 6: A microbial sample taken at 10 AM contained $1x10^5$ CFU/mL. The count reached to $1x10^{10}$ CFU/mL at 8 PM of the same day. The growth rate (h^{-1}) of the microorganism would be _______. (up to two decimal points)

[GATE Food Technology – XE-G2018]

Solution

The rate of growing of microorganisms is proportional to initial number of organisms and can be calculated as:

$$\ln\frac{N}{N_0} = kt$$

Where N_0 is the original number of microorganisms at $t = 0$ and N is the number of microorganisms at time $t = 10$ hour.

Therefore, growth rate (h^{-1}) of the microorganisms can be calculated as:

$$\ln\left(\frac{1 \times 10^{10}}{1 \times 10^{5}}\right) = k \times 10$$

$k = 1.15\ h^{-1}$

7. Pasteurization Equipment

The most common processing of heating milk known as pasteurization to enhance the shelf life and safe consumption is not only is affected by specific time-temperature combination but also largely influenced by the mode of operation and design specification of equipment. The pasteurization equipment has been modified numerous times to enhance the working of pasteurizers based on scale and speed of operation. As discussed in the previous section, pasteurization is classified into two categories *i.e.* Batch Process and Continuous process.

7.1 Batch or VAT Pasteurizer

Batch or Vat pasteurizer is the oldest and easiest method of pasteurization for processing comparatively small volumes of food products. It generally consists of a jacketed vessel tailored with an agitator, temperature sensor and a source of heat. Batch Pasteurizers are essentially very large vessels in which dairy products are pasteurized in a jacketed stainless-steel vessel, called as Vat and thus, also known as Vat Pasteurizer. The vat consists of an inner vessel, enclosed by an insulated outer casting to form a jacket with a different operational mechanism through which the heating or cooling medium is circulated. The Jacketed vessel holds the milk with an arrangement of common inlet and outlet for milk. The bottom of the vessel is designed with a slope towards the outlet to warrant total drainage at the collector end. Hot water or steam used in jacketed vessels ensures the heating of the product by maintaining specific temperature (Dhotre, 2014). The jacketed vessel is equipped with a bourdon tube pressure gauge, safety valve, temperature sensor, etc. for monitoring and safety operation purposes.

The propeller type agitatormountedon top of the tank facilitates in uniform heat transfer and prevent localized overheating of the food products. The food product is finally cooled in the vessel using chilled water and refrigerant circulated through the outer jacket of a pasteurizer.

The biggest hurdle that restricts the usage of batch pasteurizer for small volume is because of the huge amount of energy required and delayed cooling in comparison to the continuous pasteurizer. Batch pasteurizer, because of its construction is also not capable of heat regeneration and in result causes low thermal efficiency. However, because of its lower capital cost. Batch or vat pasteurization is used extensively by small scale milk processing plant for butter and ice cream production specifically.

7.2 Continuous Pasteurizer

As discussed earlier, continuous type pasteurization also known as HTST pasteurization system is operated at high temperature for a shorttime (72°C/15 sec) and instantly cooled below 5°C. Generally, any continuous (HTST) pasteurizer mainly consists of a balance tank, heat exchanger system, holding tube, flow diversion valve, booster pump, heating mechanism, chilling, mechanism and a back-pressure valve. The dairy industry uses commonly three types of heat exchangers namely plate heat exchanger, tubular heat exchanger, and scraped-surface heat exchanger that depends on the nature and scale of operation.

The primary purpose of the process control system in HTST pasteurization is to safeguard that the milk gets anticipated holding at a specified temperature. It depends on numerous factors viz., heating medium and flow rates of milk, range of operational temperature and heat transfer rate. The PHE is the most vital component in HTST pasteurizer which facilitates uniform heat transfer rate.

7.3 Tunnel Pasteurizer

Tunnel pasteurization is an alternative to flash pasteurizer functions at a lower temperature (60-65°C) for a longer duration and is used mainly for in-bottle or in-package pasteurization, carried out after pouring and sealing the product in bottles and packages. This method is used for the processing of products packed in can primarily, however also being used in bottles/packages and passed through a tunnel-like pasteurizing structure categorized in different temperature zones.

The first two or three zones are the preheating section which helps in the gradual raising of the temperature of the product and prevents the risk of package rupture due to pressure build-up.

8. Future Pasteurization Innovations

Since the inception of pasteurization techniques, pasteurization has destined a long journey in terms of scale, speed and operation. The technology of pasteurization as thermal processing has revolutionized milk processing techniques and thus the availability of milk and milk products throughout the world. Advancement in techniques like heat regeneration has further caused to prevent energy consumption over 85% and thus, fulfilling economic and environmental prospects also. In recent past few years, non-conventional techniques of thermal processing like ohmic heating, inductive heating, microwave heating, pulsed heating, high-pressure processing, etc. have been implementing, however, pasteurization methods using plate heat exchanger is anticipated to remain in fashion because of its efficiency in terms of energy and operation as well as research being carried out on de-fouling techniques or antifouling plates.

References

Davis, J.G. Adictionary of dairying, 2nded., London: Leonard Hills Books, 1955, pp. 786-810.

Dhotre, A.V. Milk pasteurization and equipment. In: P.K. Mandal and A.K. Biswas, ed., Animal Products Technology, 1sted. New Delhi: Studium Press (India) Pvt. Ltd., 2014, pp.51-78.

Morris, S.A. Processes calculations and their applications. In: Food and Package Engineering, 1st ed., John Wiley & Sons, Inc, 2011, pp. 207-252.

Pal, M. Public health hazards due to consumption of raw milk. The Ethiopian Hearald, 2012, P.10.

Pal, M. Zoonoses. 2nd ed., Satyam Publishers, Jaipur, India, 2007.

Silva, F.V.M. da and Gibbs, P.A. Principles of thermal processing: Pasteurization. In: Engineering of Thermal Food Processing, Simpson, Ricardo. (Ed), CRC Press, London, 2009, pp. 13-39.

Simpson, R. In Engineering Aspects of Thermal Food Processing. Editors: CRC, Boca Raton, USA, 1st ed., 2009, 13-48.

VonSoxhlet, F. Infant Milk and Infant Nutrition. Munich Medical weekly, 1886, 33: 256-276.

Von Soxhlet, F. ÜberKindermilch and Säuglingsernährung. Münchenermedizinische Wochenschrift. 1886, 33: 256–276.

Westhoff, D.C. Heating milk for microbial destruction: A historical outline and update. Journal of Food Protection,1978,41:122–130.

14

Cleaning in Place

Kumar Sandeep, Kumaresh Halder, Kirtiraj K. Gaikwad and Ajit Kumar Singh

1. Introduction

To maintain good quality of product & shelf life, it is very important & essential to concentrate on Cleaning & Sanitization process at plant level & to follow the CIP process properly. An attempt is made in the current chapter to present details pertaining to cleaning in fluid milk and milk products manufacturing &processing plants. This chapter will be helpful the students of Food & Dairy Technology for better understanding the process milk plant cleaning & sanitation technology.

The prime role of cleaning is to remove the milk and milk residue from different dairy equipment. The left over residue due to improper cleaning in the dairy equipment may leads to increase the bacterial load & growth and finally deteriorate the quality & shelf life of product. Secondarily, excessive deposition of soil and mineral will reduce the efficiency of the process processes where heat transfer is involved like pasteurization and scaling of the pipelines, tanks & equipment.

2. Principle and Methods of Cleaning

For all the successful method, proper procedure & methods of CIP must be followed effectively. The proper temperature, chemical concentration, time & flow rate must be good enough so that the milk residue & remaining soil can be break &removed by the physical & chemical action.The equipment should by clean without any trace of soil or detergent after post rinse & to carry away the loosened soil and detergent solution. For more effective cleaning acidify the process of cleaning after alkaline cleaning whenever applicable. By this way accumulation of alkaline residual of detergent or any minerals from rinse water. For effective cleaning, must have to adhere to

i. Flow rate, time, temperature, detergent concentration along with physical action and

ii. The step wise procedure of pre-rinsing, washing & rinsing

2.1 Manual Cleaning

This process is a primary process and used at most of the dairy plants which require dismantling of the equipment and their different parts which are of product contact and manually rinsed, brushed and cleaned. The dismantled pipes, bends etc must be washed, brushed and poured in the sanitization tank with the detergent of the desired concentrations or may be cleaned with power-driven brush. All pumps, valves, caps, tees, ells, ferrule unions, etc., must be brushed, washed & cleaned. The gaskets must be checked on the regular & frequently and replaced wherever necessary. The storage tanks, tankers & other vessels must be manually cleaned & thoroughly washedwith proper detergent solution.

The concentration of the operator who is doing the manual cleaning, detergent concentration & proper brush which is being used is important for the proper manual cleaning process and must be verified manually. The brush be of food grade material which do not leave the bristles, rust free otherwise it may lead to the food safety hazards and strong enough to resist heat & chemicals. The brush shall be cleaned easily without adhering the dirt which may leads to growth of bacteria and fungi.The various sizes of nylon brush is widely used in a dairy plant.

2.2 Cleaned-Out-Of-Place (COP)

COP referred as "cleaned-out-of-place" which requires the dismantling of the equipment and their parts after rinsing so that different sections of pipelines & all parts can be placed in a recirculation tank for physical & chemical cleaning.

2.3 Cleaned-In-Place (CIP)

Cleaned-In-Place (CIP) process isgenerally applied in dairy industry plant. It include cleaning and sanitizing of milk processing without disassembled the equipment and pipes by using acid, alkali and hot water treatment under predetermine time and temperature. Time count modules, microprocessors and computers have been attached with the CIP systems of advance dairy plant that maintain thesimilar repeatable process of functions. CIP remove the manual physical action of the brush from the cleaning process and replaces it with chemical and hydraulic physical action, so plant personnel, maintains the concentration of chemical and sanitizing solution level on a daily basis is very important because very strong chemicals are used during the process that may be give adverse effect on pipes or equipment as well as people health (Mauck et al., 2001).

3. Nature of Soils

Soil residues generally consist of milk constituents. Water maysupplydifferent concentration of soil residue. Primary organic residues of milk are fat, protein and lactose, whiledifferent milk salts are consider in inorganic residues.

Lactose is easily clean by rinse water. Fats are readily removed above their melting point of approximately 90°F. Denatured proteins by the application of pasteurization can cause residues which are difficult to remove. Effect of heat treatment on milk component and their removal treatment method has highlighted in Table 1.

Table 1: Effect of heat treatment on milk components and their removal

Milk Component	Effect of heat treatment	Treatment for removal
Fat	Toughest to clean	Surface active agents
Lactose	Toughest to clean	Water
Protein	Tougher to clean	Alkali
Mineral Salts	Tougher to clean	Acid

Milkstone formed in the plant can be identified by visual examination or by using simple chemical tests. Table 2 outlines some of the milkstone characteristics alongwith the reasons for their formation and methods for their removal.

The situation becomes even more complex in a milk UHT plant as protein will be the predominant soil at temperatures of up to 115°C whilst mineral deposits will prevail as the temperature increases further. Each type of soil will need a specific regime for removal.

4. Cleaning Parameters

The Soil & Milk Stone are binding on the surface by adhesive forces. The forces have to overcome and to loosen from the surfaces so that soil can leave the surface easily. The four key parameters are,

Mechanical Action- Mechanical action through pressure, flow rate and flow velocity.

Table 2: Characteristics of Milkstone

Visual Characteristics	Component	Cause	Method/Treatment for its removal
Blue/Rainbow blue	Protein	Usage of non-chlorinated cleaner, Inadequate pre-rinse	Usage of chlorinated alkali and/or Strong acid
White/Yellow deposits	Mineral	Deposition of mineral from water/milk	Usage of chlorinated alkali and/or Strong acid
Greasy	Fat	Low temperature or improper concentration of detergent	Alkali
White, grey, chalky deposits	Mineral from Water	Improper rinse, no acid rinse, incompatible alkali, usage of hard/untreated water, non-usage of acid detergents	Acid
Red to brown/black deposits	Iron particles from water	Water Supply	Strong acid

Temperature: Cleaning effectiveness increases with temperature.

Concentration: Cleaning effectiveness increases with concentration to a maximum, but might even decrease then.

Time- Time will be a function of temperature and chemical activity.

In order to remove the soil energy is required in a cleaning process. The energy may be in various forms like kinetic, chemical & thermal energy. The effectiveness of cleaning is determined by these three factors. All the four parameters are interconnected and if one parameter is changed, the others three need to be modify of the same or the best result. Sinner's circle (Figure 1) diagram shows which include flow, temperature, concentration and time. The circle diagram was originally constructed in 1959 by Dr Herbert Sinner, a chemist who worked for Henkel, a German detergent supplier.

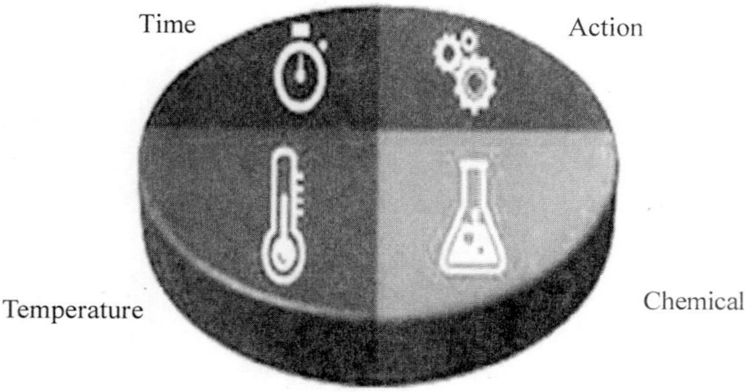

Fig. 1: Sinner's Circle

4.1 Mechanical force

The shear force created by the flow in CIP is act as a mechanical force. Compare how car is cleaned with a water hose with a nozzle or without a nozzle. Cleaning with a nozzle you will observe that the velocity of the water is more & water jet gets "harder" as restricted the area through which water is passing. In a plant level the pump is responsible for increasing the flow velocity of the cleaning chemicals. It is advisable that the flow must achieve turbulent flow with the minimum velocity of 1.5 m/s which will be adequate mechanical force for a proper CIP. The CIP flow has various purposes – act as transport for CIP chemicals to the dirty and soiled surfaces, remove the soil from the surface, remove the soil from the equipment being cleaned. The design of the plant is a Prerequisite Program (PRP) of the CIP and its mechanical forces.

4.2 Chemical force

The chemical force is the second force which uses the soil to leave a surface. Chemicals are used to clean the equipment with the flow which is a type of mechanical force. First alkaline detergents are often used. It removes the organic soils like protein, fat & sugars. Sodium Hydroxide is widely used as alkaline detergent. The pure alkaline detergent is recommended which is formulated with other cleaning components suitable for hard water and for best cleaning result. It has more wetting properties which dissolve the dirt more than other alkaline solutions.

The recommended dose of the Alkaline solution for CIP process is 0.5-2% for most of the applications, for some food applications also higher doses is used. High doses may linked with proteins and make difficult to remove from the cleaning objects. The intermediate process between the Alkaline detergents & Acidic detergents is flushing with hot water because the alkaline detergent reduce the efficiency of the acid. The next step is the acid cleaning. Acid dissolve inorganic soils like minerals. Commonly used acids are nitric acid and phosphoric acid. The detergent company also supplies formulated acid detergents for best cleaning result (Handbook, Tetra Pak, 2016). Generally 0.5-1.5% strength of nitric acid is used for CIP process. Care must be taken for higher doses of Acid as it has adverse effect on Stainless Steel & Polymer material.

4.2.1 Types and Properties of Cleaning Compounds

The neutral, alkaline or acid type of cleaning compound found in milk plant. To remove different types of soils and different methods, different types of cleaners are necessary. The alkali & acid is the main component of all formulated detergents. It may include additional component:-

i. Wetting agent or Surfactants of lower surface tension which makes the cleaning more efficiently by allowing wetting the surface more efficiently.

ii. In order to soften the water calcium and magnesium ions bind with sequestering agents.

iii. An oxidation compound boosts the cleaning effect. Sodium hypochlorite and hydrogen peroxide are few of them.

iv. Metal ion only binds with complex forming agents which is different from sequesting agents which bind with a number of metal ions.

v. Complex-forming agents can only bind one metal ion per molecule in contrast to sequestering agents, which can bind to a number of metal ions.

A good cleaning compound should have as many of the following properties as possible:

i) Economical
ii) Property of Soft Water
iii) Low toxic
iv) Less Corrosive
v) Easily Soluble
vi) Storage Stability
vii) Easy to Handle & Use
viii) Rinsing Property

Cleaning compounds mainly contain various chemicals that can be categorized as following groups,

4.2.1.1 Alkalis

pH more than 7 is referred as alkaline cleaners. These types of cleaners are commonly used in dairy industry because it alters the fat, protein & sugars, so these are converted into soap and can be removed with water. The alkalis are mainly consists of sodium hydroxide (caustic soda) potassium hydroxide (caustic potash), sodium carbonate (soda ash), and sodium silicates. Trisodium phosphate (TSP) is also grouped into the alkali as it yield hydroxide ion when react with water. Strong alkali bases like caustic soda and potassium hydroxide can cause tissue damage & corrosion. The less expensive source of alkali is soda ash. It imparts & incorporates into manual cleaning as it is mild. Hard water gets precipated with alkali and used in the hard water areas and blended with various ingredients which softens the water. Suspension of soil & inhibiting corrosion is an important property of the sodium silicate. Sodium silicates are good soil suspending agents with an additional property of inhibiting corrosion of soft metals such as aluminum. Trisodium phosphate has good emulsifying propertiesand softens water by precipitating the hardness.

4.2.1.2 Polyphosphates

These are also referred as complex phosphates. They have excellent emulsifying & soil dispersing properties which prevent scaling of equipment. Sodium tripolyphosphate and sodium hexametaphosphate are few complex phosphates.

4.2.1.3 Organic Chelating Agents

To prevent scale deposition and to control hard water these ingredients are blended into detergents. The gluconic acid is mostly used organic acid and salts of ethylene diamine tetra acetic acid.

4.2.1.4 Chlorine Compounds

These are very often added to alkaline cleaners and various source of chlorine is available. They solubilize the protein molecules by peptizing the protein & enhancing the rinsing the equipment, make it free from water spots.

4.2.1.5 Wetting Agents

They promote the soil removal and often blended with alkali & acid cleaning compounds. In neutral cleaners these are primary active ingredients. They enhance the cleaning chemicals to penetrate the soils. They rinsing the cleaning solution by emulsifying fats & suspend the soils. Some of them have the ability to defoam when used in circulation cleaning. Most of the wetting agents do not give more loads in effluent treatment as it broken down. Now a day's most of the wetting agents are biodegradable.

4.2.1.6 Acids

pH lower than 7 are referred as acid cleaners.These types of cleaners are commonly used in dairy industry because it descale the equipment & eliminate the minerals. These are also used as single steps cleaning process for heat exchangers.

Both organic and inorganic acids and salts is used with wetting agents often. The most effective & commonly used inorganic acids are phosphoric acid as it is relatively safe and less corrosive than other inorganic acids. For removing heavy scale hydrochloric acid are used sometime but care must be taken as it is very corrosive. To clean HTST pasteurizer's combination of nitric acid and phosphoric acid are also used.Citric and hydroxyaceticacids are organic acid which is milder and less corrosive but due to the high cost it is very limited.

5. CIP Systems

Consider Pump and a Pot system which act as a CIP process which is connected to a pipeline loop which utilize the heat and flow along with detergent to clean the line, object & equipment which can be controlled either manually or automatically. Pumps & valves are the controlling functions of the CIP cycle which are activated repeatedly after each cycle time. The automated system is more sophisticated which is either programmed or push button system. The system includes Logic Control Panels with adjustable timer, temperature & other parameters. Punched card programmer is also used as other types of controllers. Solid state programmable controllers are recent type with microprocessor; computers & time count module are applied in the CIP system. These controllers have additional advantages over other controllers,

i. No restriction of time limit in any steps of the programme.

ii. More reliable and better service as no moving parts.

According to the conditions, capital availability, and preferences of the brewer, a number of systems are available.

5.1 Single Use System

In this system of CIP a single tank is used for both the purpose of storage & circulation of caustic/acid in a closed loop. Both the pre-rinse & final rinse are fed from the mains which is difficult to achieve the required flow rate for turbulent flow. These units are small in size and are located adjacent to the system/equipment that is to be cleaned and sanitized. An example of single use system is a Boil Out System (Figure 2).The advantages of using this system are that it requires a low capital investment, low floor space and no additional equipment is required for performing the CIP. The disadvantages associated with this system are; higher consumption of detergent/acid/water, longer CIP duration and very difficult to monitor and validate the effectiveness of cleaning.

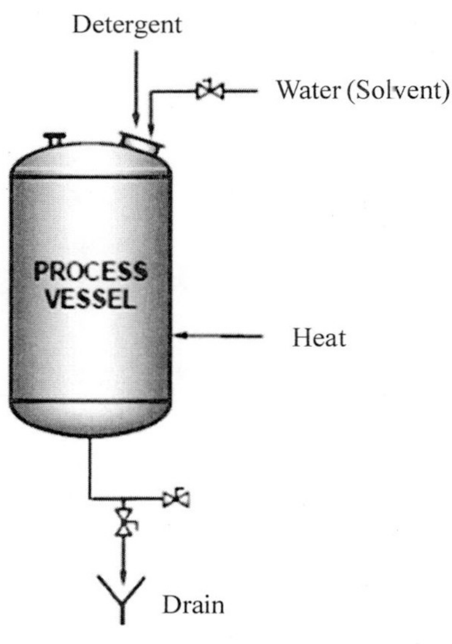

Fig. 2: Boil Out System (Source:Jeffery and Sutton, 2013)

5.2 Single Tank Total loss system

A single tank is used in this system for all rinses and the CIP chemicals are drained at the end of the every rinse cycle (Figure 3). This system have more cost of running, low capital cost and only used in the pharmaceutical industry and at critical non-contamination applications. Detergent and water usage are much higher alongwith longer cleaning time.

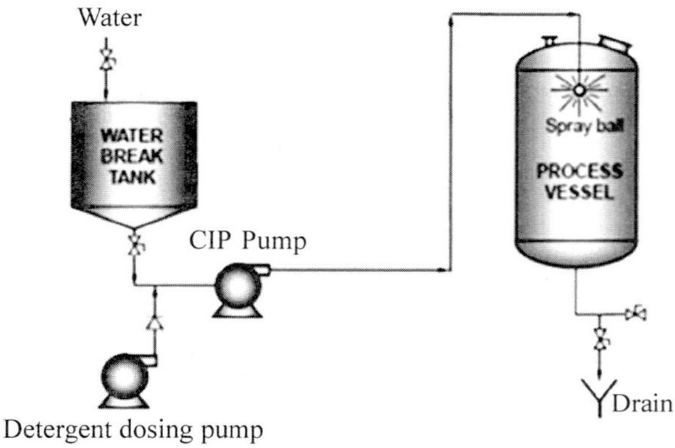

Fig. 3: Total Loss System (Source:Jeffery and Sutton, 2013)

5.3 Re-use system

In the re-use system the detergent or acid is recovered and is used as many tines as possible (Figure 4). Single Use Recirculation System are flexible in nature with a low installation cost and required a small floor space. Risk of cross contamination is reduced while using Single Use Recirculation System, but it is not suitable for large centralized systems.

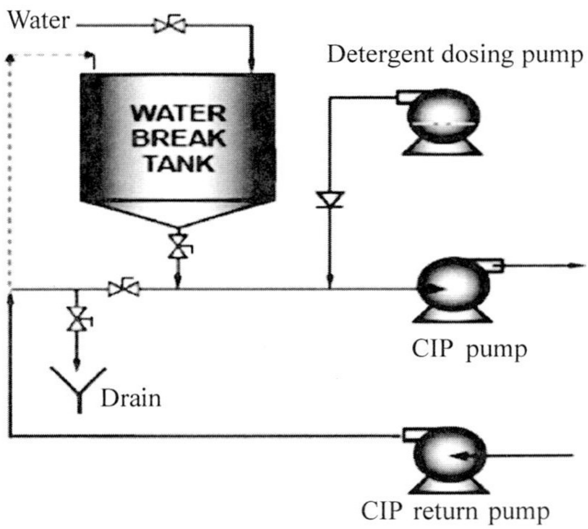

Fig. 4: Re-use System (Source:Jeffery and Sutton, 2013)

5.4 Multi-use system

Multi-use systems combines the features of both single-use and re-use systems (Figure 5) and are designed for cleaning the wet systems. Re-use systems provide the advantage of good water, detergent and acid usage and are suitable for centralized CIP systems. Drawbacks associated with re-use systems are that they are non-flexible, cost of equipment and installation is quite high as compared to other types of CIP systems. The major food safety risk associated with re-use systems is of cross contamination for dis-similar products (viz., milk and cultured products) or raw and pasteurized milk and milk products. Multi-use CIP (Figure 6) systems are comprised of rinse water break tank, recovered rinse water tank, and detergent tanks.

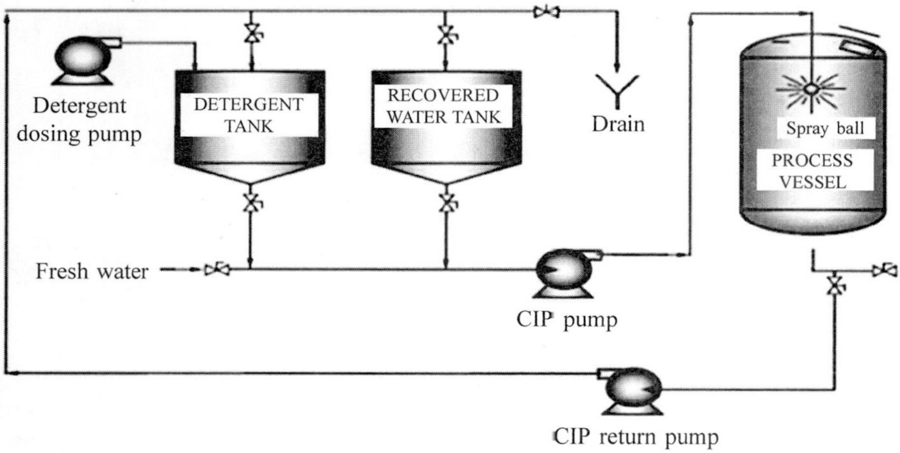

Fig. 5: Re-use System with recovery tank (Source:Jeffery and Sutton, 2013)

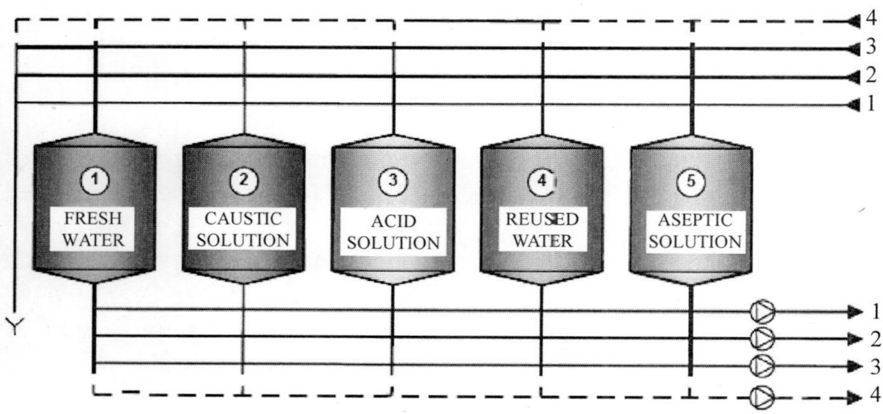

Fig. 6: Multi Channel Re-use System (Source:Jeffery and Sutton, 2013)

6. Assessment of Cleaning Efficiency

The efficiency of the CIP process is accessed by the product contact surfaces do not contaminate the food products and must be free from the residual film or soil which comes in contact. The cleaning can be controlled by validation, verification & monitoring.It may also be assessed by the following parameters:-

i) On both wet and dry surface the contaminant should not be visible in the proper lightening condition.

ii) While touching with clean fingers it should not feel like greasy or smoothness when rubbed with other fingers.

iii) It should be odorless with not any objectionable odor.

iv) No discoloration of the new white facial tissue after wiped over the surface various times.

v) The surface is completely wetted when water is draining from it.

vi) When inspected with a long wave ultraviolet light which do not give the sign of fluorescence.

vii) The product should not be come in re-infection which contacts the surface after sanitization process.

viii) The traces of any residual of CIP chemicals in the final flush water.

The common method applied widely to determine the presence of microbial load after the CIP process is assessment of final flush water. The water used for the CIP process must be potable from any source of water i.e. water supplies, bore well etc. The limits of the CFU (Colony forming units) of the potable water should not be more than 100 cfu/ml & absent of coliform and *E.Coli* in 100 ml. It is therefore very important to check the flush water at various passages throughout the plant for any increase in micro-organism.

7. Advance CIP Method

The current practices of CIP using chemical detergents are needed to be replaced for food safety enhancement. CIP process using chlorine-free detergents used in one of the study clearly indicates reduction in environmental impact factors (Sandberg et al., 2011). Apart from this, application of Electrolyzed Oxidizing (EO) water in a research carried out at the laboratory of the Pennsylvania State University is an effective method in terms of operation as well as environmental impact. Basically, EO water is produced by the electrolysis of 0.1% solution of sodium chloride. Whereas, acidic EO water is generated with chlorine and hydrochloride at the anode, alkaline EO water is generated using sodium chloride at the cathode of the electrified chamber. It has been found that the pH of acidic water is 2.6 with a free chlorine concentration up to 80 ppmand high oxidizing reduction potential (ORP) around 1100 mV whereas, the alkaline EO water has a pH value around 11.0 and an ORP about 2800 mV (Sharma and Demirci, 2003). Since, the function characteristics of alkaline and acidic EO water solutions are similar to alkaline and acid wash chemical used for traditional milking system CIP, alkaline and acidic EO water solutions are found to be recommended in place of chemical detergents in milking system CPI (Wang et al., 2019).

References

Handbook on Cleaning in place: A guide to cleaning technology inthe food processing industry, Tetra Pak, Sweden, 2016, pp. 1-40.

Jeffery, N. and Sutton, E. Design for CIP brochure, Suncombe Ltd., United Kingdom, 2013, pp. 1-46.

Mauck, J.F., Holley, R.A. and Jakubowski, J. Guidelines forcleaning and sanitizing influid milk processing plants. The Dairy Practices Council, United States, 2001, pp. 1-45.

Sandberg, M., Christiansson, A., Lindahl, C., Wahlund, L. andBirgersson, C. Cleaning effectiveness of chlorine-free detergents for use on dairy farms. Journal of DairyResearch, 2011, 78: 105-110.

Wang, X., Demirci, A., Graves, R.E. and Puri, V.M. Conventional and emerging clean-in-place methods for the milking systems. In, Raw Milkedited by Nero, L.A., De Carvalho, A.F., ed., Raw Milk, Academic Press,2019. pp. 91-115.

15

Evaporation

Subir Kumar Chakraborty and Harsh Prakash Sharma

1. Introduction

Foods available in nature and, in some cases, the products made from it thereof contain quite an amount of water. This water not only adds volume to the food products but also provides a sight for attack by the micro organisms. Also, removal of water becomes essential if this water interferes with the end product chracteristics. There are several processes in food industries which are designed to get rid of this excess water, such processes are called concentration processes. The processes which can remove water from food substances are,- evaporation, freeze concentration and membrane based processes. All these process have their limitations in rendering solid content to the final product (Fig. 1). Perhaps, these limitations can be overcome by combinig the processes.

The objectives that are fulfilled by the concentration processes can be enumerated as follows:

- The liquid food can be preconcentrated to desired levels before further processing vide a compulsive costlier process, this would ensure the economics of the process.
- The volume of the food material can be reduced, thus decreasing the overall costs incurred in storage, packaging and transport.
- The water activity (a_w) can be reduced by increasing the concentration of the soluble solids. This would lessen the chances of attack by microbes.
- The heat sensitive organoleptic properties of the food materials can be kept intact as the concentration processes can be selected as per the requirements of the product.

Evaporation is an important unit operation in food industries which is still a very cost effective means of water removal. The mechanical equipments which are used for evaporation are called evaporators. Sometimes the desirable solid content is not achieved in a single evaporator, then an array of evaporators is used, such an arrangement is called multiple effect evaporation. Out of all the concentration processes evaporation can achieve solid content up to 50% solid content. Besides the capital cost involved, the success of an evaporation process is mainly judged by the end product obtained, its solid content and its retention of properties. Depending upon the properties of the evaporating liquid, a processor has the independence to choose from a wide range of evaporators discussed in detail later in this chapter.

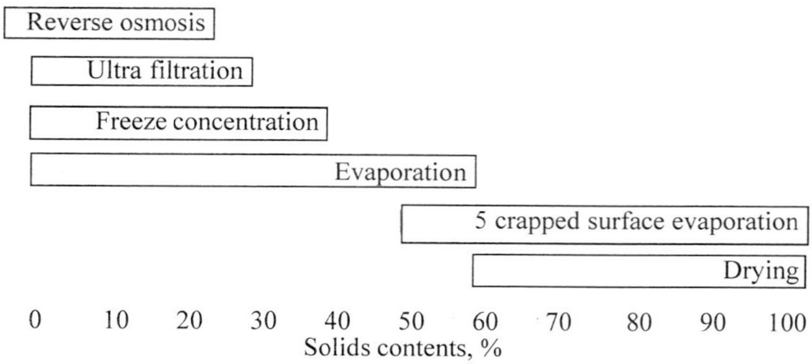

Fig. 1: Range of solid contents attainable in liquid foods by various water removal processes

2. Feed liquid properties

The problems associated with evaporation process can be to a huge extent countered if the properties of the liquid to be evaporated are well understood. These properties form the criteria for selecting and designing the evaporators. Some of these properties are, -

2.1 Concentration

The dilute feed gets concentrated upon entering the evaporator leading to change in properties. Density and viscosity change with the change in solid content of the feed. These changes register substantial variations in the heat transfer characteristics of the liquid. Certain feed soluition might also crytalise out and clog the system. The boiling point of the solution may rise considerably as the solid content increases. This phenomenon is called boiling point elevation and is governed by Duhring's rule. This empirical rule states that the boiling point of a solution is a linear function of the boiling point of the pure solvent at the same pressure. The presure inside the evaporators is less than atmospheric pressure so that boiling can take place at a lower temperature. Incase of increase in solid content leading to increase in boiling point calls for a further reduction in pressure over the boling liquid. Also, the height of the evaporator causes an additional pressure (hydrostatic head) of the liquid column over the level beneth the free surface. This additional pressure also causes an increase in the boiling point of the feed liquid. In other words, the boiling point of the liquid in an evaporator varies with depth. Therefore the heating system should be such that this additional pressure is accounted for, else overheating might result causing heat damage to the product. The boiling point at the mid point of the liquid column is considered for designing the heating system for an evaporator.

2.2 Foaming

Liquid food materials which are high in organic substances, especially proteins, foam during the evaporation process. This foam is carried away by the vapours of the liquid causing high entrainment losses. Using a suitable food grade anti-foaming agent or an efficient entrainment collection system would solve this problem.

2.3 Temperature sensitivity

Certain food substances are highly sensitive to heat. Such substances should not only be evaporated at a lower temperture but also the residence time in the evaporator should also be short. Lowering the pressure over the boiling liquid may to certain extent mitigate this problem.

2.4 Aroma loss

The aroma and flavour components in many liquid foods, fruit juices, etc. are more volatile than water. When such liquids are evaporated these components can be removed with the water vapour causing deterioration in the quality of the resulting concentrate. Volatile components are commonly recovered from the vapour as an essence by fractional distillation. The stripped liquid is then concentrated separately and reblended with the recovered essence.

2.5 Scaling / Fouling

The food substances are biological in nature. They contain a lot of higher molecualr weight organic substances (proteins and carbohydrates), which upon heating deposit scale on the heating surface. This scale not only hiders heat trasfer, but are also sights for possible accidents. Repeated cleaning with liberal use of surfactants is highly desireable to check fouling.

2.6 Construction material

Whenever possible, evaporators are made of cast iron and steel. Many solutions however, attack ferrous metals and / or are contaminated by them. Special materials such as copper, nickel, stainless steel, aluminum, impervious graphite and lead are then used. Since these materials are expensive, high heat transfer rates become especially desirable to minimize the first cost of equipment. Based on the food product to be handled by the evaporator the choice of material of construction is presented in **Table 1**.

Table 1: Construction material of evaporators for different types of food products

Products	Material of construction
Most Dairy And Food Products	304/316 Stainless Steel
Most Fruit Juices	316 Stainless Steel
Sugar Products	Carbon Steel (304/316)
Foods Containing High Salt (Nacl)	Titanium / MonelHigh Alloy Stainless SteelsDuplex Stainless Steels
Caustic Soda Low Concentration (<40%)	Stress Relieved Carbon Steel
Caustic Soda High Concentration (>40%)	Nickle

2.6 Other characteristics

Such as specific heat, heat of vaporization, freezing point, gas liberation on boiling, toxicity, explosion hazards, radioactivity, and necessity for sterile operation must be considered by the designer of an evaporator.

3. Material and energy balance in evaporators

An evaporator's basic components are a heat exchanger (calendria) for heating the liquid feed, and a covered chamber for evaporation to take place (Fig. 2). Steam is fed into the heat exchanger for heating the liquid feed in the evaporation chamber. Steam loses it heat to the liquid feed and condenses, the condensate is removed from the heat exchanger and is used for preheating the liquid feed about to enter the evaporation chamber. Inside the evaporator the liquid feed takes up heat from the condensing steam and water in it gets vapourised. The vapours are removed from the evaporator through an entrainment capturing unit to arrest the accompanying solids and other volatiles. The liquid feed which is now increased in solid content is removed as the concentrated product from the evpaorator.

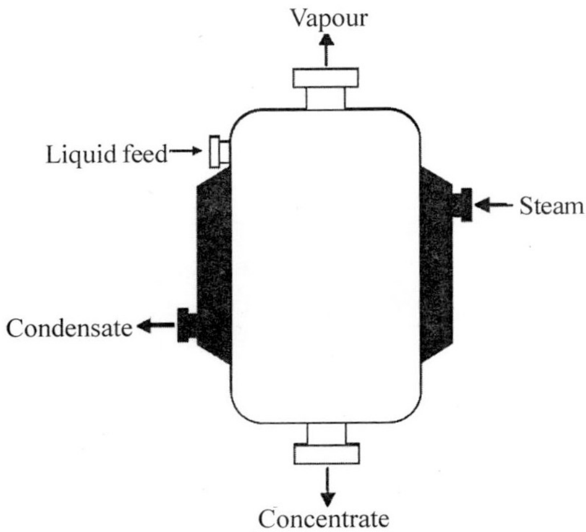

Fig. 2: Basic construction of an evaporator

3.1 Mass balance in evaporator

The mass balance equations can therefore be written as,-

$F = P + V$ (1)

$F\,(x_f) = P\,(x_p)$ (2)

where,

F =Mass flow rate of liquid feed, *kg per h*

P =Mass flow rate of concentrate , *kg per h*

V =Mass flow rate of vapour , *kg per h*

x_f=Mass fraction of solids in liquid feed

x_f=Mass fraction of solids in concentrate

3.2 Energy balance in the evaporator

Heat Entering The Evaporator	=	Heat Leaving The Evaporator	
Heat Content of Liquid Feed + Heat Content of Steam	=	Heat Content of Concentrate + Heat Content of Condensate + Heat Content of Vapour	
$Fh_f + Sh_s$	=	$Ph_p + Sh_c + Vh_v$	(3)
Heat Given Away By Steam	=	Heat Required for Sensible Heating of Liquid Feed to Bring it to Boil	
$S\lambda_s$	=	$UA(T_f - T_v)$	(4)

Where,

S = mass flow rate of steam, kg per h

S, enthalpy of steam entering at some temperature (T_s) is h_s, latent heat of steam at T_s is λ_s, condesate is leaving at temperature, T_s, at this temperature the enthalpy is h_c, overall heat transfer coefficient of the heat exchanger, *U*, area of the heat exchange surface, *A*, temperature at which boling of the liquid feed is taking place, T_v, temperature of the lquid feed, T_f, at this temperature the enthalpy is h_f, vapour of the liquid feed is leaving at the temperature at which boiling is taking place, T_v, and has enthalpy, h_v.

The assumptions that are made before carrying out the energy balance are,-

(i) steam is entering at temperature, T_s and the condensate is also being collected at T_s, which means that steam is only changing phase inside the heat exchanger or that the steam is only giving its latent heat, alone;

(ii) all the heat given out during the phase change is utilized in raising the tempearture (T_f) of the liquid feed to its boiling tempearture (T_v), i.e. the latent heat of steam is utilized for sensible heating (if $T_f < T_v$) and vapourising water from the liquid feed. Evaporation will take place in a flash, called 'Flash evaporation' if $T_f > T_v$;

(iii) the *U* is constant throughout the energy balance. In actuality, with time as the solid content of the feed increases, this would alter its heat transfer characteristics. Moreover, the latent heat of the feed would also change during boiling leading to an elevation in the boling point, in that case the vapour leaving would be superheated.

Evaporator capacity is defined as the amount of water evaporated from the liquid food per unit time. It is expressed as kilogram of water evaporated per hour.

$$E = \frac{V}{S} \tag{5}$$

where,

E= steam economy, unit

S=amount of steam consumed, kg per h

V amount of water evaporated, kg per h

4. Types of Evaporators

4.1 Once - through and circulation type evaporators

Evaporators may be operated either as once - through or as circulation type. In once - through operation the liquid feed passes through heat exchanger tubes only once, release the vapour, and leave the unit as concentrate. The evaporation is accomplished in a single pass. The ratio of concentrate to liquid feed is limited in an once through evaporator. These evaporators are therefore well suited to multieffect evaporation process, where total amount of concentration can be spread over several effects. Agitated film, falling film and rising film evaporators are always operated once through. Once - through evaporators are more suited for heat sensitive materials as high vacuum over the evaporation space allows boiling of the liquid at lower temperature

In circulation evaporators a pool of liquid feed is held within the evaporator. Fresh liquid feed also falls into this pool and the mixture is passed through the tubes. While some part of the concetrated liquid feed leaves the evaporator, rest falls back to the pool only to be resend to undergo the same process. This evaporation process often leads to over heating of the feed, therefore, it is not suited for heat sensitive food products.

Incoming feed mixed with the liquid from the pool, and the mixture passes through the tubes. Unevaporated liquid discharged from the tubes returns to the pool, so that only part of the total evaporation occurs in one pass. All the liquor in the pool must therefore be at the maximum concentration. Since the liquid entering the tubes may contain several parts of concentrate from the last part of feed, its concentration, density, viscosity, and boiling point are nearly at the maximum. Accordingly the heat transfer coefficient tends to be low. Prolonged heating of even a small part of a heat sensitive material like a food can ruin the entire product. Circulating evaporators, however, can operate over a wide range of concentration between feed and thick liquor in a single unit, and are well adapted to single effect evaporation. They may operate either with natural circulation, with the flow through the tubes induced by density differences, or with forced circulation, with flow provided by a pump.

4.2 Batch evaporator

These are the simplest evaporators. They are nothing but a jacketed vessel that is heated with a vapour or liquid heating medium (Fig. 2). The liquid feed is heated inside this jacketed vessel till required amount of concnetration is achieved. Once the desired concentration has been achieved the concentrated liquor is discharged and fresh feed again introduced. The technology is quite crude, it does not suit to the heat sensitive products. The heat transfer coefficients are very low as there is no scope of agitation or motion of the liquid feed. There is always a static head of liquid over the boiling liquid, increasing the boling point of the liquid which is at the bottom of the evaporation vessel.

4.3 Horizontal short tube evaporator

This is the oldest form of evaporator and still finds widespread commercial use. There are a bundle of tube horizontally placed inside the evaporation chamber (Fig. 3). The overall heat transfer coefficent is low owing to the less circulation space for the liquid feed. Genearly used for low viscosity liquids that do not scale upon heating.

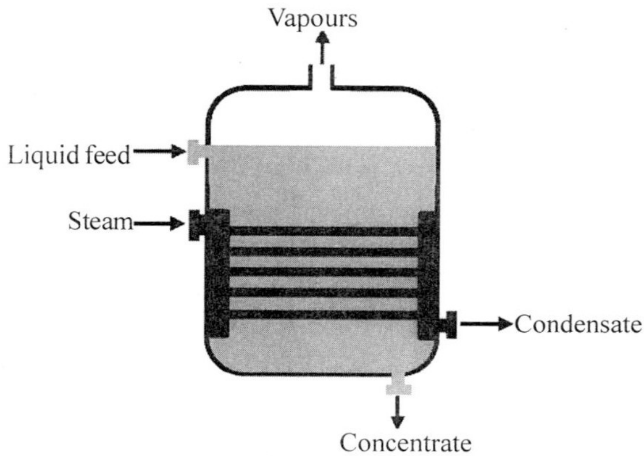

Fig. 3: Horizontal short tube evaporator

4.4 Vertical short tube evaporator

The heating tubes are arranged vertically (Fig. 4). The length of the tubes vary between 0.5 to 2 m and the diameter ranges between 2.5 and 7.5 cm. A channel is left open at the central part (25 to 40 per cent) of the arrangement of tubes for allowing a natural circualtion on the basis of specific gravity between the hot and the cold liquid feed. These evaporators are used for concentrating sugar cane juice, beet root juice and malt extracts.

The main advantages are:

- Short head space is required
- Withstand liquid feed that scales moderately
- The heat transfer coefficients are quite high for thin liquid (5 – 10 centipoise)
- Relatively inexpensive and maintenace cost is also very less

Limitations

- Temperature sensitive materials and crystaline products cannot be used.

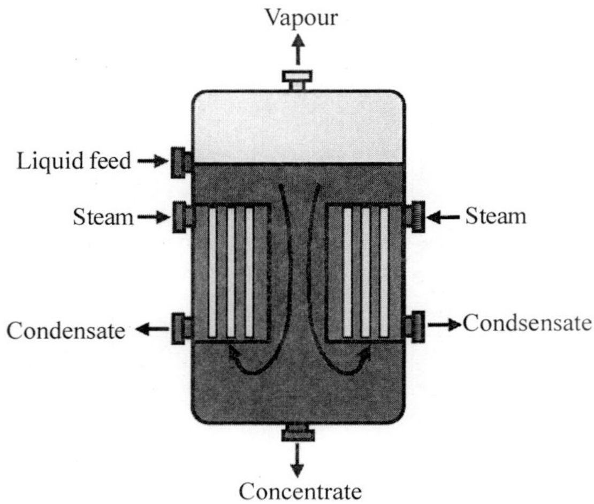

Fig. 2.4: Vertical short tube evaporator.

4.5 Rising film evaporator

The rising film evaporator (Fig. 5) was the first tc be introduced in the long tube evaporators' category. Length of the evaporato: varies from 3 to 12 m while diameter of tubes from 25 to 50 mm. The main objective was to reduce the residence time in comparison to the short tube evaporators. Liquid is preheated to near boiling before it is introduced at the bottom of the tube assembly. Boiling commences a short distance up the tubes. The vertical tubes are heated from the outside by steam. Expansion on vaporization causes high velocity vapour bubbles and slugs of liquid to rise rapidly up the tubes. The residence time is short and the heat transfer coefficient is good. A major difference from falling film evaporator with repsect to product characteristics is that this evaporator an handle highly viscous liquids, like, - honey, tomato paste and other saturated solutions.

The boiling in the tubes is complex process. As vapourization progresses, convective heat transfer takes place with vapour bubbles dispersed in the liquid, as the liquid column further rises, these bubbles coalesce. The bubbles form slugs which later collapse to give an annular flow regime in which there is vapour only in the central core and the liquid climbs as a thin film carried up along the wall. All told, condition of annular flow exists in the tube. As the liquid climbs this thin film gets thinner, eventually the film collapse leading to dry wall conditions. There are nucleate boiling with forced convection boiling conditon arising in the tube due to high velocity liquid – vapour mixture.

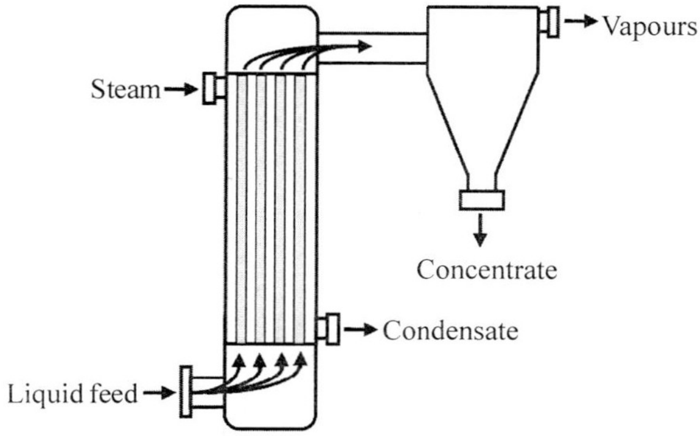

Fig. 5: Rising film evaporator

4.6 Falling film evaporator

It is a very common evaporator of food industries. Liquid feed enters the top of the evaporator at about boiling temperature and is incident on the distributors (Fig. 6). The distributors cause the liquid to be evenly distributed into the tubes. The liquid falls down through the tubes as thin film due to gravity and the residence time is very low. The tubes are 5 to 15 meters long. These tubes are heated from outside by steam and evaporation takes place under vacuum. The distribution is a very critical aspect of falling film evaporator. The heat transfer coefficient of this evaporator is very high, it can operate with liquid upto viscosity of about 200 m Pa.s. A key operating limitation is that the tubes inner surface should be kept wet once the process is on, else the feed entering may get burnt.

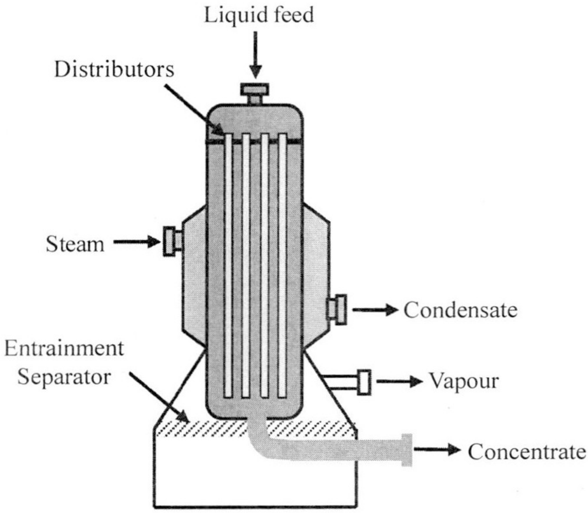

Fig. 6: Falling film evaporator

4.7 Rising / Falling film evaporator

This evaporator is deviced to combine the features of rising film evaporator and falling film evaporator, possessing the advantages of both (Fig. 7). The tubes in the bundle are divided into two sections; the first, functioning as the rising film and the second, functioning as the falling film. Liquid feed enters the first section from the bottom, boiling proceeds as the liquid rises through the tube. The mixture of vapours and concetrated liquid is made to fall through the pipes of the falling film evaporator. The heat transfer coefficient here is higher than normal falling film evaporator because the vapours aid in distribution and the fall of the liquor. This type of evaporators can handle large volumes of clear fluids and foamy liquids.

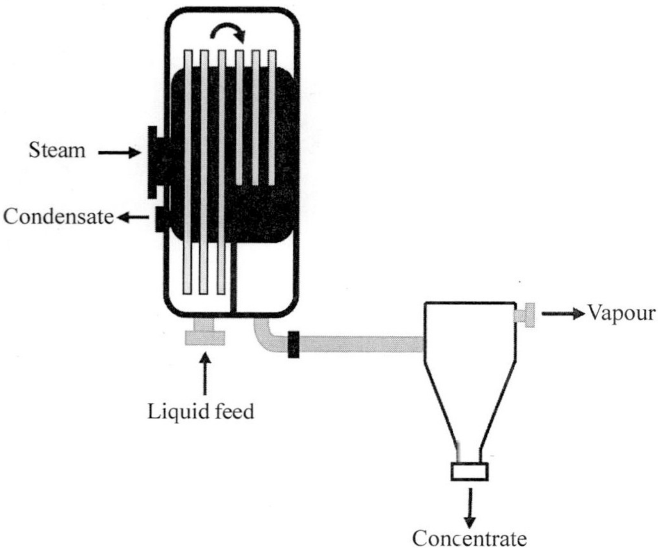

Fig. 7: Rising / Falling film evaporator

4.8 Forced circulation evaporator

In this evaporator the liquid feed is forced through the heat exchanger by a pump (Fig. 8). For low viscosity liquid feeds centrifugal pumps are used, while for higher viscosities positive displacement pumps are used. The hot liquid from the heat exchanger is released inside a flash chamber, where vapour is separated with the help of an impact plate. High heat transfer coefficients are achieved in this heat exchanger because of the velocity (2 – 6 m/s) given to the liquid feed by the pump. Viscous and particulates can be handled by this type of evaporator. Tomato pastes as well as crystalization of sugar can be carried out in this type of evaporators.

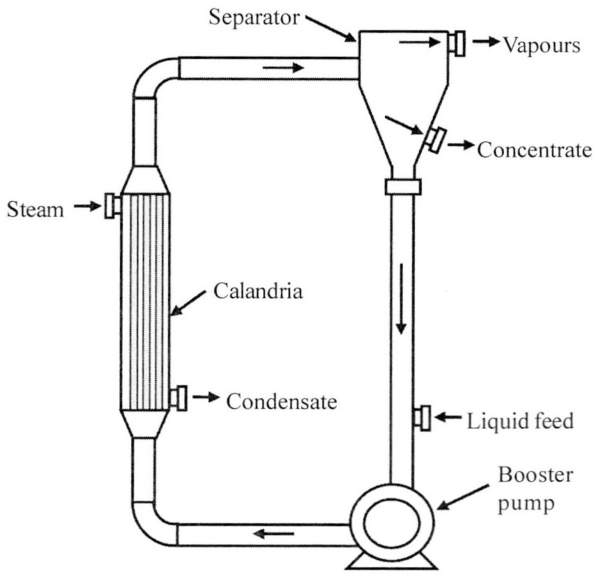

Fig. 8: Forced circulation evaporator

4.9 Gasketted plate evaporator

The operating characteristics are similar to plate heat exchangers. The plates are spaced 3 to 6 mm apart, nitride or butyl rubber gaskets are used to separate the plates. The plates being in contact with each other at some points so that pressure (working pressure can be 15 kg/cm^2 gauge) loading might not cause distortions. Turbulence is introduced because of the corrugations on the plate surface resulting in a high heat transfer rate at low Reynold's number. Liquid feed is directed through the adjacent layers between the plates. The first passage of the liquid feed is a rising film type and the second is a falling film type. The steam plates are arranged alternately between each product passage (Fig. 9). Depending on the gasket, the flow can be parallel or counter current. Normally the liquid feed is introduced at temperature slightly higher than the evaporating temperature within the plate annuli so that evaporation taked place in a flash. Ensuing flash distributes the feed uniformly all over the plate. The gaskets confine the fluids and prevent the vapours from escaping to the atmosphere. The mixture of vapour and partially concentrated liquid feed now falls by gravity to the falling film annuli. At the bottom of the falling film annulus a rectangular duct connects all the plate units and transfer the concentrated liquid feed into a separating device. The velocity of liquid feed falling down is high, hence the heat transfer coefficents are high and the residence times are short. Ergo, they are suited for concentrating heat sensitive fluids. Dismantling and assembling this evaporators is easy and it requires less space. Plate evaporators are used to concentrate coffee, soups, marmalades and citric juices.

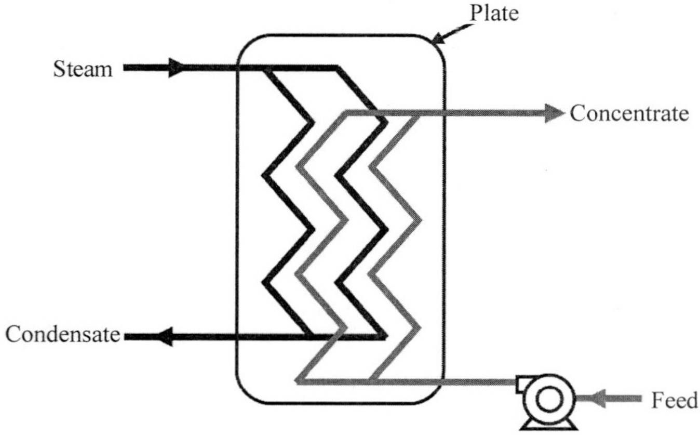

Fig. 9: Flow pattern on either side of a gasketted plate evaporator

4.10 Agitated thin film evaporator

This evaporator successfully counters the problems associated with most of the food materials. The volatiles are not lost, the product temperature is also favourable as separation is carried out under vacuum. The agitated thin film or wiped film evaporator comprises,- a cylindrical wall with hot surface and a rotor housed inside. The cylinder wall is kept warm by a jacket. The cylinder – rotor assembly can be horizontal or the more common vertical. Liquid feed enters from the top, above the heating zone; the rotor distributes the feed over the inner circumference of the cylindrical body. The liquid feed spirals down the inner wall while the rotor generates turbulence, increasing the heat transfer. The vapour and volatiles of the feed are trapped from the top of the evaporator, the non volatile components are collected from the bottom. Also, the rotor checks the fouling as it continuously washes the wall by the bow waves generated by it. The overall residence time of the liquid feed in the evaporator is very less. The evaporator provides short residence time, high turbulence and surface cleansing as it key features, thus, this type of evaporator is able to handle heat sensitive, viscous and fouling causing substances. The viscosity range that this evaporator can handle ranges from 1 – 50000cP. In food industries agitated thin film evaporators are used for concentrating lecithins, enzymes, fruits and vegetables purees, fermentation broths and plant extracts.

4.11 Scrapped surface evaporator

As the name suggests, there is a scrapper inside the evaporation vessel. This scrapper rotates as the evpaoration is carried out.The axis of rotation of the blades is same as that of the jacketted cylinder (Fig. 2.10). The scrapper is so designed that its' blades do not allow deposition of evaporating material to the surface of the evaporator, while is felicitaties the movement of less concentrated substances towards the surface. These evaporators are used for highly viscous (upto 20 Pa.s), heat sensitive and fouling substances. They are used for tomato pastes, gelatine crystalization of whey.

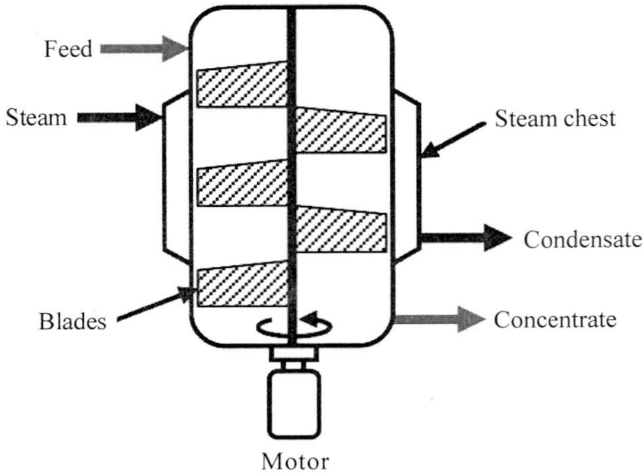

Fig. 10: Scrapped surface evaporator

4.12 Vapour recompression evaporator

During the evaporation process in the evaporators, a lot of vapour is generated from the liquid feed. In vapour recompression evaporators, the vapour is compressed in an compressor and is sent back to the calandria to be reused as a heating medium. The use of vapour recompression for improving the economy of an evaporation process has been practiced for many years. These systems can result in considerable savings in overall energy requirements when compared with multiple-effect evaporation.

There are two types of vapour recompression evaporators:

(i) Thermal vapour recompression (TVR), in which a steam jet compressor is used to compress the vapour. When high pressure (> 100 psi) steam is available, a portion of the water vapour from the liquid feed is recompressed by steam jet venturi and returned to the steam chest of the evaporator for heating purposes. TVR is an inexpensive way of improving the economy of the evaporation. This method can also be used in conjunction with multiple effect evaporators. TVR is not suited for evaporators handling liquid feed that foul a lot.

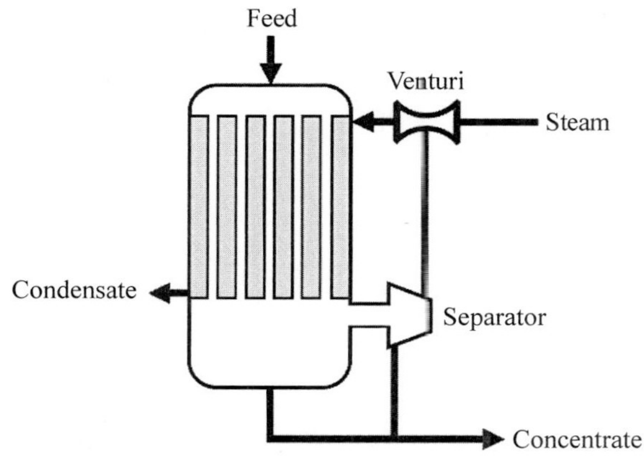

Fig. 11: Thermal Vapour Recompression Evaporator

(ii) Mechanical vapour recompression (MVR), where a mechanical pump is employed to compress the vapour generating from the evaporator, this high pressure vapour is then send to the steam chest for using its heating value (Fig. 12). MVR requires energy to compress the vapour only, latent heat is there with the vapour, an MVR evaporator is theoretically as efficient as 100 effects. Inefficiencies in the compression process decreases the efficiency of the MVR to 30 to 55 effects. Energy supplied to the compressor can be derived from an electric motor, steam turbine, gas turbine or an internal combustion engine. The cost of the compressor increases the overall cost of these evaporators, but in the long run it proves to be economical.

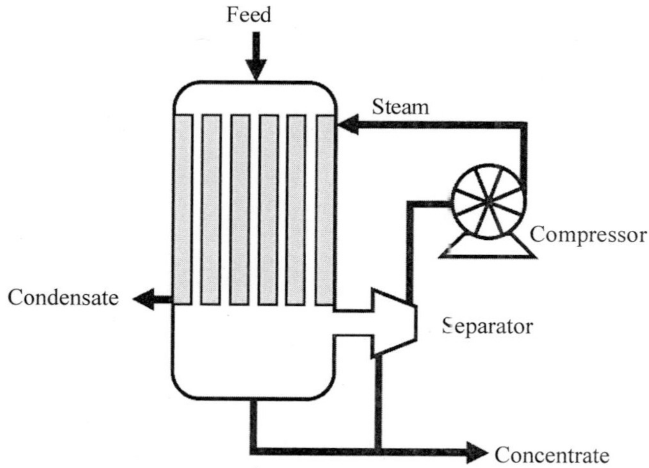

Fig. 12: Mechanical Vapour Recompression Evaporator

The TVR and MVR is not suitable for liquids that foul a lot and for liquid there is an appreciable boling point rise.

5. Multiple Effect Evaporators

Single effect evaporators are those evaporators where the amount of steam required is just equal to the amount of water to be evaporated. The vapour generated during the evaporation is sent to the condenser. However, it is always not possible to carry out the removal of desired amount water from the feed in a single effect. In multiple effect evaporators there are a number of effects. Removal of water from the feed takes place in all the effects till the required concentration level is achieved, the vapours produced from first effect is used to provide heat to the next effect. As water is removed, two things are increased,- solid content and boiling point of the concentrating liquid. To make the concentrating liquid boil in the subsequent effects, the pressure over the free surface is decreased, i.e. the vacuum is increased. The decreased pressure creates a suction effect and acts as a driving force for the concentrating liquid to move from one effect to the another.

In a two effect evaporator, for every one kilogram of steam supplied two kilogram of vapour can be generated. As the number of effects increase, the steam economy increases. Again, increasing the effects can harm the economics of the process, so one has to be able to weigh the options and decide accordingly. Thumb rule is that if water has to be evaporated at a rate higher than 1350 kg/h then multiple effect evaporators must be used. Since the product characteristics vary in terms of heat sensitivity, viscosity, fouling, crystalization, etc., so in multiple effect evaporators different feeding arrangements are adopted to counter the detrimental effects, if any, that are initiated during the evaporation process.The multiple effect evaporators can be classsified on the basis of the feeding arrangement into,- forward feed, backward feed, mixed feed and parallel feed. All these arrangements have their own sets of advantages and limitations (Table 2).

5.1 Forward feed

Fresh liquid feed enters the first effect at about boiling point for the pressure conditions in the evaporator (Fig. 13). In this feeding arrangement the best steam is utilized by the fresh liquid feed, thus the chances of heat damage to the product is very less. The vapour generated is utilized as the heating source for the subsequent evaporators.

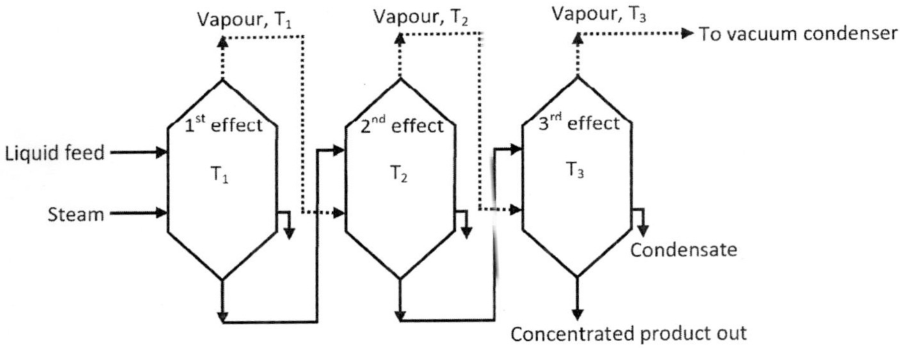

Fig. 13: Forward feed multiple effect evaporator

5.2 Backward feed

It is also called reverse feeding arrangement. The best steam gets to handle the most difficult to evaporate part of the concentrating liquid (Fig. 14). Since the concentrating liquid is moving against the pressure gradient, so mechnical pumps are used to carry the liquid towards the last effect.

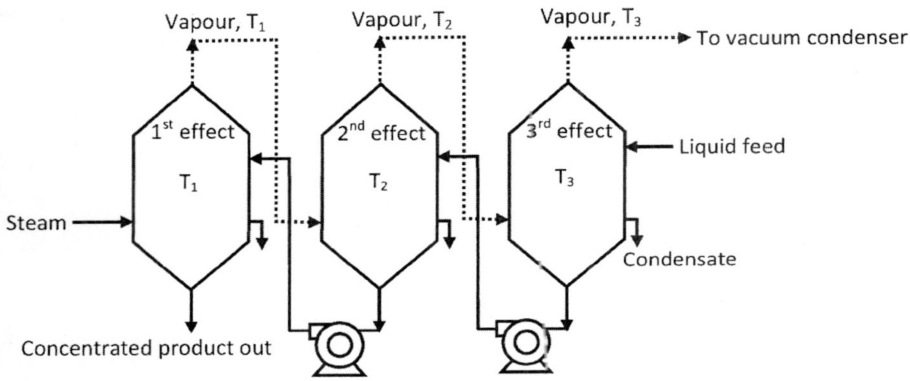

Fig. 14: Backward feed multiple effect evaporator

5.3 Mixed feed

In mixed feed arrangement, forward feed and backward feed arrangements are combined i.e. product is initially fed in second effect, from there it goes to third effect. After third effect it comes to first effect. However, heating medium flows in forward direction only (Fig. 15).

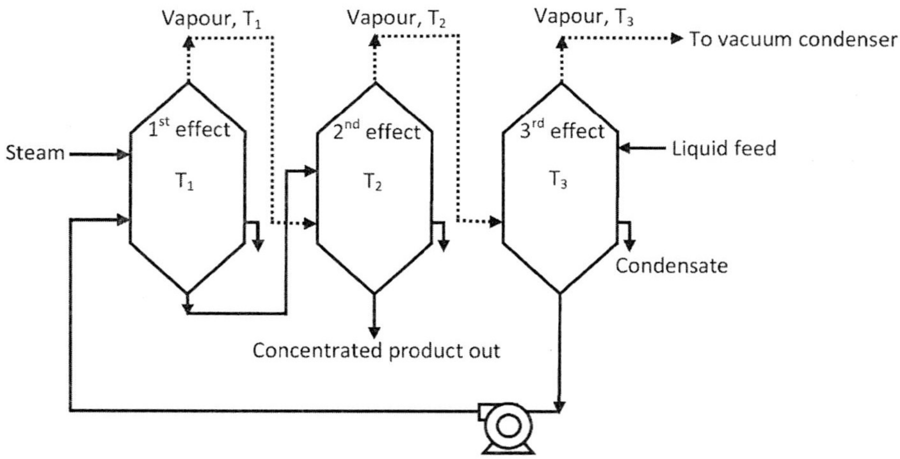

Fig. 15: Mixed feed multiple effect evaporator

5.4 Parallel feed

It involves introduction of fresh feed and collection of concentrated product from each effect (Fig. 16). There is no transfer of liquid feed from one effect to another. This feeding arrangement of multi effect evaporator is used for liquid feeds that crystalize when concentrated.

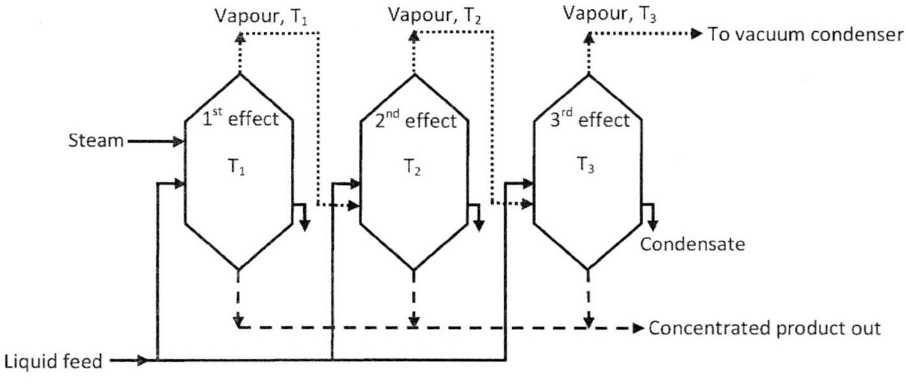

Fig. 16: Parallel feed multiple effect evaporator

6. SELECTION OF EVAPORATORS

Evaporators come with different features. The choice of an evaporator depends upon the liquid feed that has to be handled and product specifications of the concentrated product. The choice of evaporator is guided by the throughput, viscosity that can be handled, solid content, fouling and foaming tendencies of the product. Residence time of the food in evaporators is a key parameter of the evaporation process. Foods have a lot of heat sensitive attributes, nutritional and organoleptic; the residence time should be such that these properties of food might not get withered during evaporation. Also, evaporation leads to an increase in viscosity (as the solid content rises), severely affecting the flow characteristics. So, options have to be weighed with due consideration to liquid feed characteristcs while choosing an evaporator. A comparitive of advantages and limitations possessed by the different evaporators has been reported in Table 3 and in Table 4 the criteria for selection of different evaporators on the basis of liquid feed characteristics has been summarised.

Table 2: Comparitive advantages and limitations of various feeding arrangements in multiple effect evaporator.

Feeding Arrangement	Advantages	Limitations
Forward Feed	• Simple to Operate. • No Pumps are Required. • Chances of Heat Damage to the Product is Not There.	• Rate of Evaporation Falls With Each Effect. • Best Quality Steam Used on Initial Feed which is Easiest to Evaporate. • To Gather Only Latent Heat from Steam the Liquid Feed Has to be Introduced at Boiling Point.
Backward Feed	• Best Quality Steam used on the Most Difficult Material to Concentrate. • Better Economy and Heat Transfer Rate as Effects Are Not Subject to Variation in Feed Temperature • Feed Meets Hotter Surfaces as it Becomes more Concentrated Thus Partially Offsetting Increased Viscosity Rease in Viscosity	• Inter Stage Pumps Necessary. • Most Viscous Product is Subjected to the Highest Temperature, Hence, Higher Risk of Heat Damage. Changes of Fouling is More.
Mixed Feed	• This System Has the Simplicity of Forward Feed and the Economy of Backward Feed. • It Especially Suits Very Viscous Foods.	• The Arrangement is Complex and is Expensive to Maintain.
Parallel Feed	• Is Exclusive to Crystal Production Upon Evaporation. • The Crystallization Process Can Also be Controlled. • Prevents the Need to Pump Crystal Slurries.	• Extraction Pumps Required for Each Effect. • Most Complex and Expensive of all the Arrangements.

Table 3: Advantages and limitations of various types of evaporators

Type of Evaporator	Advantages	Limitations
Batch	• Oldest form of Evaporators. • Very Simple In Construction. • Present Day use is Limited to Products that are not Heat Sensitive. • Heat Transfer Coefficients Can be Improved by Introducing Agitation.	• Obsolete In All Respects.
Horizontal Short Tube	• Head Room Required is Less • Well Suited For Processes where the Final Product is Liquid.	• Not Suitable of Liquids that Scales
Vertical Short Tube	• Head Room Required Is Less • High Heat Trasfer Coefficients With thin Liquids (Upto 5 – 10 Cp) • Inexpensive to Manufacture.	• Not Suitable for Heat Sensitive Materials. • If there is no Agitation then it Cannot be used for Crystaline Products.
Rising Film	• Short Residence Times • Good Heat Transfer • Can Handle High Viscosity Liquids and Products with Suspended Solids	• Cannot Operate with Low Temperature Differences. A Significant Part of the Temperature Difference is Required to Provide the Pressure Difference to Carry the Liquid Upwards.
Falling Film	• Large Heat Transfer Area. • Can Handle Heat Sensitive Products. • They Can Operate with Temperature Differences as Low As 2°c. • High Energy Efficiency.	• Cannot Handle High Viscosity Liquids and Products with Suspended Solids. • They Can Operate with Liquids Up to A Viscosity of About 0.2 Pa.S.
Rising / Falling Film	• Low Residence Time. • Low Cost. • Less Floor Space Requirements. • High Heat Transfer Rates.	• High Head Room is Required. • Required Concentration is Not Obtained in one Go, Recirculation is Almost Mandatory. • Not Suited for Materials that Scale as it wil Choke the Narrow Tubes of the Evaporator.
Forced Circulation	• High Recirculation Rates Through the Heat Exchanger. • High Heat Transfer Coefficients. • Can Handle Viscous Liquids and Particulates. • Very Less Fouling, as Evaporation	• High Residence Time. • Complexities Involved Because of More Moving Parts.

(Contd.)

Type of Evaporator	Advantages	Limitations
	Takes Place in the Flash Chamber.	
Gasketed Plate	• Adapted To Handle Heat Sensitive, Viscous And Foamy Materials. • It Is Very Compact, Requires Less Floor Space and Head Room. • Easy To Clean as all the Parts Are Accessible. • Capacity Can be Varied Easily Altering the Number of Plates.	• Any Leakage in the Gasket Can Cause A Huge Trouble. • Evaporation Capacity is Less, Not More than 40000 Litres Per Hour.
Agitated Thin Film	• Short Residence Time In Heating Zone. • High Heat Transfe Coefficient Because of the the Turbulence Created by the Rotor. • Less Back Mixing. • Ability to Handle Thick Substances (High in Solid Content / Viscous) • Higher Recovery of the Product as the Rotor Cleans the Heating Walls.	• High Costs Are Involved.
Scrapped Surface	• Residence Time is Short. • Very High Viscosities (Upto 20 Pa.S) Can Be Handles.	• Heat Transfer Area is Less. • Power Requirements Are High. • Costly Process.

Table :

Type of Evaporator	Liquid Feed Characteristics						
	Clean	High Capacity	Solids / Crystals	Scaling	Foamy Substances	Thermal Sensitivity	Viscous
Batch	√						√
Horizontal Tube	√						
Vertical Short Tube	√						
Rising Film	√	√					
Falling Film	√	√	√			√	
Rising / Falling Film	√	√			√		
Forced Circulation	√	√	√		√		
Gasketed Plate	√	√	√	√	√	√	√
Agitated Thin Film	√	√	√	√	√	√	√
Scrapped Surface	√	√	√	√	√		√

16

Drying

Rachna Sehrawat, Somya Tewari, Prabhat K. Nema and Khursheed A. Khan

1 Introduction

Drying is a thermo physical and physicochemical process, which involves removal of excess water from a solid, semi-solid or liquid food material by evaporating it into vapour phase on applying heat, to reduce growth of micro-organism or to completely eliminate their growth. Hence, reducing the water activity increases shelf life of food products. Drying medium for carrying evaporated moisture can be air or steam. Benefits of drying include increase in shelf life, easy to transport, handle food material and reduce packaging cost as it reduces bulk weight. Dried food material can be utilized during off season and in the areas where it cannot be grown. Transformation of particular product likes grapes to raisin or we can say creating a new food, chewiness and crispiness are some of the desirable feature of dried products. Early harvest of horticultural produce is also possible so that losses due to shattering could be minimized using drying. But, there are certain limitations in conventional driers (solar, tray, drum and spray driers) like structure damage of food products, loss of nutritional quality (vitamins and antioxidants) and difficulty to regain the original shape after rehydration. However with advancement in expansion of knowledge of drier, loses in nutritional quality, changes in structure are minimized with better rehydration property (freeze drier, Vacuum drier superheated steam drier) (Kaur et al., 2014)

2 Principle of drying

Drying is a complex unit operation governed by heat and mass transfer principles along with physico-chemical transformations. To increase product temperature and to vaporise the liquid present, latent heat is to be supplied. Heat may be supplied to food material continuously or intermittently and can be transferred by conduction (direct or indirect driers), convection (direct driers), by putting in microwave or electromagnetic field or some drying medium like steam, air and radiation can be used. Reducing moisture content from food material involves:

- Vaporisation of water present over the surface of food material
- Migration of water from inside the food material over its surface

Migration or transport of water inside the food material may occur by following mechanism (Jangam and Mujumdar 2010):

- Liquid diffusion: It generally occur if temperature of wet material is below or less than boiling point of water
- Surface diffusion
- Vapour diffusion: It occurs if water vaporises inside the food material
- Hydrostatic pressure difference: It occur if rate of vaporisation inside the material exceed
- Knudsen diffusion: When drying is carried out very low pressure and temperature Knudsen diffusion occur (freeze drying)
- Above mechanisms can occur in combination also.

Depending upon mode of heat transfer drying process can be classified as:

- Air drying: Air is used as heating medium and is in direct contact with food materialand evaporated moisture becomes part of surrounding air. More than 85% of dryers in industries are based on convective heating.
- Vacuum drying: It is carried out at under vacuum and low temperature mode of heat transfer may be radiation or conduction
- Freeze drying: Principle behind freeze drying is sublimation (transition from solid to gas phase directly without passing through liquid phase)
- Steam drying: In place of air steam is used as drying medium in indirect contact with food medium, for direct contact it should be superheated (adding heat to the saturated steam to make it superheated).

Factors responsible for controlling the moisture removal by air from a food depends upon:

- Amount of water vapour present in the air
- Temperature of air
- Amount of air passing over the food commodity.

Apart from air temperature and humidity, air velocity also plays key role in controlling drying rate.

3 Basic terms encountered in drying

3.1 Moisture content

It represents the amount of water available in the moist food products. It can be expressed on wet basis and dry basis also.

Moisture content on wet basis (MC_{wb}): It is expressed as amount of water present per unit mass of moist food material

$$MC_{wb}(\%) = \frac{Weight\ of\ water}{Weight\ of\ moist\ product} \times 100$$

$$MC_{wb}(\%) = \frac{Weight\ of\ water}{Weight\ of\ water + Weight\ of\ dry\ matter} \times 100$$

Moisture content on dry basis (MC_{db}): It is expressed as amount of water present per unit mass of dry food material

$$MC_{db}(\%) = \frac{Weight\ of\ water}{Weight\ of\ dry\ matter} \times 100$$

Moisture content on dry basis may be more than 100%as amount of water present in food material may be more than the dry matter where Moisture content on wet basis lies between 0 to 100%.

Relationship between MC_{wb} and MC_{db}

Dividing numerator and denominator of equation 2 with weight of dry matter

$$MC_{wb}(\%) = \frac{Weight\ of\ water(ww)/\ weight\ of\ dry\ matter}{Weight\ of\ water(ww)/\ weight\ of\ dry\ matter + 1} \times 100$$

$$MC_{wb}(\%) = \frac{MC_{db}(\%)}{MC_{db}(\%) + 1}$$

Similarly moisture content on dry basis can be calculated as

$$MC_{db}(\%) = \frac{MC_{wb}(\%)}{1 - MC_{wb}(\%)}$$

Water present in food material can be in following distinct form as shown in Fig. 1:

Unbound water: Moisture in food material is hold loosely by solid matrix and vapour pressure exerted is equal to saturated vapour pressure of pure water, when temperature is same.

Bound water: Moisture in food material is hold tightly (physically or chemically) by solid matrix (cell wall and other cell constituent like cellulose, starches, proteins). Vapour pressure exerted is lower than vapour pressure of pure water, when temperature is same. The bound water found in biological materials and the water in colloidal system may exist as occuluded water, capillary water, osmotic water, colloidal water bound by physical forces and chemically bound water

Free water: It includes both unbound as well as bound water present in large capillaries. It acts as dispersing medium or solvent for colloidal particles. Water activity of free water is unity.

3.2 Psychrometry

Mostly air is used as heating medium as well carry away evaporated moisture with exception in freeze drying, Vacuum drying (used for heat sensitive material) and superheated steam drying. Knowledge of moist air, which is mixture of dry air-water vapour, is vital for designing of driers. Subject which involves study of thermodynamic properties of moist air is known as psychrometry.

Moist air: Moist air is a mixture of water vapour and dry air.

Saturated air: Maximum amount of water vapour that air can hold at a given temperature. However when saturated air is cooled some of water vapour will condense.

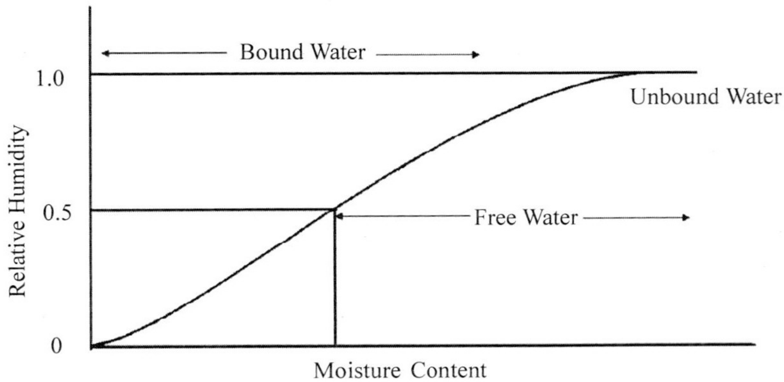

Fig. 1: Different moisture content with relative humidity at constant temperature

Dry bulb temperature (T_{db}): It is indicated by dry thermometer, when immersed in vapour gas mixture. It represents actual temperature of moist air.

Wet bulb temperature (T_{wb}): It is indicated when bulb of thermometer is wrapped in a wet cloth or wick and give indirect measurement of dryness in the air. Difference between wet bulb and dry bulb temperature gives wet bulb depression. But, depression at 100% is zero as

Dew point temperature (T_{dp}): It is the point at which condensation occur. Cooling of moist air at constant humidity and pressure, then at a particular temperature it will become saturated and water vapour will condense. This temperature will be the dew point temperature.

Absolute humidity: Mass of water vapour present in unit mass of dry air is known as absolute humidity or humidity ratio or specific humidity.

$$Absolute\ humidity = \frac{Weight\ of\ water\ vapour}{Weight\ of\ dry\ air} \times 100$$

Relative humidity (RH): It represents amount of water vapour in the air to the maximum amount it can hold. It is having 6kg of water vapour and it can hold 30 kg of water vapour so relative humidity is 20 % (RH = 6kg × 100/ 30kg). It is also defined as ratio of partial pressure of water vapour in air to the partial pressure of water vapour in saturated air at same temperature.

$$RH = \frac{Weight\ of\ water\ vapour\ in\ air}{Weight\ of\ water\ vapour\ required\ to\ saturate\ air} = \frac{p_w}{p_{ws}}$$

Humid heat (C_p): It indicates heat capacity of moist air. It can be defined as quantity of heat required to raise the temperature of unit mass of moist air (dry air and water vapour) by unit degree rise in temperature at constant pressure. It can be expressed as:

C_p= Specific heat of dry air + Specific heat of water × SH

C_p= 1.005 kJ/kg °C + 1.888 kJ/kg °C × SH

Specific Humidity (SH): It is the amout of water vapour unit wights up dry air can hold. It is expressed in kg/kg of dry air.

Humid volume (V_H): Humid volume or specific volume of air is defined as volume occupied by unit mass of moist air. It has inverse relationship with density.

$$V_H = (0.082T + 22.4)\left(\frac{1}{29} + \frac{H}{18}\right)$$

T (°K) denotes temperature of air

Psychrometric chart: A psychrometric chart depicts thermal and physical properties of air as shown in Fig. 2. It's a plot between dry bulb temperature on abscissa and humidity ratio on ordinate axis. Properties of air (dry air and water vapour mix) are very useful in designing the drier for particular food commodity and other processes. Calculations of these properties using equation are very tedious and time consuming, but with the help of psychrometric chart they can be easily found. If we know any two properties of air other can be easily found from psychrometric chart.

How to find different properties of air on psychrometric chart (Fig 2a):

- RH lines are hyperbolic lines which are drawn asymptotically corresponding to saturation curve.
- From the abscissa T_{db} can be determined.
- From the ordinate axis absolute humidity can be determined.
- From a steady point follow the line inclined to the horizontally and intersects 100% RH i.e saturation temperature boundary.
- To determine T_{dp} move horizontally to left along with constant humidity ratio until reaches 100% RH.
- Specific enthalpy lines are represented by slope from upper left towards the lower left side.
- Specific volume is represented by parallel family of lines which are equally spaced.

Using the psychometric chart find the following

- To find absolute humidity of air when RH and a dry-bulb temperature: First find intersection of RH (A) and T_{db}(B) in Fig.2.b i.e point C then follow the chart horizontally right to read off the absolute humidity i.e point D.
- To find the T_{wb} from RH and dry-bulb temperature: From the intersection of the T_{db} and RH lines i.e point C in Fig.2.c extrapolate towards left inclined parallel lines to the saturation line to read off the wet-bulb temperature at point E and F will give the T_{wb}.

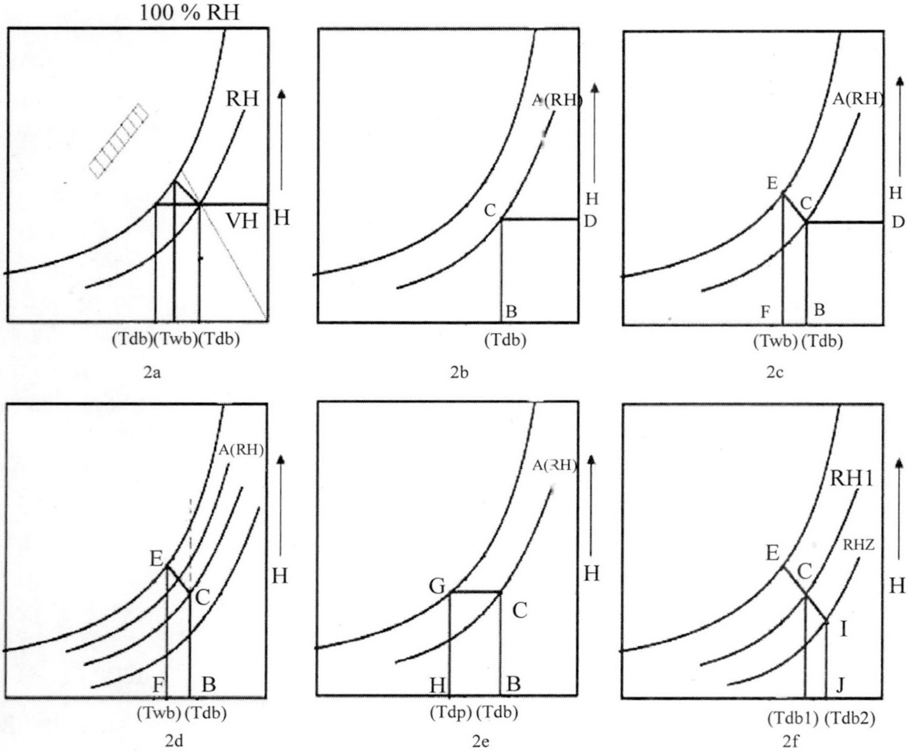

Fig. 2: Psychrometric chart for air-water system

- To find the RH of air using T_{wb} and a T_{db}: Intersection of T_{wb} lines and T_{db} lines will give point C than hyperbolic line i.e A will give the RH.
- To find the T_{dp} of air cooled adiabatically from a T_{db} and RH: From point of intersection of T_{db} and RH i.e A in fig 2.e follow the absolute humidity lines towards left until reaches RH 100% curve i.e point G and H will give the T_{dp}.
- To find the change in RH of air havingT_{wb}, heated from a dry bulb temperature of 30°C (T_{db1}) to a dry-bulb temperature 70°C (T_{db2}): Find the intersection of the T_{wb} lines and the T_{db1} lines i.e point C as shown in Fig 2.f, and follow the T_{wb} line to the intersection with T_{db2} lines i.e point I ; read the sloping RH line at each intersection. Difference between two relative humidity will give the change in RH at same $T_{wb.}$

3.3 Water activity (a_w)

Relationship between water content of food and shelf life has been recognized long back, although it is imperfect. It has been observed that different food products with same water content have different shelf life. So shelf life depends not only on water content in food but also on intensity of its association with other constituents. Weaker the association of water with other constituent more it will be available and will support the growth of microbial population, hydrolytic chemical reactions and enzymatic activities hence lower will be the shelf life. Thus availability of water present in food commodities indicates water activity. This is reason that drying preserves food as it reduces water activity of product below which microbial growth can be prevented and minimize other chemical reaction rate. Water activity (a_w) can be defines as ratio of partial water vapour pressure (p) exerted in headspace of food to the vapour pressure of pure water (p^o) is known as water activity at same temperature. It is measured as the equilibrium relative humidity (ERH), the per cent relative humidity (RH) of an atmosphere in contact with a product at the equilibrium water content.

a_w = ERH/100= p/p^o

So water activity can be used in food product design and food safety

Food product design: It will help the food designer in formulating shelf stable food products as they will keep the a_w below which undesired chemical and microbial growth will be prevented. Simultaneously it will be helpful in limiting migration of moisture within food commodities made with multiple ingredients. If rice flakes with low water activity are packed with high water activity of raisins then with time, migration of water from raisin to bran flakes occur which will result in soggy flakes and hard resins.

Food safety: In many cases a_w is considered as critical control point for Hazard Analysis and Critical Control Points (HACCP) programs.

Different a_w is required by different microbes for their growth into food products below which they could not grow. If a_w is below 0.90 bacteria cannot survive, yeast below 0.80 are incapable to grow and activity stops below 0.60 for molds. Apart from microbial growth other activities like deteriorative and enzymatic reaction ceases below 0.3 a_w (Fig. 3)

Based on colligative properties of solutions water activity in food can be estimated by measuring:

- Vapour pressure
- Freezing point depression of a liquid
- Boiling point elevation
- Osmotic pressure
- Dew point and wet bulb depression
- Bithermal equilibrium
- Suction potential, or by using the isopiestic method
- Electric hygrometers
- Hair hygrometers
- Equilibrium relative humidity of a liquid or solid

Product with $a_w > 0.85$ are categorised into high moisture food like fruits, meat, milk etc as shown in Table 1. They require refrigeration condition throughout their storage. Food like cakes, flour, jams are intermediate moisture foods and a_w lie between 0.60-0.85. These do not require refrigeration storage but chances of spoilage by yeast and moulds are there so have shorter shelf life than low moisture food. Food products having $a_w < 0.60$ are categorised as low moisture foods (cereals and milk powder). These have longer shelf life even without refrigeration.

Table 1: Water activity of different food commodities

Water activity	Product
High moisture foods	
Bacon	0.968
Meat	0.97
Eggs	0.97
Evaporated Milk	0.983
Apples	0.988
Bananas	0.979
Tomatoes	0.998
Peas	0.99-0.98
Vegetables	0.97
Milk	0.995-0.983
Intermediate moisture food	
Cake	0.75-0.72
Jams	0.80-942

Jelly (grape)	0.80
Low moisture foods	
Malted milk powder	0.19
Chocolate powder	0.258-0.251
Rice	0.591-0.531
Walnut	0.58
Tea instant	0.130-0.26

3.4 Equilibrium Moisture Content (EMC) and Sorption Isotherms

Equilibrium Moisture Content

EMC is the moisture content at which vapor pressure of the water present in food is equal to the vapor pressure of moisture content present in the surroundings.

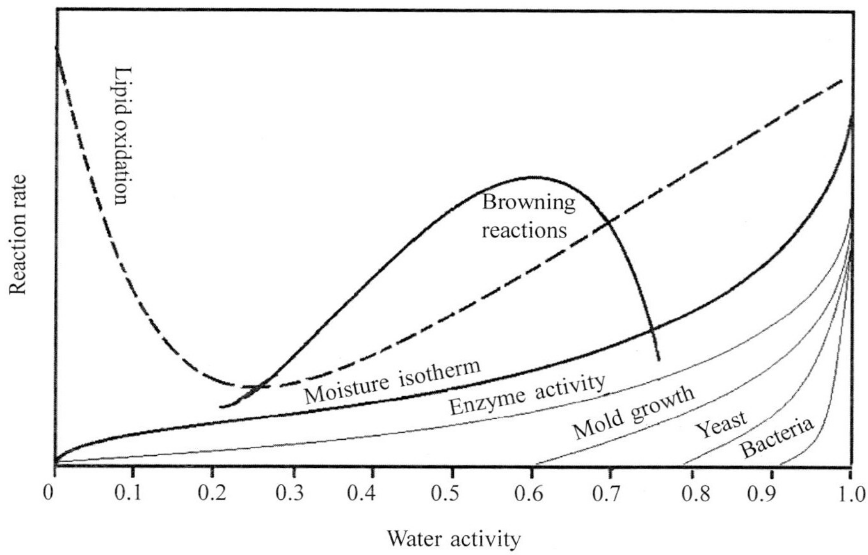

Fig. 3: Relationship between water activity and other chemical or biochemical reaction rate

Sorption Isotherms

Sorption Isotherm is the graph obtained after plotting EMC versus relative humidity at a given temperature or in other words EMC as a function of water activity. When the isotherm is obtained for the food product that attains EMC by losing moisture to the surroundings, it is called desorption isotherm, and when it is obtained for the food product that attains EMC by gaining moisture content from the surroundings, it is called adsorption isotherm. With drying, the former is more important as the product loses moisture content. The extent of sorption and desorption is dependent on the vapor pressure of the surroundings and the product itself. The difference between desorption and adsorption is known as hysteresis.

The shape of isotherm may vary from product to product depending on the differences in capillary effect, colligative effects of dissolved components such as salt and sugars and surface interactions. The shape of moisture sorption isotherm (MSI) is usually is sigmoidal, showing a J-type shape for foods containing large amounts of sugar and small soluble molecules, not rich in polymeric materials.

Fig 4 shows the general shape of MSI. Typically it can be divided into three distinct zones A, B and C, indicating different water binding mechanisms at individual sites on the solid matrix. Region A corresponds to the site where water is tightly bound and is unavailable for reaction, depicting monolayer adsorption of water vapor with no distinction between adsorption and desorption isotherms. In Region B, the water molecules are loosely bound to the sites, confined to smaller capillaries as compared to region A. In Region C water is freely available for reactions, as it is loosely held in voids, large capillaries and crevices.

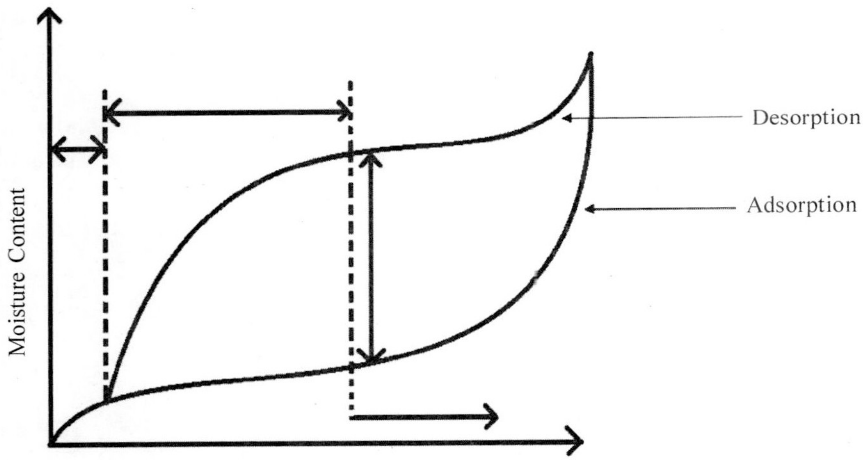

Fig. 4: Moisture Sorption Isotherm

Measuring the Isotherms

The isotherms can be measured using various mathematical models derived theoretically or by modification of existing models. A few practical techniques used to measure sorption isotherms are classified as gravimetric, manometricand hygrometric. While Gravimetric method measures mass change either continuously or periodically, manometric employs manometers to measure vapour pressure of water in equilibrium with food products, and hygrometric method measures relative humidity surrounding the food at particular moisture content.

4. Drying Kinetics

Drying kinetics mainly comprises of calculating the rate of drying which is a function of moisture content of the product. For a wet solid under fixed drying conditions, drying process can be roughly divided into three phases i.e. first falling rate period, second falling rate period and constant falling rate period. Also, the Drying Rate (R) for it is given by

$$R = -\frac{m_s dX_f}{dt}$$

where

$X_f = (X-X^*)$

X = moisture content at dry basis

X^*= equilibrium moisture content

t= time

Fig 5 demonstrates a drying curve (obtained under constant drying conditions), which is a plot of R versus X. Drying curves are difficult to obtain in actual drying processes as the conditions are not constant in actuality but can be obtained by extrapolating or interpolating limited data over a wider range of operating conditions.

Depending on the rate of evaporation the drying curve is divided into three periods (Fig 5):

- The constant rate period
- The first falling rate period
- The second falling rate period

In the constant rate period AB drying rate is fully governed by the rate of external heat and mass transfer and the rate at which water evaporates from the surface of food is equal to the moisture migrating from the interior of the food to its surface, thus maintaining free water availability on the surface. The drying period remains independent of the product being dried.

The temperature of food product is usually at the wet bulb temperature of drying air while the drying rate depends on:

- Temperature difference of air and wetted surface at constant air velocity and relative humidity
- Humidity difference between air stream and wet surface at constant air velocity and temperature and,
- Air velocity at constant temperature and relative humidity

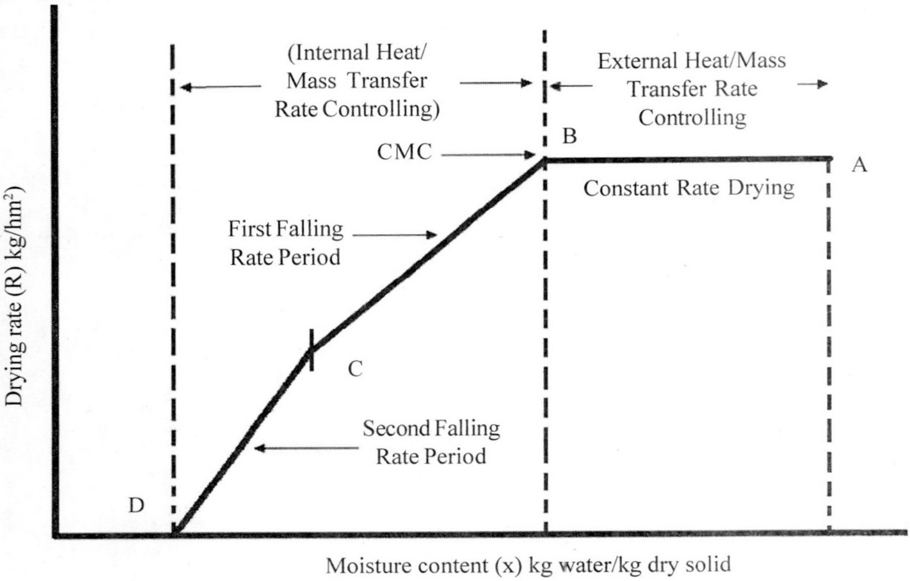

Fig. 5: A typical drying curve

The constant drying rate period continues till Critical Moisture Content (CMC) is reached. CMC depends on the type of product and its thickness.

In the first falling rate period (BC) the rate of drying depends on removal of moisture from the external surface after the moisture diffusion from the internal surface to the external free surface. It depicts a decrease in the drying rate as the wet surface area decreases since the rate of moisture evaporation from the external surface is more than the moisture migration from inside to outside due to internal transport and drying conditions. The first falling rate period ends at the reduction of wet surface to zero. Mostly agricultural products are dried in falling rate drying period.

The second falling rate period is mostly sub surface evaporation and continues till the equilibrium moisture content is reached, during which the drying surface first attains partial unsaturation and then complete unsaturation.

Sometimes materials may display more than one CMC thus influencing the shape of graph obtained, which occurs due to structural and chemical changes in the materials influencing their drying mechanisms.

Mechanisms for capillary diffusion in porous products

Drying time calculation

The time required to reduce the moisture content of food from initial to final is called drying time, and can be calculated by using initial moisture content and critical moisture content values.

The drying rate (R) (water per kg dry solid per unit time) is given as

$$R = -m_s \frac{dX}{dt}$$

Where dX stands for change of moisture content on dry basis, and negative sign stands for decrease in moisture content.

Now for constant rate period R is constant, if we integrate within limits of initial (X_i) and critical (X_c) moisture content then the time for constant rate drying period (t_c) is given as

$$\int_0^{t_c} dt = -m_s \int_{Xi}^{Xc} dx / \mathrm{Rc}$$

Hence, $t_c = \dfrac{m_s (X_i - X_c)}{R_c}$

However, for falling rate period drying rate can be given by

$R_f = aX+b$

because diffusion and drying rate govern the drying process.

$R_{f1} = aX_1+b$ and $R_{f2} = aX_2+b$

where a is the slope of line b its intercept X_1 initial moisture content X_2 final moisture content for first falling rate period

For BC in the drying curve drying rate period can be given by

$$\int_0^{t_f} dt = -m_s \int_{X1}^{X2} \frac{dX}{aX+b}$$

$$t_f = m_s \ln\left[\frac{aX_1+b}{aX_2+b}\right]$$

also the slope a will be

$$a = \frac{R_{f1} - R_{f_2}}{X_1 - X_2}$$

Substituting the value of a in $t_f = m_s \ln\left[\frac{aX_1+b}{aX_2+b}\right]$

we get

$$t_f = m_s \frac{X_1 - X_2}{Rf_1 - R_{f2}} In\left[\frac{R_{f1}}{R_{f2}}\right]$$

As the falling rate period can be represented by a straight line from Xc to Xe,(equilibrium moisture content) and where $X1 < Xc$ and $X2 < Xc$, then the drying rate and gradient at any moisture content X are given as

$$R_f = a(X - X_e)$$

$$a = \frac{R_f}{Xc - Xe}$$

$$R_f = \frac{Rc(X - Xe)}{Xc - Xe}$$

$$t_f = m_s \frac{(Xc - Xe)}{(R1 - R2)} In\frac{R_{f1}}{R_{f2}}$$

$$t_f = \frac{ms}{Rc}(X_c - X_e) \ln\frac{(X_1 - X_e)}{(X_2 - X_e)}$$

Now total drying time

$t = t_c + t_f$

$$t = [(X_1 - X_c) + (X_c - X_e) \ln \frac{(X1 - Xe)}{(X2 - Xe)}]$$

4. Classification of driers

Drying has been always fascinating and innovative area for continuing research due to variation in commodity to be dried, modes of heat transfer and medium of heating air (Table 3).

Some of distinctive features of drying are given below

- The commodity to be dried can be very viscous or solid, liquid and is dried as either a sheet or as a bed of particles (Fig. 6).
- Mode of heat transfer can be conduction, convection and radiation. Mostly dryers use convective type of drier but radiation can also be utilized as additional source of heat (Fig. 7).
- The medium of heating can be air or superheated stream.

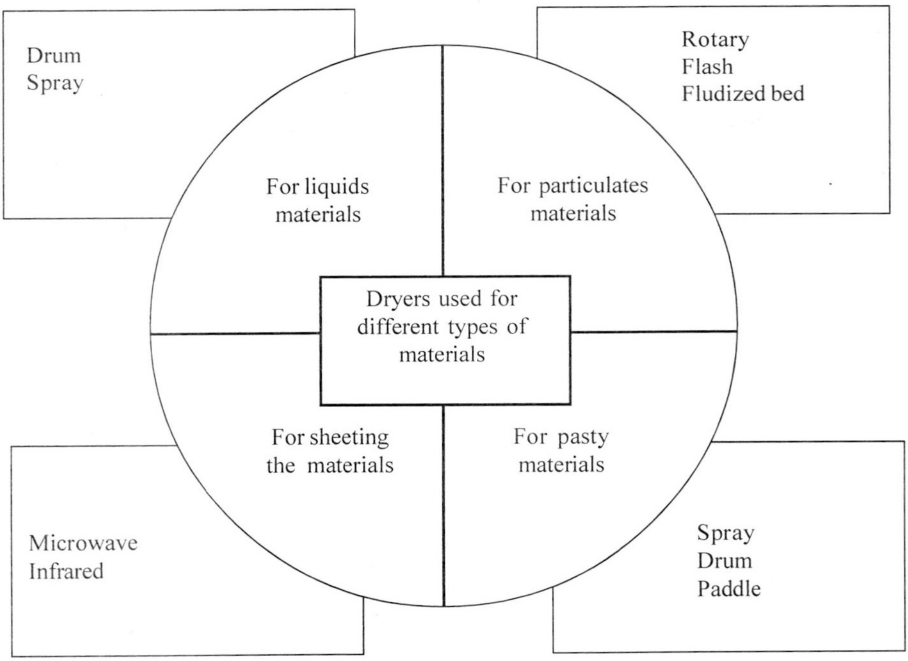

Fig. 6: Dryers used for different types of material

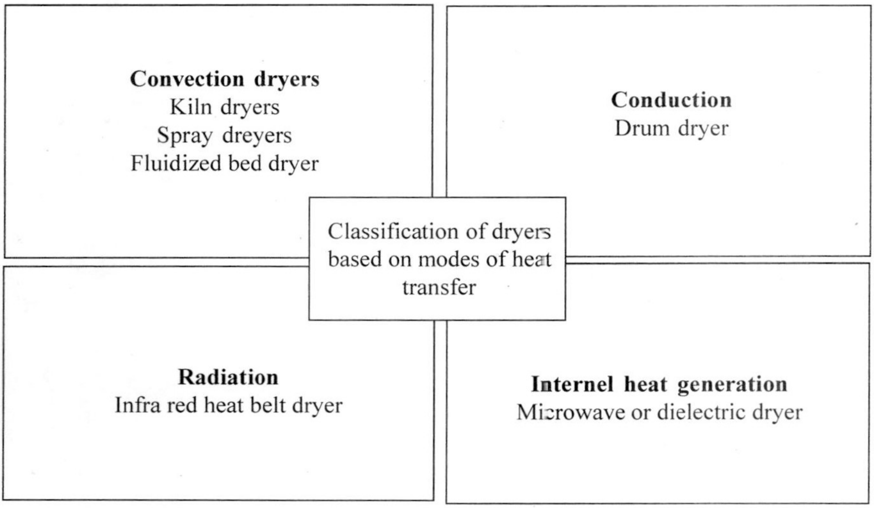

Fig. 7: Classification of dryers based on modes of heat transfer

- The commodity may be carried on conveyors, belts, trays or pneumatically by the air.
- They may be operated in batch or continuous process
- They can be operated under atmosphere or under vacuum.

Drying being energy intensive process needs improvement to make it energy efficient. Around 20-25% of energy consumed in Germany and Denmark, 10-15% in Canada, USA, UK and France on industrial drying (Mujumdar and Devahastin, 2000). The major cost is attributed by operating cost rather than the capital cost of drier. Conventional dryers gave non uniform product may be due to over or under drying of food material, harder texture accompanied by loss of several important nutrients. So drying system needs continuous research and developments to reduce the operating cost as well minimize the nutritional losses and provide good quality of dried food products. Various key areas for improvements in conventional dryers have been reported by Mujumdar A. S. and Huang (2007), Jangam et al (2011). Also recent advances in drying like superheated steam drying, heat pump-assisted drying and hybrid drying techniques are also discussed by Jangam et al (2011).

The kinetics of drying, product quality and energy consumption are important issues in the drying of food materials

Table 3: Variation while drying using different driers

Characteristics	Variation
Product size	Microns to tens of centimetre
Product porosity	Zero to 99.9%
Time	Fraction of seconds to 5 months (wood)
Production capacity	0.10kg/h to 100t/h
Product speed	Zero (stationary) to 2000m/s (tissue paper)
Drying temperature	Below triple point to above critical point of liquid
Operating pressure	Millibar to 25 atmosphere (low pressure, atmospheric and high pressure)
Mode of heat	Continuous or intermittent
Source of heat	Convection, conduction, radiation, microwave or radio frequency electromagnetic fields
Medium of applying heat	Superheated steam, hot air, sun, supercritical fluid, heat pump, inert medium
Phase of material	Solid, semi-solid, liquid

References

Jangam SV and Mujumdar AS (2010). Basic concepts and definition. In: SV Jangam, CL Law, AS Mujumdar (Eds.), Drying of Foods, Vegetables and Fruits, vol.1, ISBN - 978-981-08-6759, Published in Singapore, pp. 1-30.

Jangam SV (2011). An overview of recent developments and some r&d challenges related to drying of foods. Drying Technology: An International Journal. 29(12): 1343-1357.

Kaur BP, Sharanagat VS and Nema PK (2014). Fundamental principles and basics of drying. In: PK Nema, BP kaur and AS Mujumdar. Drying Technologies for Foods: Fundamentals and Applications, ISBN 9789383305841, 9383305843, New India Publishing Agency, New Delhi, India, pp 1-22

Mujumdar A S and Devahastin (2000). Fundamental principles of drying. In Mujumdar's practical guide to industrial drying / edited by Sakamon Devahastin Montreal: Exergex Corporation, c2000pp 1-22.

Mujumdar AS and Huang LX (2007). Global R&D Needs in Drying. Drying Technology: An International Journal 25:4, 647-658.

Singh V, Hedayetullah M, Zaman P and Meher, J (2014). Postharvest technology of fruits and vegetables: an overview. Journal of Postharvest Technology 2: 124-135.

17

Freeze Drying

R.C. Verma and G.P. Sharma

Important Terms

Collapse: A phenomenon causing collapse of the structural integrity of a freeze dried product due to too high a temperature at the drying front.

Collapse Temperature: The temperature above which collapse occurs.

Collector: A cold trap designed to condense the water vapour flowing from a product undergoing freeze drying.

Eutectics: Areas of solute concentration that freeze at a lower temperature than the surrounding water. Eutectics can occur at several different temperatures depending on the complexity of the product.

Eutectic Temperature: The temperature at which all areas of concentrated solute are frozen.

Freeze Drying: The process of drying a frozen product by creating conditions for sublimation of ice directly to water vapour.

Isothermal Desorption: The process of desorbing water from a freeze dried product by applying heat under vacuum.

Lyophilization: The freeze drying process.

Prefreezing: The process of cooling a product to below its eutectic temperature prior to freeze drying.

Primary Drying: The process of removing all unbound water that has formed ice crystals in a product undergoing freeze drying.

Secondary Drying: The process of reducing the amount of bound water in a freeze dried product after primary drying is complete. During secondary drying, heat is applied to the product under very low pressures.

Sublimation: The conversion of water from the solid state (ice) directly to the gaseous state (water vapour) without going through the liquid state.

Vapour Pressure: The pressure of the vapour in equilibrium with the sample.

Triple point: At a particular temperature and pressure a point (Fig. 1, point B) where three phases viz. solid, liquid and vapour coexists.

1. Introduction

Most plant materials are composed of very high percent of water (as high as 95%). Because of presence of high amount of water they are susceptible to microbial activity which spoils the products leaving it to useless for human consumption. It not only changes its taste and colour but may develop toxins into it making it unsafe. There are various methods for preserving them but to make them shelf stable, removal of water from foods keep them from spoilage for a longer period. Drying is the age old practice widely used to preserve various foods.

In conventional drying methods such as mechanical tray drying or solar drying or any other traditional drying method heat energy is used to remove water in order to reduce water activity. Longer exposure of foods to high drying air temperatures not only reduces their sensory and nutritional quality but also the food structure collapses leading to case hardening which prevents the food material to regain its shape to original.

In freeze drying similar preservation effect is achieved by reducing water activity without heating the product. The food quality can be retained to its original at the most. As it is clear that this drying takes place at low temperature after freezing, prevents the nutrition losses from the products.

2. Principle of freeze drying

In freeze drying the solid (frozen moisture present in the form of ice) is directly converted to gaseous form (water vapour) without transforming into liquid phase. The process is mainly followed by three steps, firstly, temperature of the food material is lowered converting it to frozen state, secondly evacuating the space surrounding the frozen food material or lowering the air pressure in the chamber and lastly by supplying heat to the frozen food removes the moisture (ice) from the frozen material directly in the form of water vapour due to low pressure. The process of freeze drying will be discussed in detail in the following sections.

3. Phase diagram of water

The phase diagram represents the physical state of matter at different pressures and temperatures (Fig. 1). Here the figure depicts the ICE/WATER/WATER VAPOUR system. The curve A-B is termed as sublimation curve, at which the solid and vapour are in equilibrium. Similarly curve B-C termed as the melting curve and curve B-D termed as the evaporation curve where solid and liquid and the liquid and vapour respectively are in equilibrium. The curve B to D, terminates at point Cr, which is called the critical point. At critical point, there is no clear distinction of the phases and the vapour cannot be condensed by increasing the pressure, rather highly compressed vapour will result. If the atmosphere pressure is higher than 4.58 mmHg, water shows all three states (solid, liquid, gas) by changing the temperature. In general water exists in liquid state and if the temperature is lowered below 0°C (Freezing arrow), it turns to solid form (ice) and if temperature is raised beyond 100°C, it converts to gaseous phase called water vapour (evaporation arrow). A little change in temperature or pressure may cause the phase to change abruptly from one state to other. At 4.58 mmHg the melting pressure curve (B to C), vapour pressure curve (B to D) and sublimation pressure curve (A to B) meet at one point B. The point (B) where all the three phases coexists (a point at which all the three curves meet) is called triple point. The triple point has a unique value for pressure and temperature, and is a reference point, for which the Celsius temperature scale is defined.

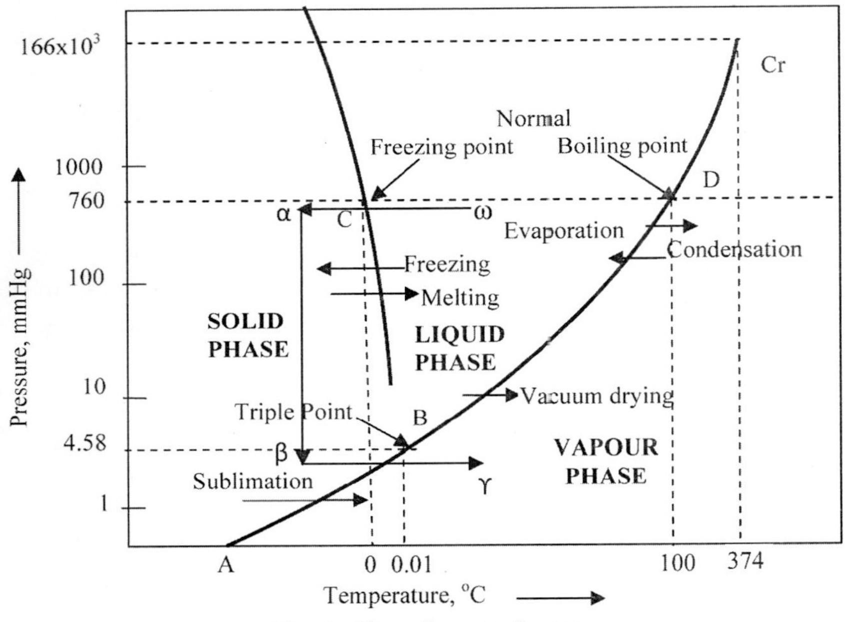

Fig. 1: Phase diagram of water

4. Sublimation

The phase diagram of water (Fig. 1) indicates that at pressures more than 4.58 mmHg (0.006 atm), the ice will change to water while crossing the B-C curve and on further heating it may cross the curve B-D and the water will be converted to water vapour. In conventional drying the B-D curve is crossed for evaporation of moisture. Ice can also be directly changed to vapour phase if the vapour pressure and temperature are below those of the triple point of water, i.e. below 4.58 mm Hg (611.73 Pa) and 0.01°C respectively. It should now be apparent that for any pressure in a system below the triple point pressure (4.8 mmHg), we can change water from the solid to vapour phase crossing curve A-B without passing through the liquid phase is called sublimation. Hence sublimation is the direct transition from solid state to gaseous state without melting. Sublimation occurs at a definite range of temperatures and pressures, depending on the substance in question.

5. Fundamental steps of freeze drying

The freeze drying method (process) involves the following three stages: (a) the pre-freezing stage, (b) the primary drying stage, and (c) the secondary drying stage.

5.1 Pre-freezing

Since freezing drying is a change of state from solids phase to gaseous (vapour) phase, the food material to be dried is to be frozen adequately in the beginning (process ω-α, Fig. 1) before the drying starts.

i) *Freezing rates* : Freezing rates during pre-freezing stage affects the quality of the final product. Freezing rates can be defined as rapid freezing and slow freezing. Rapid freezing/cooling results in large number of smaller size ice crystals while slow freezing/cooling results in large size ice crystals less in numbers. Though small size crystals help in preserving the product structure microscopically which is more difficult to dry while large size crystals in slow cooling rates results less restrictive channels in the food during the drying process.

The freezing rate depends on type and condition of the product to be freeze dried and the type of freezer used, can be classified as-

- slow pre-freezing (0.2 cm/h)
- quick pre-freezing (0.5 – 3 cm/h)
- rapid pre-freezing (5 – 10 cm/h)
- ultra rapid pre-freezing (10 – 100 cm/h)

(ii) *Freezing point depression:*Most of the food products to be freeze dried consists of primarily water as solvent and the material suspended or dissolved in it are solutes. Presence of these solutes depresses the freezing point of the suspension then the freezing point of pure water. When this aqueous suspension is cooled, water separates as crystallization of ice proceeds creating more concentration of solute in the remaining solution. This increase in concentration of the solutes in remaining solution further lowers the freezing point. Although the product may appear to be frozen because of ice present, it has unfrozen suspension yet to be frozen. The mixture of various concentrations of solutes within the solvent constitutes the eutectic of the suspensions. Only when all of the eutectic mixture is frozen, the product is called properly frozen. The temperature at which eutectic solution freezes is called *eutectic temperature.*

It is thus very important to freeze the product to be freeze dried below the eutectic temperature (usually to values ranging from "40 to "50°C) before the commencement of freeze drying process to start. Some of the suspensions become more viscous while lowering the temperature instead of forming eutectics. The food suspension thus undergoes glass formation during the freezing process. The product freezes at *glass transition point* forming vitreous solid. Such products are extremely difficult to freeze dry.

5.2 Primary Drying

After the product is frozen, it is placed in the drying cavity where the ice can be removed via sublimation (process β - γ, Fig. 1). The drying cavity is connected to the condenser where solvent vapour is condensed on the cold surface. The desired level of vacuum is maintained by connecting it with a vacuum pump. Sublimation requires that the pressure of the system (freeze dryer) at which the product is dried must be less than or near to the equilibrium vapor pressure of the frozen solvent. If, for instance, frozen pure water (ice) is processed, then sublimation of pure water at or near 0°C and at an absolute pressure of 4.58 mmHg could occur.

The freeze drying (sublimation) rate depends on the vapour pressure difference between the ice front in the frozen product and the ice at the condenser. Since vapour pressure is related to the temperature, it is desired that the condenser temperature should be less than the product temperature preferably a minimum of 10 degree difference is better. Too high product temperature may alter the structure of the product.

In order to sublimate ice from the frozen product, vapour pressure difference is to be maintained. The chamber pressure and temperature are carefully controlled so the entire process of drying takes place by sublimation. The pressure in the chamber is maintained with the help of a vacuum pump (Fig. 2). The vacuum speeds up the sublimation.

In practice, however, freeze drying is carried out at very low total pressure (typically 10 to 50 Pa). At such low pressure, water vapour has a very large specific volume. For example

- 1.0 g of ice at 1.0 mbar transforms to a volume of 1 m^3 vapour
- 0.1 mbar transforms to a volume of 10 m^3 vapour
- 0.01 mbar transforms to a volume of 100 m^3 vapour

In order to remove such large volumes of vapour in gaseous state, the vacuum pump must have unrealistically large displacement capacity. To overcome this problem, the vapours are condensed as ice crystals on the surface of condensers kept at extremely low temperature (typically – 40°C or less).

The condenser provides the cold surface to the water vapour to adhere and freeze to ice. The condenser also helps the vacuum pump to handle less vapour which is high energy consuming operation. The condenser is used to collect the condensable gases from the product by condensing over it and the vacuum pump removes the non-condensable gases from the system.

During sublimation the crystalline water is removed from the frozen product. About 95% of the product moisture may be reduced in this phase. Since vapour pressure is related to the temperature, the product temperature is carefully raised and maintained higher than the cold trap (ice condenser) temperature. The product is heated by radiation heating or conduction through shelf heating (process β - γ, Fig. 1).Almost ten times as much energy is required to sublime a gram of water from ice to vapour as is required to freeze it. Heating of the frozen product to be followed carefully as it can affect the structure of the frozen product at the drying front. The product should not be raised above the eutectic temperature in case of crystallized product and collapse temperature of glass transition temperature in case of amorphous products. Molecules of vapour migrate due to difference in pressure from frozen product to the condenser (cold trap). The process continues for many hours (even days) while the material gradually dries out.

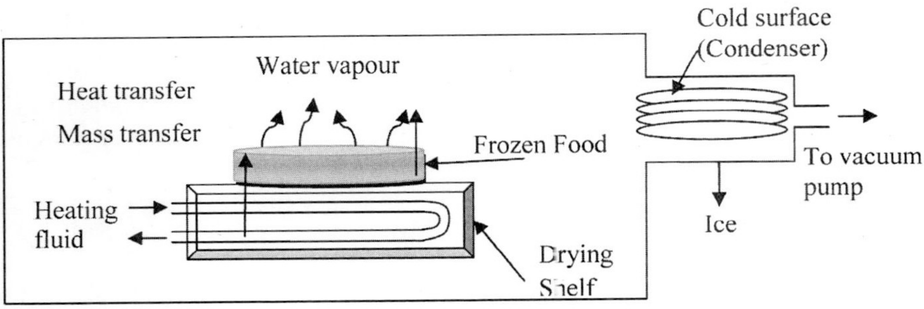

Fig. 2: Freeze drying System

5.3 Secondary Drying

As the drying of frozen material occurs in two stages, in the first stage, i.e. primary drying, sublimation of the frozen water (ice crystals) occurs which normally removes most of the water from the food. Though at the end of primary drying, the product appears to be dried some residual moisture that did not freeze still left in the product in the form of ionically-bound water molecules. This is termed as sorbed or bound water in the product which may be as high as 7-8%. It is necessary to remove this water, hence the continued drying is necessary at slightly higher temperature keeping the other conditions same (vacuum and the condenser temperature) to remove this residual moisture to optimum values. The drying temperature can be a little above the atmospheric temperature seeing the sensitivity of the product. The second stage drying is due to desorption during which most of the water adsorbed on the solid matrix is removed. The product temperature is increased to desorbe the bound moisture during secondary drying. The bonds between the moisture and product are broken by higher temperature to facilitate the drying in this phase. The desorbed water vapor is transported through the pores of the material that is dried. Typically, freeze drying is carried to a final moisture content of 1–3%.

By their nature, freeze dried materials are hygroscopic and exposure to moisture during storage can destabilize the product. Packaging used for freeze dried materials must be impermeable to atmospheric moisture. Storing products in low humidity environments can reduce the risk of degradation by exposure to moisture. Oxygen is also detrimental to the stability of most freeze dried material so the packaging used must also be impermeable to air. The detrimental effects of oxygen and moisture are temperature dependent. The higher the storage temperature, the faster a product degrades. Most freeze dried products can be maintained at refrigerator temperatures, i.e. 4-8°C. Placing freeze dried products at lower temperatures extends their shelf life. The shelf life of a freeze dried product can be predicted by measuring the rate of degradation of the product at an elevated temperature. This is called accelerated storage. By choosing the proper time and temperature relationships at elevated temperatures, the rate of product degradation can be predicted at lower storage temperatures.

Once the material is dried sufficiently, it's sealed in a moisture-free package, often with an oxygen-absorbing material. As long as the package is secure, the material can sit on a shelf for years and years without degrading, until it's restored to its original form with a bit of water (a very small amount of moisture remains, so the material will eventually spoil).

Generally the process is completed in 24 to 48 hours to reduce the moisture content of the product to 1 to 3%. The process stabilizes the product so that it can be stored for many years without affecting the nutritional quality.

Since the operation is slower than conventional dehydration, energy costs for refrigeration are high and, in freeze drying, the production of a high vacuum is an additional expense. This, together with a relatively high capital investment, results in high production costs for freeze-dried foods.

6. Advantages and disadvantages of Freeze drying

Advantages

1. Low temperature drying helps in preserving loss of nutrition
2. Retention of original shape and size of the product helps in restoration of original quality of product.
3. Freeze dried products have a very high surface area which enables them to be reconstituted quickly and easily with the re-introduction of the solvent (usually water).
4. Low temperature drying also helps in retention of sensory qualities such as colour, taste and volatile components of the product, thus preserving its flavour too.
5. It avoids the collapse of cellular structure thus case hardening is absent in dried products thus helping in better rehydration.
6. It helps in prolonging storage life.
7. Product storage does not need refrigeration or can be stored at ambient temperature.

Dis-advantages

1. The initial cost as well as operational cost is high
2. Long process time (typically 24 h drying cycle)

3. Limited capacity
4. Needs specialized equipments
5. Many freeze dried products are hygroscopic and need to keep in closed containers for preservation.
6. High Energy cost (2 to 3 times more than other methods)
7. High capital cost of equipment (about 3 times more than others)

7. Factors affecting the drying process

1. Surface area and thickness of food product: The drying will be longer if the product thickness is more. Generally, the larger the surface area of the frozen material, the faster the rate of lyophilization, and, conversely, the thicker the frozen material, the slower the rate of lyophilization. Sample thickness affects the ability of a sample to absorb and transfer heat to the surface undergoing sublimation. Because water vapor must pass through dried material, the rate of lyophilization in thick samples is slower, especially if the dried material collapses onto the surface of the frozen material. Shell freezing minimizes collapsing by increasing the surface area.
2. Vapour pressure difference or condenser temperature: In the drying process vapour pressure difference between the food surface and the ice surface on the condenser is the driving force. Greater the driving force, faster is the vapours driven off from the food surface.
3. Thermal conductivity: Poor thermal conductivity of solid will slow the drying process as the heat transfer will be slow from the frozen material to the food surface.
4. Solute concentration: More solids present in the food will resist the escape of vapour from the material.
5. Ice crystal size: Size of ice crystals will depend on the rate of freezing. Faster freezing will help in smaller size ice crystal formation. With the small size crystals, the drying will be faster.
6. Eutectic temperature: Eutectic temperature is the most important factor determining how much sample can be lyophilized at one time. Vapor pressure decreases as the eutectic temperature lowers, but the rate of heat absorption by the sample remains the same and may cause melting. Diluting the sample with water prior to freezing can prevent melting.
7. Instrument factors

8. Vacuum: A vacuum of at least 133 x 10^{-3} mbar is required for lyophilization.
9. The condenser must be 10-15°C colder than the eutectic temperature of the material being frozen.

Besides these factors following factors are must to be considered to be critical in freeze drying process.

7.1 Critical point

A 'critical point' occurs at the end of a phase line where the properties of the two phases become indistinguishable from each other. At temperatures above the critical temperature a gas cannot be liquefied. Critical points are usually found at the high temperature end of the liquid-gas phase line.

7.2 Collapse temperature

During sublimation, the drying temperature at surface must be kept low enough to prevent pore structure collapse (also called shrinkage) caused by partial melting. When collapse occurs, vapor escape is cutoff, the product foams, and drying rapidly fails. The collapse temperatures are different for different products e.g. collapse will occur at surface temperature greater than -20°C for coffee extract (Table 1) or as low as -46°C grape juice.

Collapse can occur if flow-pressure drop is excessive or vapour pressure difference is too large, but it can be prevented, if need be, by reducing the condenser operating temperature or by reducing the sublimation rate.

As explained before, not all the water in the frozen material is in the form of ice. The non-frozen portion of the water and the water vapour adsorbed on the matrix may act as plasticizer and soften the dry layer, if the temperature is not sufficiently low. Such softening results in the collapse of the porous structure of the dry layer and in loss of all the advantages of freeze drying. The glass transition properties (explained later) of the dry layer have a profound influence on the tendency of the material to undergo collapse. Knowledge of the temperature at which collapse occurs is essential for the proper control of the process.

Table 1: Collapse temperatures* for selected foods in freeze drying

Food	Collapse temperature °C
Coffee extract (25%)	-20
Apple juice (22%)	-41.5
Grape juice (16%)	-46
Tomato	-41
Sweet corn	-8 to -15
Potato	-12
Ice cream	-31 to -33
Cheddar cheese	-24
Fish	-6 to -12
Beef	-12

* Above *Collapse temperatures* there is an instantaneous irreversible collapse of the food structure, which restricts the rate of vapour transfer and effectively ends the drying operation. Usually lower than freezing point.

The majority of products that are subjected to freeze drying consist primarily of water, the solvent, and the materials dissolved or suspended in the water, the solute. Products usually freeze in two ways, depending on the makeup of the product.

7.3 Eutectic temperature

Most samples that are to be freeze dried are eutectics which are a mixture of substances that freeze at lower temperatures than the surrounding water. When the aqueous suspension is cooled, changes occur in the solute concentrations of the product matrix. As cooling proceeds, the water gets separated from the solutes as it changes to ice, creating more concentrated areas of solute. These pockets of concentrated materials have a lower freezing temperature than the water. As temperature drops, individual solute reaches saturation point and crystallizes out. Hence the temperature at which a crystal of an individual solute exists in equilibrium with the unfrozen liquor and ice is its eutectic temperature. The eutectic temperatures for glucose is -5°C; sucrose -14°C, NaCl -21.13°C and for $CaCl_2$ -55°C. However it is difficult to identify individual eutectic temperatures in the complex mixture of solutes in foods and the final eutectic temperature is therefore used. This is the lowest eutectic temperature of the solutes in the food (e.g. ice cream -55°C; meat -50 to -60°C; bread -70°C) (Fellow). It is very important in freeze drying to prefreeze the product to below the eutectic temperature before beginning the freeze drying process.

The eutectic temperature is very important to determine the drying temperature and vacuum. During the drying process the product temperature is only controlled by the vacuum, not by the shelf temperature. At the beginning of the drying process, the product temperature is proportional to the vacuum, in accordance with the water vapour pressure above ice. The product temperature is independent from the shelf temperature. During the drying process it is very important that the product does not melt, therefore the drying temperature should be at least 10°C below the eutectic temperature. It is possible to determine the drying vacuum according to vapour pressure above the surface of ice.

Example:

Eutectic temperature	-10°C
Drying temperature	-20 °C
Drying vacuum	1,030 mbar

Hence prior knowledge of eutectic temperature of the product is important to know the temperature to which the product is to be frozen.

7.4 Glass Transition

The second type of frozen product is a suspension that undergoes glass formation during the freezing process. Instead of forming eutectics, the entire suspension becomes increasingly viscous as the temperature is lowered. Finally the product freezes at the glass transition point forming a vitreous solid. This type of product is extremely difficult to freeze dry.

8. Energy transport

The heat transfer through the dried layers of the product to the non-dried centre of the material does not remain as good as it was at the beginning of drying, so that the temperature of the ice core can be increased. If the temperature of the ice core is higher than the eutectic point the product will melt. So it is very important to control the vacuum and the product temperature during the drying process.

9. Heat and Mass Transfer in Freeze Drying

Like any other dehydration process, freeze-drying is also a simultaneous heat transfer and mass transfer process. For hot air drying the drying rates are primarily limited by mass transfer and the temperature gradients are minimal, while moisture content gradients are steep. In freeze-drying situations, mass transfer occurs through porous layers with high effective diffusivity and very low thermal conductivity. The most widely used freeze-drying process is based upon the heat of sublimation being supplied from the surrounding gases to the sample surface. Then the heat is transferred by conduction through the dried material to the ice surface.

The distribution of moisture content through the material undergoing lyophilization is different from that observed in other dehydration processes. Ideally, two distinct zones, separated by a fairly sharp interface are observed: a 'frozen' zone and a 'dry' zone. The frozen zone is at the original moisture content. The dry zone is devoid of ice crystals and the only moisture in it is water adsorbed on the solid matrix (Figure 3). The proportion of adsorbed moisture is usually small and the water content distribution across the material can be idealized as a step function. The interface between the two zones is known as the 'sublimation front'. Heat can be delivered to the sublimation front by a number of mechanisms. Two modes of heat transfer are the most commonly applied as shown in Fig. 2.

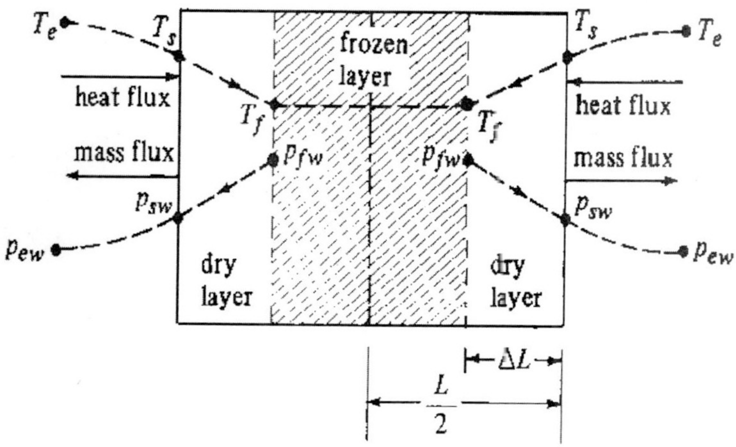

Fig. 3: Radiation heat transfer process

1. Radiation from hot surfaces: in this case, heat is delivered to the exposed surface of the dry zone by radiation and then travels through the dry zone to the sublimation front, mainly by conduction. The water vapour released travels also through the dry zone, in the opposite direction, from the sublimation front to the surface and from there to the condenser.

2. Contact with a hot surface: in a common embodiment of this mode, trays containing the frozen material are placed on heated shelves. Heat is transferred to the sublimation zone by conduction through the frozen layer. Water vapour is transferred through the dry layer as before.

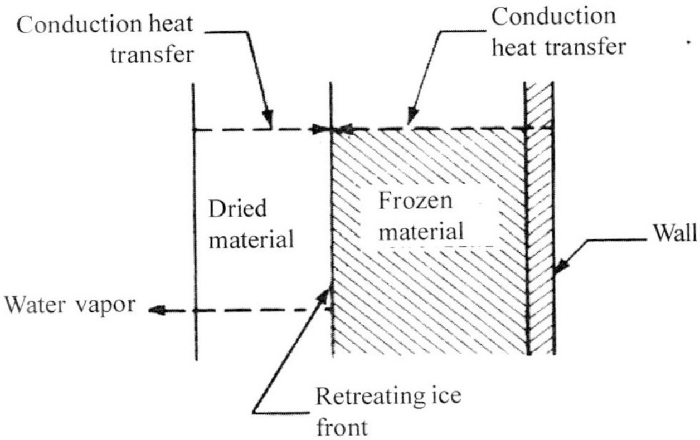

Fig. 4: Conduction heat transfer process

Equation for Heat-transfer Flux in Freeze Drying

Consider the case, which is simplified but is nevertheless typical of most vacuum freeze-drying operations, where the material to be dried is heated by applying radiation to the dry surface, and its internal frozen layer temperature is determined by the balance between heat and mass transfer. For the sake of simplicity, we shall consider a slab geometry with negligible end effects, as shown in Fig. 3. We shall assume that the maximum allowable surface temperature T_s is reached instantaneously and maintained constant throughout the drying cycle. We also assume that the partial pressure of water vapour at the drying surface, p_s, is constant and that all of the heat is used for sublimation of water vapour.

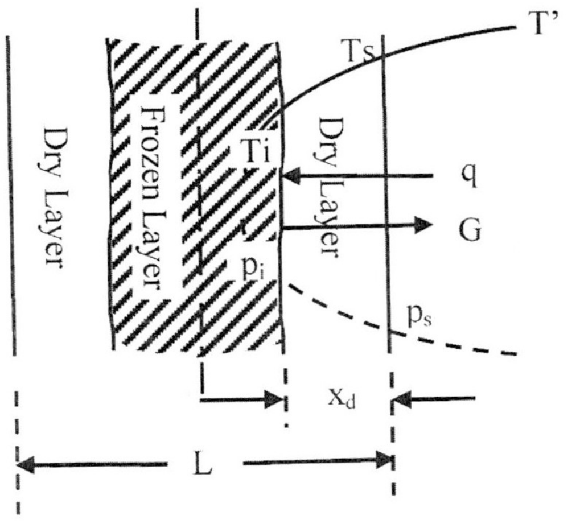

Fig. 5: Schematic representation of heat and mass transfer in a freeze drying of slab

Under these conditions, the heat transfer flux to the surface is equal to that conducted through the dry solid assuming a pseudo steady state at any instant is given by

$$q = Ak_d \frac{(T_s - T_i)}{x_d} \tag{1}$$

Where

k_d is the thermal conductivity of the dried layer, W/mK;

x_d is the thickness of dried layer, m;

L is the thickness of the slab, m;

T_s is the surface temperature, °C;

T_i is the temperature at the sublimation front, °C;

A is the surface area of the slab, m^2.

In the similar manner the mass transfer flux of the water vapour G from the sublimation front can also be expressed as

$$\frac{q}{\lambda_s} = G = -\frac{dW}{dt} = Ab\frac{(p_i - p_s)}{x_d} \tag{2}$$

where

dW/dt is the rate of moisture removed from the dry layer, kg/s;

b is the permeability of the dry layer, kg/s m Pa;

p_s is the partial pressure of water vapour at dry surface, Pa;

p_i is the partial pressure of water vapour at sublimation front, Pa;

λ_s is the latent heat of sublimation J/kg (roughly 2836 kJ/kg ice).

In practice the lowest economical chamber pressure is approximately 13Pa and the lowest condenser temperature is approximately -35°C

If the initial moisture content of the ice-dry layer m_o drops to final moisture content m_f, given slab geometry, we obtain a relation between loss of weight and the rate of recession of the interface.

$$-\frac{dW}{dt} = A\rho_d\left(m_o - m_f\right)\frac{dx_d}{dt} \tag{3}$$

where

ρ_d is the bulk density of solids in the drying slab, kg/m^3; m_o is initial moisture content, kg water/kg solids; m_f is final moisture content, kg water/kg solids.

By combining Eqs. (2) and (3) we obtain

$$x_d dx_d = \left[\frac{b}{\rho_d\left(m_o - m_f\right)}\right]\left(p_i - p_s\right)dt \quad (4)$$

We can make the assumption that the rate of heat transfer is equal to the rate of sublimation multiplied by latent heat of sublimation, λ_s, hence

$$Ak_d\frac{\left(T_s - T_i\right)}{x_d} = Ab\frac{\left(p_i - p_s\right)}{x_d}\lambda_s \quad (5)$$

During drying the moisture content falls from the initial high level in the frozen zone to a lower level in the dried layer, which depends on the water vapour pressure in the drying cabinet. When heat is transferred through the dry layer the relation between pressure in the cabinet and pressure at the ice surface is obtained by simplifying equation 5 as

$$p_i = p_s + \frac{k_d}{b\lambda_s}\left(T_s - T_i\right) \quad (6)$$

Since we assume that p_s, b, λ_s, k_d, and T_s are all constant, we have a linear equation relating p_i with T_i. This means that if the assumptions inherent in our analysis are true, the frozen layer temperature Ti must remain constant throughout the drying cycle.
As a consequence, Eq. (4) can be readily integrated to calculate the freeze drying time, because it contains only the two variables x_d and t_d as

$$\int_o^{L/2} x_d dx_d = \frac{b\left(p_i - p_s\right)}{\rho_d\left(m_o - m_f\right)}\int_o^{t_d} dt \quad (7)$$

and the freeze drying time, t_d is given by

$$t_d = \frac{L^2\rho_d\left(m_o - m_f\right)}{8b\left(p_i - p_s\right)} \quad (8)$$

Where L is thickness of the slab (m), t_d is drying time

Because of the equivalence of heat and mass transfer shown in Eq. (5) we can equally well integrate the heat-transfer equation, Eq. (2), and have a corresponding equation for freeze drying time, Eq. (9)

$$t = \frac{L^2 \rho_d \left(m_o - m_f\right) \lambda_s}{8k_d \left(T_s - T_i\right)} \qquad 9$$

We see therefore that drying time depends on the following variables:

1. maximum permissible surface temperature (T_s)
2. initial and final moisture contents (m_o, m_f)
3. bulk density of solids (r_d)
4. latent heat of sublimation (1_s)
5. thickness of the slab (L)
6. thermal conductivity of the dry layer (k_d)
7. permeability of the dry layer (b)

Frozen layer temperature depends neither on the overall thickness of the drying material nor on the thickness of the layer already dried.

18

Membrane Separation Process in Food Industry

Prodyut Kumar Paul

1. Introduction

Membranes separation processes have gained widespread applications in food industry. High selectivity of membrane has been exploited for separating a particular component from complex liquid food mixture and for controlled release of active ingredients like preservatives or flavor compounds. Today, a number of chemical processes utilizing the unique properties of membrane have been developed for commercial application in food industry with specific advantage of low energy requirement. Membrane separation/molecular sieving is one among those food engineering principals which has found broad range of applications including concentration of heat labile compounds, recovery of solvent, downstream processing, flavor volatile recovery, deacidification, desalination, waste water treatment etc.

Membrane technology is an emerging and evolving separation technology and because of its multidisciplinary characters, it can be used to perform a large number of separations in food processing. The primary motivations for using membrane technology to replace traditional thermal evaporation are to substantially reduce the thermal damage to food components. The second reason for using membrane technology is energy saving. However, many membrane processes incur high capital and operational costs related to membrane replacement and cleaning so the overall economics should be considered when evaluating membrane technology. Additionally, some membrane processes cannot match the performance of thermal evaporation in terms of one-pass product concentration (low retention or selectivity) or have low permeation rates (low permeability).

2. General aspects of Membrane separation

2.1 Membrane separation system

A membrane is a semipermeable or selectively permeable barrier which allows some of the molecules or ions to pass through it while retaining others. In membrane separation, the feed stream is pumped onto the membrane surface maintaining a pressure gradient across the membrane. A portion of the feed stream, known as permeate, passes through the membrane while the other constituents are rejected or retained by the membrane called retentate (Fig.1).

The semipermeability or selectivity of the membrane arises from its ability to recognize the difference either in molecular diameters or chemical potentials or vapour pressures or ionic charges of mixture components. The essential components of a membrane separation system consist of a suitable membrane; a membrane module to house the membrane and an appropriate system to generate required driving force for transport of material across the membrane. The membrane can either be nonporous, microporous, mesoporous or macroporous. By IUPAC classification, pores between 20 Å and 500 Å are called mesoporous and pores larger than 500 Å are called macroporous. A membrane can also either be a solid or a liquid. It can be a polymeric material, inorganic refractory compound or a metal or their composites.

Normally a chemical potential gradient provides the driving force required for transport of material from one side of the membrane to other. In most application, a pressure gradient across the membrane is created by pumping the feed at high pressure or by maintaining vacuum at the permeate side of the membrane. Thus the feed side of the membrane is always at higher pressure than the permeate side to provide the required driving force for mass transfer. In some other application like electrodialysis, the driving force is provided by the electrical potential gradient across the membrane.

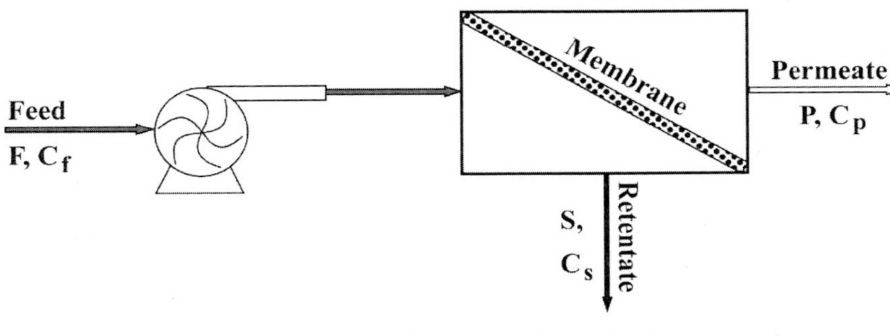

Fig. 1: Schematic of Membrane Separation Process

2.2 Stages of Mass Transfer

The components transferred through the semipermeable membrane in the separation processes should have to overcome a series of resistance as it passes from concentrated solution at feed side to the diluted solution at permeate side. Therefore, a series of successive stages occurs during the global mass transfer that can be summarized as follows (Figure 2):

Stage 1: Transfer from the bulk feed solution to the membrane surface.

Stage 2: Passage of the component from the solution at membrane surface into the membrane.

Stage 3: Transfer through the different layers of the membrane, from the wall in contact with the feed solution to the wall in contact with the permeate solution.

Stage 4: Transfer from the membrane wall to the permeate solution.

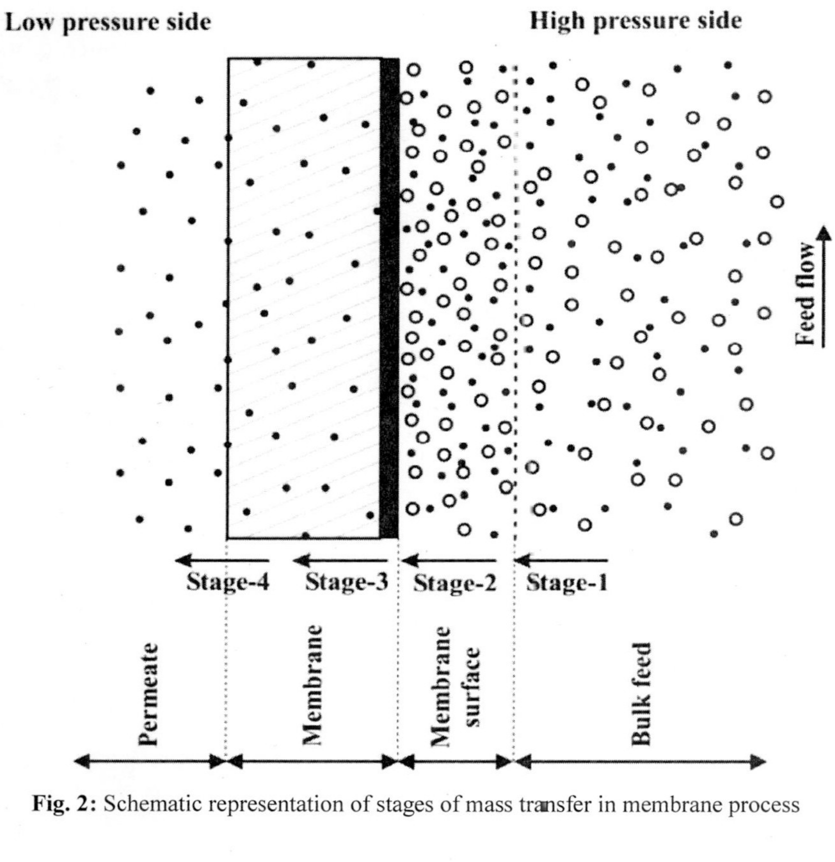

Fig. 2: Schematic representation of stages of mass transfer in membrane process

3. Classification of membrane separation

Each separation process using membrane can be tailor-made to suit the feed characteristics, separation requirement and throughput requirement. This can be achieved by varying the membrane type, selecting a suitable module and selecting a driving force depending on the feed characteristics. Accordingly, a wide variety of system can be made available to effect an efficient separation. Therefore, a comprehensive classification of membrane separation process will help to select a suitable system for efficient separation. Membrane separation process can be classified depending upon-

a) Direction of mass transfer with respect to the membrane surface

b) Driving force for mass transfer

3.1 Direction of Mass transfer with respect to membrane surface

Membrane process can be achieved either in dead-end or cross flow mode (Sablani, 2007):-

3.1.1. *Dead-end filtration*: In dean-end filtration, the feed stream moves perpendicular to membrane surface and it passes through the membrane as filtrate. Particulates and aggregates accumulate on the membrane surface and form filter cake which reduces the transmembrane flux and increases the feed pressure with time (Fig. 3).

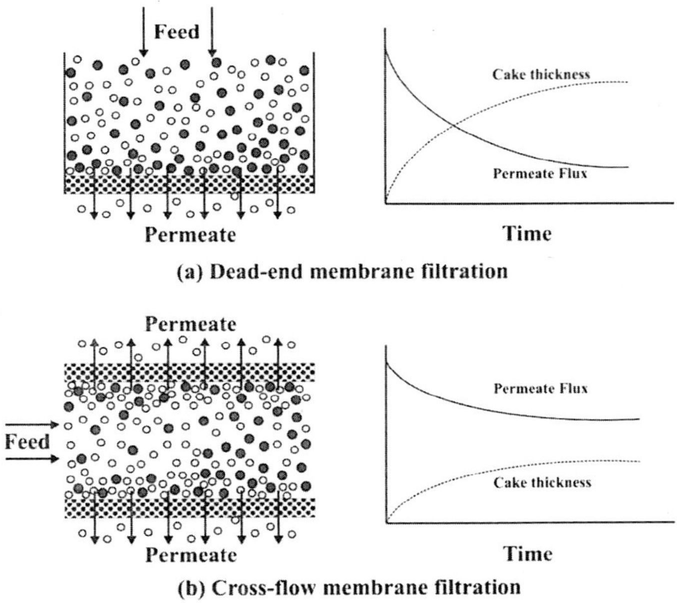

Fig. 3: Schematics of (a) Dead-end and (b) Cross flow membrane filtration

3.1.2 ***Cross flow or tangential flow filtration***: In cross flow or tangential flow filtration, the feed stream moves parallel to the membrane surface. Some portion of the stream passes through the membrane as permeate while the remainder of the feed stream becomes retentate. This retentate may be recirculated back as feed stream or may undergo further processing. The tangential feed stream continuously sweeps across the membrane surface which prevents accumulation of particulates and aggregates and thus maintains a more steady permeate flux and transmembrane pressure. Most large scale industrial membrane filtration processes operate in cross-flow mode (Fig. 3).

3.2 Driving force for Mass transfer

In membrane filtration, separation can be achieved due to difference in pressure, concentration, temperature and electrical potential across the membrane. Accordingly, membrane processes can be classified as given in Table 1:

Table 1: Classification of membrane processes based on driving forces

Driving Force	Membrane Processes
Pressure Difference (ΔP)	Microfiltration (MF), Ultrafiltration (UF), Nanofiltration (NF), Reverse Osmosis (RO)
Concentration Difference (Δc)	Pervaporation (PV), Vapour permeation (VP), Gas separation (GS), Dialysis (D)
Temperature Difference (ΔT)	Membrane Distillation (MD), Thermo-osmosis
Electrical potential difference (ΔV)	Electrodialysis (ED), Electro-osmosis, Membrane electrolysis

3.2.1 ***Pressure driven membrane processes***: Majority of membrane filtration processes employ trans-membrane pressure gradient to achieve separation. Pressure driven membrane separation processes include Microfiltration (MF), Ultrafiltration (UF), Nanofiltration (NF) and Reverse Osmosis (RO). Comparative characteristics of pressure driven membrane processes (Wang and Zhou, 2013) are presented in the Table 2. A filtration spectrum of different pressure driven membrane process is represented in Fig. 4.

3.2.1.1 Microfiltration (MF)

Microfiltration is defined as a membrane separation process using membranes with a pore size of approximately 0.1 to 10 microns, and a relatively low feed water operating pressure of approximately 70 to 350 kPa (10 to 50 psi). The mechanism of separation in microfiltration is size exclusion. Microfiltration membranes are selected by their ability to retain components of specific sizes dissolved in the feed stream. This is referred to as the molecular weight cut-off (MWCO), which is defined as the smallest molecular weight at which at least

90% of the solute is retained by the membrane. Microfiltration membranes are characterized by a MWCO values greater than 10^4 kDa.

Microfiltration is primarily used to separate particles like sand, silt clay etc., and microorganism like yeast, fungi, bacteria etc. from other smaller particles/solutes. Effectiveness of Microfiltration in removing microorganism is enhanced when used in combination with disinfection. By physically removing the pathogens, membrane filtration can significantly reduce the use of chemical in food processing.

Another application for the technology is for removal of natural synthetic organic matter to reduce fouling potential. As such, MF can be used as a pretreatment to Nanofiltration (NF) and Reverse Osmosis (RO) to reduce fouling potential. Various applications of pressure driven membrane processes (Wang and Zhou, 2013) is presented in Table 3.

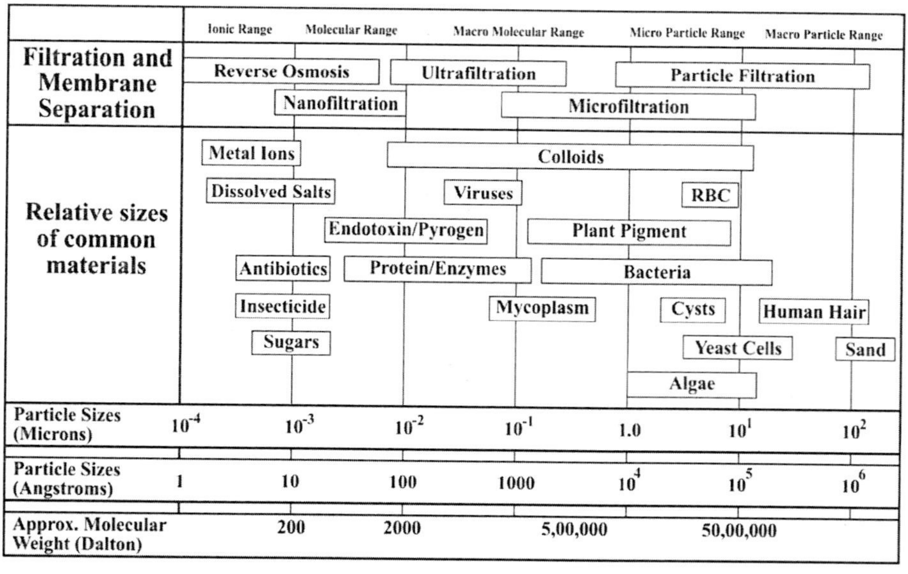

Fig. 4: Filtration spectrum of different pressure driven membrane processes

3.2.1.2 Ultrafiltration (UF)

Ultrafiltration has a pore size of approximately 0.002 to 0.1 microns, an MWCO of approximately 1,000 to 100,000 daltons, and an operating pressure of approximately 100 to 700 kPa (15 to 100 psi). In ultrafiltration, the mechanism of separation is size exclusion as like microfiltration. UF membranes are commonly used to retain all microbiological species, as well as some viruses and humic materials. UF membrane can also retain suspended protein and starches while allow salts and smaller dissolved organic compounds to permeate.

Ultrafiltration processes are widely used in biopharmaceutical protein separation, virus clarification and whey protein concentration and isolation in dairy industry.

3.2.1.3 Nanofiltration (NF)

Nanofiltration lies between the separation characteristics of reverse osmosis (RO) and UF process which is widely used for several applications such as water softening and wastewater treatment. The pore size of the NF is in the range of 0.5–1 nm. Nanofiltration removes multivalent ions (e.g., sulfate ions) and small molecules (e.g., salts, sugar etc.) with a characteristic MWCO of 1000 Da. Monovalent ions are commonly pass through NF membranes. In nanofiltration, salts with divalent anions have rejection rates in the range of 90% to more than 99%, while salts with monovalent anions (e.g., sodium chloride) have a rejection rate of 20-80%.

The nanofiltration membrane process is governed by the solute-diffusion transport mechanism, where the premeating species first dissolve into the membrane and diffuse through it. Because of the nearly nonporous nature of the membranes, NF processes are operated at significantly high pressure than the more-porous UF and MF membrane. The operating pressures of NF processes are typically in the range of 350-1500 kPa (50-225 psi).

3.2.1.4 Reverse Osmosis (RO)

Osmosis is a naturally occurring phenomenon and one of the most important processes in nature. It is a process where a less concentrated solution will tend to migrate to a more concentrated solution through a semipermeable membrane.

In the Fig. 5, two arms of the U-tube contain two different solution separated by a semipermeable membrane that allow only water molecule to pass through. The arm in the right contains a more concentrated solution (hypertonic) while the arm in the left contains pure water (hypotonic). According to Van't Hoff equation, the hypertonic solution will have more negative osmotic potential while the pure water will have zero osmotic potential. In order to achieve equilibrium in osmotic potential, the water molecules will tend to move towards hypertonic solution from hypotonic solution (pure water). This movement of water molecules results in rising of solution level in the right arm of the U-tube while the pure water level in the left arm will come down. This will results in a difference in head in the two arms. The movement of water molecule will be opposed by this head difference and will finally stop. This difference in head at this stage is defined as the osmotic pressure of the hypertonic solution.

If a pressure more than the osmotic pressure of the hypertonic solution is applied in the right arm of the U-tube, the water molecules will start moving in the

Table 2: Characteristics of pressure driven membrane filtration process (Wang and Zhou, 2013)

	Microfiltration (MF)	Ultrafiltration (UF)	Nanofiltration (NF)	Reverse Osmosis (RO)
Membrane type	Asymmetric	Symmetric, Asymmetric	Asymmetric, thin film composite	Asymmetric, thin film composite
Pore size	0.1 to 10 μm	0.002 to 0.1 μm	Almost nonporous, 0.5 to1 nm	Nonporous
MWCO	10,000 kDa	1-100 kDa	1 kDa	0.1 kDa
Total thickness	10-150 μm	150-250 μm	150-250 μm	150 μm
Composite film thickness	-	-	1 ìm or less	1 μm or less
Rejected components	Particulates, clay, bacteria	Macromolecules, protein, polysaccharides, viruses	High molecular weight compounds, monosaccharides, disaccharides, oligosaccharides, multivalent ions	High molecular weight compounds, Low molecular weight compounds, sodium chloride, glucose, amino acids
Membrane materials	Polymeric, ceramic	Polymeric, ceramic	Polymeric (thin-film composite and integrally skinned)	Polymeric (thin-film composite and integrally skinned)
Membrane module	Hollow fiber	Spiral wound, Hollow fiber, Plate and frame	Spiral wound, Plate and frame	Spiral wound, Plate and frame
Operating pressure	70 to 350 kPa (10 to 50 psi)	100 to 700 kPa (15 to 100 psi)	350-1500 kPa (50-225 psi)	1 and 7 MPa (150-1000 psi)

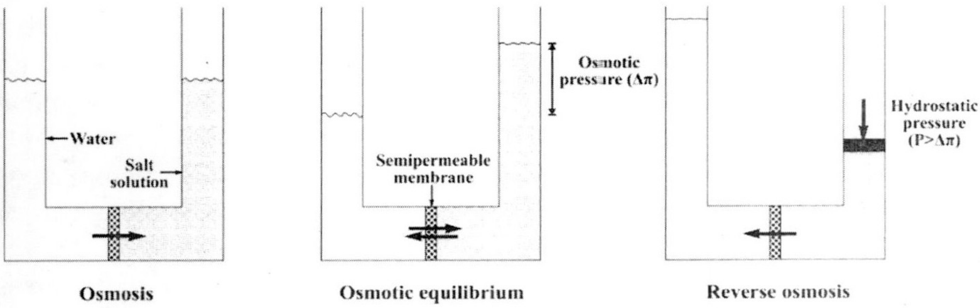

Fig. 5: Schematic diagram of osmosis and reverse osmosis

reverse direction i.e., from the right arm towards the left arm. This movement of water molecule from a hypertonic solution to hypotonic solution overcoming the osmotic pressure of the hypertonic solution is called reverse osmosis.

Reverse Osmosis employs dense non-porous membrane for liquid separation. Dissolved salts, inorganic and organic solutes with a molecular weight greater than approximately 100 Da are rejected while water molecules are able to pass through RO membrane. It uses pressures between 1 and 7 MPa (150-1000 psi).

3.2.2 *Membrane process driven by transmembrane concentration difference:* A nonporous selectively permeable membrane can be used for organic solvent removal from aqueous stream, ethanol or butanol recovery from fermentation broths, solvent dehydration to meet dryness specifications, and organic-organic separations. Pervaporation and vapor permeation are two promising membrane based technologies that can serve as alternative to technologies conventionally used for such separation. Unlike membrane filtration processes, which rely on an applied liquid pressure gradient and size sieving to accomplish a separation, pervaporation and vapor permeation separate compounds based on a chemical activity driving force and the sorption and diffusion of the compounds through the membrane.

3.2.2.1 Pervaporation

Pervaporation is recognized as a separation process in which a binary or multicomponent liquid mixture is separated by partial vaporization through a dense non-porous selectively permeable membrane. The process results in a vapour permeate and a liquid retentate. In the process, the feed is first heated up to the operating temperature and then brought into contact with the active (feed) side of the pervaporation membrane. The better permeating component preferentially passes through the membrane and is continuously removed in the form of vapour from the permeate side of the membrane. The continuous removal of the vaporous permeate, by reducing the total pressure on the permeate side of the membrane by using a vacuum pump system or by sweeping an inert gas on the permeate side of the membrane, creates a concentration gradient across the pervaporation membrane. This concentration gradient acts as a driving force for the process. The mass transfer across a pervaporation membrane is carried out in three steps. Initially, the permeating components are absorbed into the membrane at feed side. These absorbed components are then transported across the membrane by diffusion according to Fick's law. Finally, these component are removed in a vapour state from the permeate side into a vacuum or sweeping gas (Fig. 6).

3.2.2.2 Vapour permeation

Vapor permeation is similar in principle to pervaporation. In vapour permeation, the feed is a mixture of vapors or vapors and gases while in pervaporation the feed is a mixture of liquid (Fig. 6). As in pervaporation, the permeate partial pressure is maintained by use of a vacuum or an inert sweep gas. There is no change of phase involved in its operation. Thus, compared to pervaporation, the addition of heat equivalent to the enthalpy of vaporization of the permeating component is not required in the membrane unit and there is no temperature drop along the membrane. Operation in the vapor phase also eliminates the effect of the concentration polarization prevalent in liquid phase separations, such as pervaporation.

Fig. 6: The process of (a) Pervaporation and (b) Vapour Permeation

3.2.3 ***Membrane process driven by transmembrane temperature difference*:** Inherent limitation of membrane filtration like inability to obtain high solute concentrations and concentration polarization phenomena can be easily overcome by creating a temperature gradient across the membrane. One such promising technology driven by transmembrane temperature difference is membrane distillation.

Table 3: Commercial application of pressure driven membrane filtration (Wang and Zhou, 2013).

	Feed	Permeate	Concentrate
Microfiltration	Water	Clarified process water	Waste product
	Fruit juice	Clarified fruit juice	Waste product (suspended solids, microorganism, undesirable protein)
	Wine	Clear wine	Waste products (fine fruit particles, spent yeast, bacteria, soil, debris and fining agents)
	Therapeutic protein	High value products	Waste product
	Amino acid	Clarified fermentation broth	Waste product
Ultrafiltration	Water	Clarified process water	Waste product
	Oil emulsion	Oil free water	Highly concentrated oil emulsion
	Enzyme	Waste products	High value products
	Washing effluent	Clarified water	Dirty waste water
	Fermentation broth	Broth with desired product	Microorganism for recycling
	Crude vegetable oil	Degummed oil	Waste products
	Milk	Lactose solution	Protein concentrates for cheese production.
Nanofiltration	Water	Softened water	Waste products
	Antibiotics	Salty waste products	Desalted, concentrated antibiotics
	Whey	Salty waste water	Desalted whey concentrate
	Dyeing effluent	Clean, salty water	BOD/COD, Colour
	Fruit juices	Salty water	Concentrated juice
	Regular beer	Alcohol rich beer	Alcohol free, flavour enriched beer
	Regular red wine	Alcohol rich wine	Dealcoholized, flavour and resveratrol enriched wine.
	Vegetable Oil extraction micelle	Recovered solvent	Solvent free crude oil
	Degummed vegetable oil	Free fatty acids	Deacidified oil
Reverse Osmosis	Water	Low salinity water	Salty water
	Whey	Low BOD permeate	Whey concentrate
	Dyeing effluent	Clean water	BOD, salts, chemicals, waste products

Table 4: Comparative Characteristics of pervaporation and vapour permeation (Kujawski, 2000)

Membrane process	Feed phase	Permeate phase	Driving force	Membrane type	Main application
Pervaporation	Liquid	Vapour	Concentration gradient	Dense, mostly lipophilic	Separation of liquid mixture including azeotropes. Dealcoholization of beer/wine.Stripping of flavor volatiles from fruit juices before pasteurization.
Vapor permeation	Vapour	Vapour	Concentration gradient	Dense, mostly lipophilic depending upon the properties of permeating vapour	Separation of vapour mixtures or vapour from gases. Removal of volatile organic compounds (VOCs) from air.

3.2.3.1 Membrane distillation

Membrane distillation can be defined as the transfer of a component, in vapour phase, between two solutions subjected to different temperatures and separated by a hydrophobic membrane. The system consists of a porous hydrophobic membrane, separating two aqueous solutions of a non-volatile component maintained at different temperatures. This process takes place at atmospheric pressure and at temperatures which may be much lower than the boiling point of the solutions. The hydrophobicity of the membrane prevents the transport of the liquid phase across the pores of the partition while water vapor can be transported across them from the warm side, condensing at the cold surface. The driving force is the vapor pressure difference at the two solution membranes interfaces (Onsekizoglu, 2012).

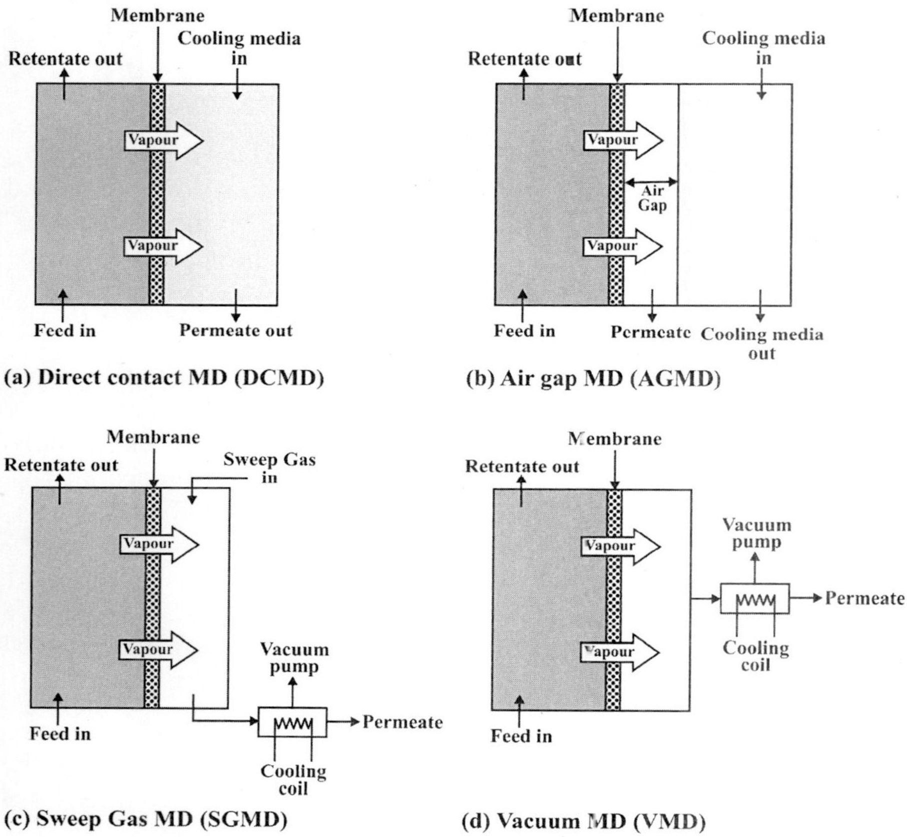

Fig. 7: Various configuration of membrane distillation (MD)

Membrane distillation can be carried on four configurations namely direct contact, sweep gas, vacuum MD and the difference between these configurations depends on the way in which the vapour is condensed and/or removed from the membrane module (Fig. 7).

The major applications of membrane distillation are the desalination and production of high purity water from brackish water and seawater. Other application of membrane distillation in food industry are fruit juice concentration, remove ethanol and the other volatile metabolites from the fermentation broth, concentration of natural food colorants, dealcoholization of wine and concentration of herbal and plant extracts.

3.2.4 ***Membrane process driven by electrical potential difference*****:** Separation of ionic species from feed stream using ion exchange membrane under the influence of electrical potential difference has long been in practice for desalination of water. Electrodialysis is one such technology which gained commercial application in large scale.

3.2.4.1 Electrodialysis

Electrodialysis (ED) is a membrane process, during which ions are transported through semi permeable membrane, under the influence of an electric potential. In the basic system, alternating cation and anion selective membranes are placed in an electric field. The cation-selective membranes permit only the transport of cations, and anion-selective membranes only anion.

When an aqueous salt solution is passed through the channel between cation and anion selective membrane, the cations are passed through the cation-selective membrane, under the influence of cathode, to reach the next channel. Similarly, the anions are passed through the anion-selective membrane to the next channel under the influence of anode. Thus, the aqueous salt solution in the channel is depleted of ions and exits the channel as desalted water. At the same time, in the next channel, ions anions and cations are accumulated which is swept away by passing a pick up solution through the channel. Thus, by placing multiple membranes in a row, which alternately allow positively or negatively charged ions to flow through, the ions can be removed from wastewater (Figure 8).

Electrodialysis has been used for producing desalinated potable water and desalting of cheese whey. Bipolar membrane electrodialysis uses a composite membrane to adjust pH of liquid foods and finds application in inhibiting polyphenol oxidase (PPO) activity in cloudy apple juice.

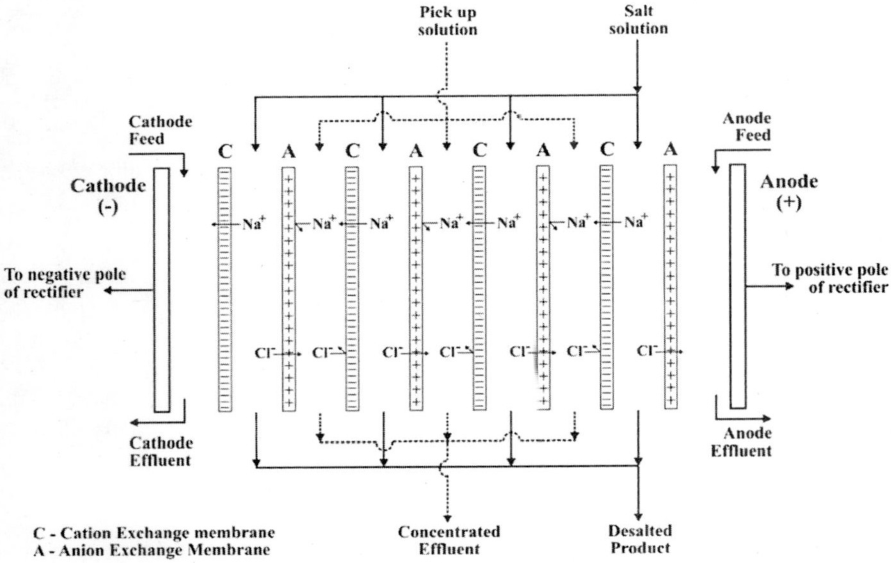

Fig. 8: Schematic diagram of electrodialysis unit

4. Membrane

The heart of the membrane process is the membrane itself. The primary role of a membrane is to act as a selective barrier. Selectivity of a membrane is due to its pore size, ability to solubilize certain components, hydrophilicity or hydrophobicity and electrical charge. Selection of a membrane based upon feed characteristics is, therefore, the most important design criteria for achieving efficient separation.

4.1 Types of membrane

Depending upon membrane structure, it can be classified into isotropic (symmetric) and anisotropic (asymmetric) membrane (Fig. 9). Isotropic membranes have a uniform composition and pore geometry throughout; such membranes can be porous or dense. Anisotropic (or asymmetric) membranes, on the other hand, consist of a number of layers each with different structures and permeabilities.

Isotropic microporous membranes are normally used in microfiltration and ultrafiltration and separation is achieved principally by size exclusion. In isotropic dense membrane, the separation of various components of a mixture is determined by their diffusivity and solubility in the membrane material. An important property of dense membrane is that even permeates of similar size may be separated when their solubility in the membrane material differs

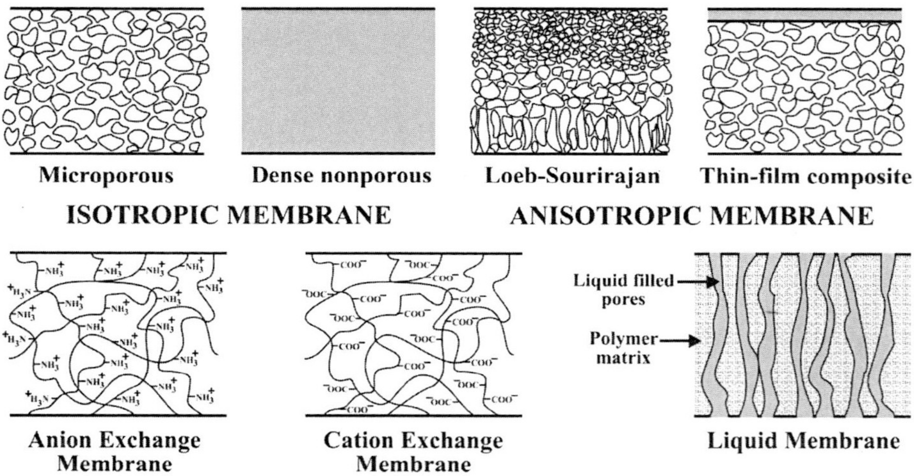

Fig. 9: Schematic presentation of different types of membrane

significantly. Most gas separation, pervaporation and reverse osmosis membrane use dense membrane to perform separation. However, these membranes usually have an asymmetric structure to improve flux.

A typical anisotropic membrane has a relatively dense, thin surface layer (0.1-0.5μm) supported on a porous, much thicker (50-150μm) microporous substructure. The surface layer performs the separation and is the principal barrier to flow through the membrane. The porous support layer provides mechanical strength. The surface layer and the support layer may be formed in a single operation or formed separately. When the dense barrier layer and the support layer are formed separately and then glued together, it is called composite membrane.

Another category of membrane is electrically charged membrane or ion exchange membrane. To make the membrane selective to cation or anion, the pores of these membranes are fixed with positively or negatively charged ions and referred to as an anion-exchange membrane or a cation-exchange membrane, respectively. Separation with charged membranes is achieved mainly by exclusion of ions of the same charge as fixed ions of the membrane structure, and to a much lesser extent by the pore size.

Carrier facilitated transport is another phenomenon that can be utilized to improvise selectivity in membranes. Liquid membranes are one such kind where a liquid carrier phase is held by capillarity within the pores of a polymeric microporous structure. The selective barrier in these membranes is the liquid carrier phase that selectively reacts with a specific permeate component to enhance its transport rate through the membrane.

4.2. Membrane material

The materials used for the preparation of membranes cover a wide range from organic polymer to inorganic materials like ceramics, glass, and metals.

4.2.1 Polymeric membranes: The majority of the membrane process is based on natural or synthetic polymers. These membranes are often called organic membrane. Natural polymers like wool, rubber (polyisoprene) and cellulose and synthetic polymers like polyamide, polystyrene and polytetrafluoroethylene (Teflon) belong to this category. Synthetic polymers are produced by polymerisation of a monomer (condensation or addition) or by the co-polymerisation of 2 different monomers. The resulting polymer is categorised as having (*a*) a long, linear chain, such as linear polyethylene; (*b*) a branched chain, such as polybutadiene; (*c*) a three-dimensional highly cross-linked structure, such as phenol-formaldehyde and (*d*) moderately cross-linked structure, such as butyl rubber. Characteristics of some of the industrially important polymeric membranes are given in Table 5.

4.2.2 Inorganic membrane: Inorganic membranes refer to membranes made of materials such as ceramic, carbon, silica, zeolite and metals such as palladium, silver and their alloys. Inorganic membranes can be both porous and dense. Microporous inorganic membranes can be obtained by coating over a porous support with a colloidal solution, called sol. The sol can consist of either dense spherical particles (colloids of oxides such as Al_2O_3, SiO_2 or ZrO_2) or polymeric macromolecules.

Inorganic membranes are more expensive than organic polymeric membranes; however, they possess advantage of temperature stability, resistance towards solvents, well-defined stable pore structure, and the possibility for sterilization.

Dense, metal membranes are being considered for the separation of hydrogen from gas mixtures. Palladium (Pd) and its alloys are the dominant material used, due to its high solubility and permeability for hydrogen. Palladium being expensive, a thin layer of it is coated on a tantalum or vanadium support film, which are also quite permeable to hydrogen.

Table 5: Characteristics of some of the industrially important polymeric membranes (Ulbricht, 2006)

Polymer	Morphology			Membrane process
	Barrier type	Cross-section	Barrier thickness (μm)	
Cellulose acetates	Nonporous	Anisotropic	~0.1	GS, RO
	Mesoporous	Anisotropic	~0.1	UF
	Macroporous	Isotropic	50–300	MF
Cellulose nitrate	Macroporous	Isotropic	100–500	MF
Cellulose, regenerated	Mesoporous	Anisotropic	~0.1	UF, D
Perfluorosulfonic acid polymer	Nonporous	Isotropic	50–500	ED
Polyacrylonitrile	Mesoporous	Anisotropic	~0.1	UF
Polyetherimides	Mesoporous	Anisotropic	~0.1	UF
Polyethersulfones	Mesoporous	Anisotropic	~0.1	UF
	Macroporous	Isotropic	50–300	MF
Polyethylene terephthalate	Macroporous	Isotropic	6–35	MF
Polyphenylene oxide	Nonporous	Anisotropic	~0.1	GS
Poly(styrene-co-divinylbenzene), sulfonated or aminated	Nonporous	Isotropic	100–500	ED
Polytetrafluoroethylene	Macroporous	Isotropic	50–500	MF
	Nonporous		~0.1	GS
Polyamide, aliphatic	Macroporous	Isotropic	100–500	MF
Polyamide, aromatic	Mesoporous	Anisotropic	~0.1	UF
Polyamide, aromatic, in situ synthesized	Nonporous	Anisotropic/ composite	~0.05	RO, NF
Polycarbonates, aromatic	Nonporous	Anisotropic	~0.1	GS
	Macroporous	Isotropic	6–35	MF
Polyether, aliphatic crosslinked, in situ synthesized	Nonporous	Anisotropic/ composite	~0.05	RO, NF
Polyethylene	Macroporous	Isotropic	50–500	MF
Polyimides	Nonporous	Anisotropic	~0.1	GS, NF
Polypropylene	Macroporous	Isotropic	50–500	MF
Polysiloxanes	Nonporous	Anisotropic/ composite	~0.1	GS PV, NF
Polysulfones	Nonporous	Anisotropic	~0.1	GS
	Mesoporous	Anisotropic	~0.1	UF
Polyvinyl alcohol, crosslinked	Nonporous	Anisotropic/ composite	<1-10	PV (hydrophilic)
Polyvinylidenefluoride	Mesoporous	Anisotropic	~0.1	UF
	Macroporous	Isotropic	50–300	MF

5. Membrane Configuration/ Geometry/ Module

Commercial operations of membrane processes often require hundreds to thousands of square meters of membrane active surface to perform efficient separation with desired throughput. Therefore, large areas of membrane required to be packed economically and efficiently into a smaller housing, called module.

Membranes are manufactured in three basic geometries namely; flat sheet membrane, tubular membrane and hollow fibre membrane. Flat sheet membranes look like filter paper in the form of filter discs or rectangular sheets. The most common type of tubular membrane looks like a single hollow tube of circular cross-section, the wall of the tube functioning as the membrane. Monolith tubular membranes look like cylindrical blocks with large numbers of parallel tubes within them. These tubes typically have diameter in the range 0.5 cm to 2 cm. Hollow fibres are also tube-like in appearance. However, these membranes have much smaller diameters of the order of 1 mm. If the inner wall of hollow fibre acts as the membrane, it is called inside-out type and if the outer wall acts as the membrane, it is referred to as the outside-in type. Double skinned hollow fibre membranes which can function both as inside-out and outside-in membranes are also available.

These membranes are arranged or packed into a membrane module. Main design considerations for module are packing density i.e., active surface area of membrane per unit volume of module, hydrodynamics to manage flow regime of the feed, feed characteristics especially viscosity of the feed, operating pressure, frequency of cleaning and cost. There are different modules available for commercial operation and principal types are- Plate and frame, tubular, hollow fibre and spiral wound module. Comparative evaluation of these modules is presented in Table 6.

5.1 Plate and Frame module

In plate-and-frame module, a flat membrane sheets with a feed spacer in one side and a permeate spacer in the other side are layered together between two end plates. The three layers i.e., the feed spacer, membrane and the permeate spacer are sealed to the end plates, forming a membrane cassette. Many such membrane cassette can be stacked together either vertically or horizontally to make up a module (Fig. 10). The membrane performs the actual filtration. The presence of spacer introduces turbulence in the feed stream, which would otherwise be laminar flow due to long and narrow flow channel. The flow turbulence promotes mixing of the feed at membrane surface for effective mass transfer and to avoid concentration polarization.

Plate-and-frame module is considerably flexible in use. Individual membrane can be replaced when damaged and capacity can be increased simply by adding additional membrane cassette. However, the main disadvantages seem to be the relatively high initial capital cost and complexity in maintaining leak-free seals under the high pressures encountered in RO.

5.2 Tubular module

The tubular module is perhaps the earliest design of industrial scale equipment using synthetic membrane. The membrane is either inserted into or coated onto the inner surface of a porous support tube designed to withstand high operating pressure. The individual tubes may be bunched together and are housed within one pressure vessel in a shell-and-tube arrangement (Fig. 10). The usual flow is through the bore of the tubes which are usually connected in series with the appropriate end-cap. The permeate collects in the shell side from where it is removed. Most manufacturers use tubes of 12.5 mm internal diameter and lengths of 100-600 cm. The number of tubes per module varies from 7 to 30, with 18 to 19 being most common.

Tubular modules have several advantages like high turbulence flow regime leading to higher flux; (ability to handle suspended matter with particle sizes as large as 1 to 1.5 mm; and easy to clean. Its use is mostly limited by low packing density, high capital cost and high pumping energy cost.

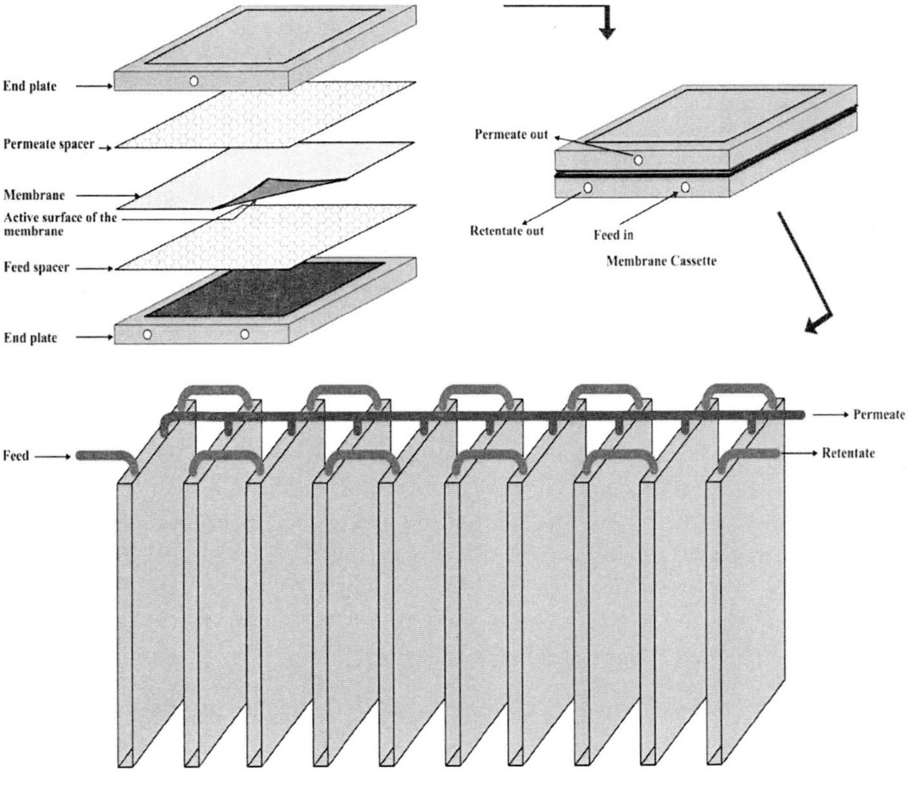

Fig. 10: Arrangement of membrane and membrane cassette in Plate and frame module

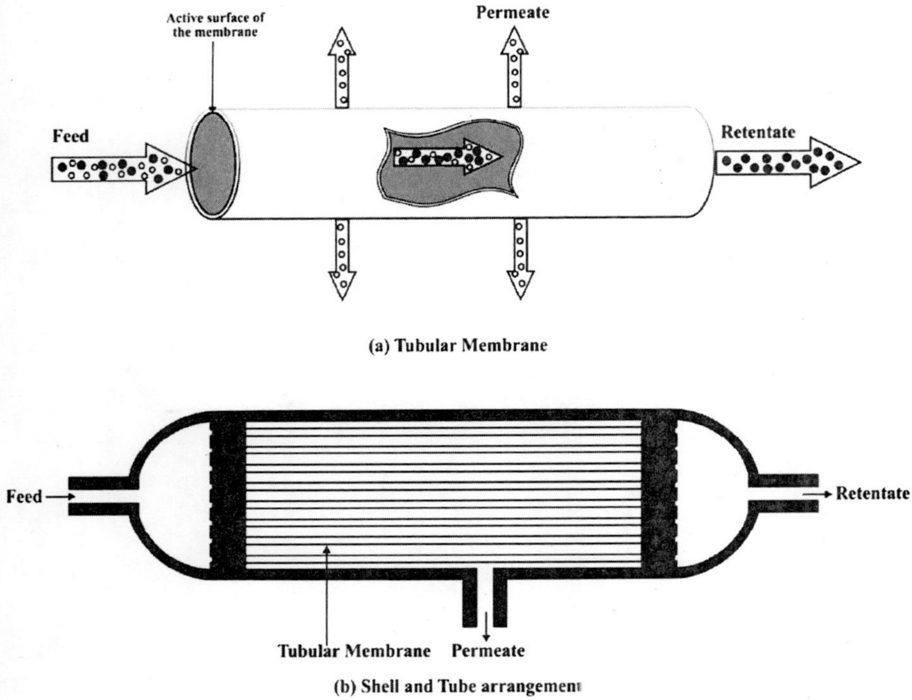

Fig. 11: Schematic diagram Tubular module

5.3 Hollow fiber module

Hollow fiber module is similar in design as tubular module. In this module, many tiny hollow fibers are bundled together and encased in a tube (shell). The hollow (fine) fibers have an internal diameter of 40–50 μm, an outer diameter of 85 μm, and a wall thickness of around 20-25 μm. The fibers are self-supporting and can withstand high pressure. Two different types of hollow fiber modules are used for RO and UF application. Shell-side feed modules are generally used for RO where high pressure upto 1000 psi (7 MPa) is applied. Fouling on the feed side of the membrane can be a problem with this design and pretreatment of the feed stream to remove particulates is required. Another module which is used for UF application is called bore-side feed where module the feed is pumped through the tube side and the permeate stream is withdrawn from the shell side (Fig. 12). This module, generally used for medium pressure feed streams upto 150 psi (1 MPa).

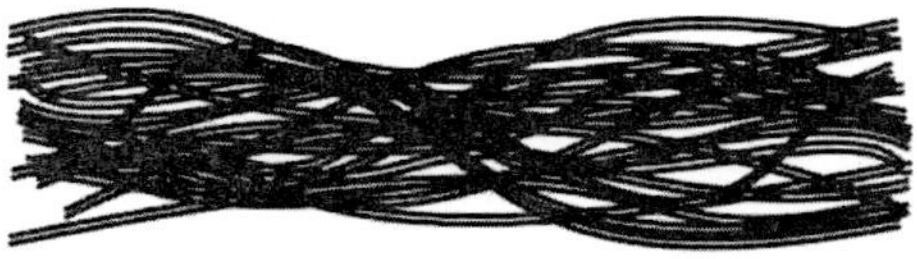

Tuft of Hollow fiber

Hollow fiber catridge

Fig. 12: Two module of hollow fibre membrane

5.4 Spiral wound module

This is one of the most compact and inexpensive designs available today. A simplified sketch of the basic concept is shown in Fig. 13. Two flat sheets of membrane are placed together with their active surfaces facing away from each other. They are separated from each other by a thin, mesh-like spacer materials called permeate collection material. Three edges of the two flat membrane sheets are glued to form a membrane envelope. Thus, the outer side of the envelope is actually the active surface of the membranes. The fourth open side of the envelope is connected to a central perforated tube. Another mesh-like feed spacer of ~25 mm thickness is placed on one side of this envelope. The whole membrane envelope-feed spacer composite is rolled into a spiral configuration around the central permeation collection tube. The membranes wound is placed inside a tubular pressure vessel forming the module which is also known as spiral wound cartridge. The number of membrane envelope in a module may vary from 4-30 depending on the diameter of the module.

Feed flows longitudinally in the feed channel, while permeate flows spirally through the permeate collection material to the permeate collection tube. Several individual spiral modules may be placed in series in a housing with the permeate tubes interconnected.

The feed channel spacers keep the wound membranes apart, forming the feed flow channel and are designed to promote turbulence, thus help to reduce concentration polarization at the membrane surface.

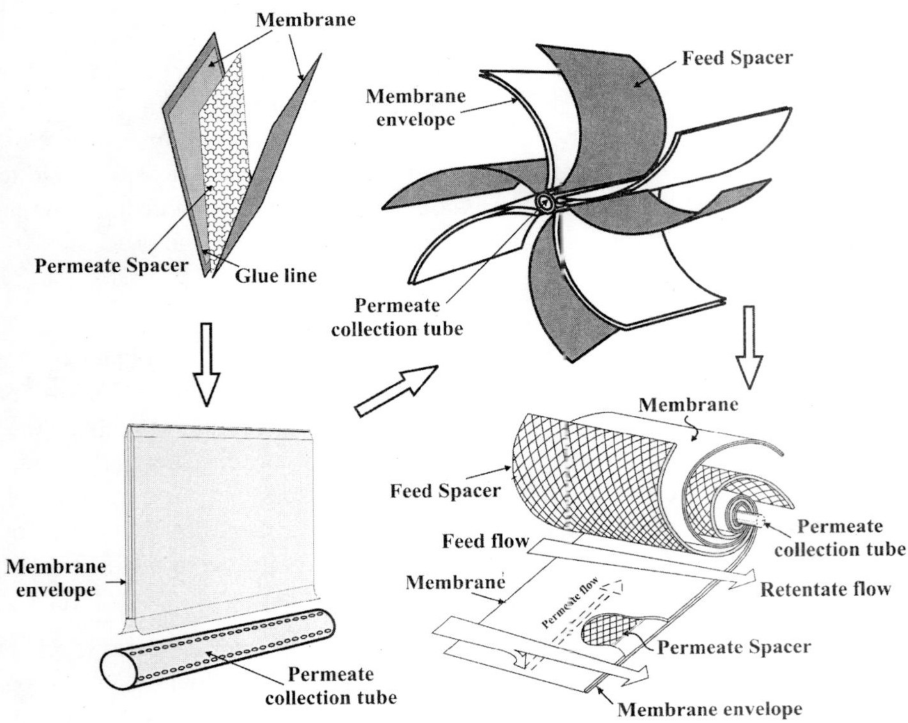

Fig. 13: Schematics of Spiral wound module

Table 6: Features of different membrane modules (Baker, 2004; Wagner, 2001)

Parameters	Plate-and-Frame	Tubular	Hollow fiber	Spiral wound
Prefiltration	Yes/No	No	No	Yes
Spacers	Yes/No	No	No	Yes
Flow pattern	Laminar	Turbulent	Laminar	Turbulent
Flux	Moderate	High	Moderate	low
Packing Density	Moderate	Low	High	High
Energy cost/Flux	Moderate	High	Moderate	Low
Resistance to fouling	Good	Very good	Very poor	Moderate
Ease of cleaning	Good	Excellent	Poor	Fair
Permeate side pressure drop	Low	Low	High	Moderate
Suitability to high pressure operation	Yes	Marginal	Yes	Yes
Limited to specific types of membrane materials	No	No	Yes	No

6. Concentration Polarization and Membrane Fouling

In membrane processes, a common problem encountered is a decline in flux through the membrane. The phenomena responsible for a decline in flux are concentration polarization and membrane fouling. These phenomena are most commonly encountered in pressure-driven membrane processes and found to affect the selectivity of the processes too. Managing this two phenomena is, therefore, very important for efficient separation by membrane processes.

6.1 Concentration Polarization

During membrane filtration, the feed solution is brought in contact of the membrane active surface by convective transport facilitated by application of pressure. The permeate (solvent) passes through the membrane leaving behind the retentate (solutes) at the membrane surface. This results in a higher local concentration of the solute at the membrane surface compared to the bulk feed, regardless of whether the solute are partially or completely rejected by the membrane. This phenomenon of increased concentration at the membrane surface is called "Concentration Polarization" (Fig. 14).

When the feed solution contains a larger proportion of hydrocolloids (such as proteins, gums etc.), concentration polarization creates an additional complication. Accumulation of these compounds in higher concentration at the membrane surface results in formation of a viscous and gelatinous layer over the membrane. This layer offers an additional resistance to the flow of permeates through the membrane.

In reverse osmosis, where the solute particles are relatively smaller and are osmotically active, accumulation of solutes increase the osmotic pressure of the solution at membrane surface. As the driving force in RO is the difference of applied pressure and osmotic pressure of solution ("P-"π), it is implied that there will be a drop in the driving force leading to reduction of flux.

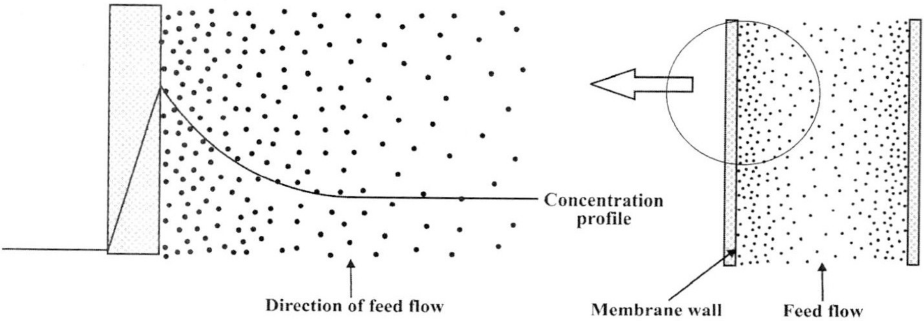

Fig. 14: Development of concentration polarization layer

6.2 Membrane Fouling

Membrane fouling is a process whereby a retentate particle is deposited on a membrane surface or in membrane pores. Membrane fouling can cause severe flux decline and affect the performance of a membrane process. Fouling may be due to adsorption of retentate particles by specific interactions with the membrane; closure of pores; formation of cake by deposition of particles in layers or due to gel layer formation at membrane surface (Fig. 15).

Fouling due to closure of pore, by the entry of particles in the tortuous pore channel, results in reduction of active membrane surface. However, a cake layer or gel layer on the surface of a membrane is an additional resistance that is in series with the membrane resistance.

Extent of fouling is depends upon several parameters such as nature and concentration of solutes and solvents; pore-size distribution and pore channel geometry; surface characteristics of membranes and its compatibility with the feed solution; and hydrodynamics of the membrane module i.e., flow regime of the feed. In addition, operational conditions such as pH, temperature, flow rate and pressure also greatly affect fouling.

Fouling can be divided into reversible and irreversible fouling based on the attachment strength of particles to the membrane surface. Reversible fouling can be removed by a strong shear force or backwashing. Formation of a strong matrix of fouling layer with the solute during a continuous filtration process will result in reversible fouling being transformed into an irreversible fouling layer. Irreversible fouling is the strong attachment of particles which cannot be removed by physical cleaning

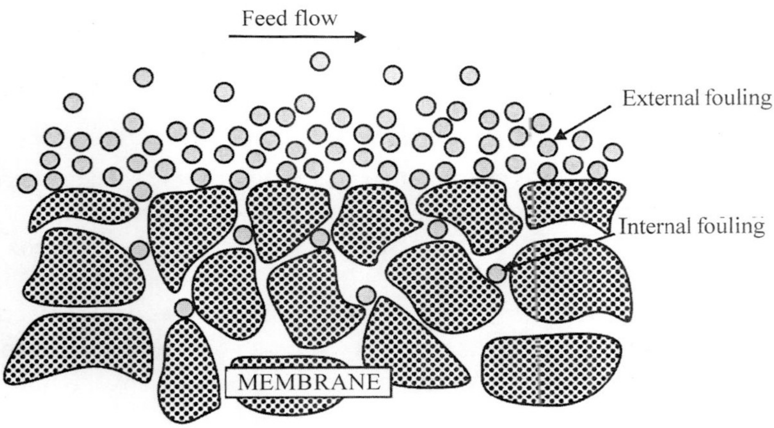

Fig. 15: External and internal Fouling of membrane

6.3 Managing Concentration polarization and membrane fouling

The major focus of equipment designers and users is to reduce the effect of concentration polarization and membrane fouling. The principal strategy to reduce the phenomenon of concentration polarization is to create turbulence at the membrane surface. Placement of wire-mesh as feed spacer, and introduction of scouring balls and static kenics mixing elements in large tubular units are some techniques to enhance turbulence at membrane surface.

Even though membrane fouling is inevitable during the filtration process, it can be controlled and alleviated. Fouling remediation can be done through pre-treatment of the feed to limit its fouling propensity, improving the antifouling properties of the membrane, membrane cleaning and backwash conditions and optimization of the operating conditions. Different approaches to prevent and reduce concentration polarization and membrane fouling are presented in Table 7.

Table 7: Approaches to prevent and reduce fouling (Field, 2010)

Direct methods	Indirect methods
• Turbulence promoters (e.g., modified membrane spacers)	• Pretreatment by filtration
• Pulsed or reverse flow	• Treatment of the membrane surface
• Rotating or vibrating membranes	• Preparation of more hydrophilic membranes
• Stirred cells with rotating blades close to the membrane	• Selection of appropriate operating mode
• Ultrasonic enhancement	• Selection of optimum operating conditions
• Periodic maintenance cleaning ✓ Chemical cleaning ✓ Hydraulic cleaning ✓ Mechanical cleaning	
• Periodic backwash with permeate or gas	
• Generation of a dynamic membrane layer	

7. Performance of membrane separation

Schematically, a single stage membrane separation process is represented in Fig. 16. A feed stream (F) with a concentration C_f of certain component is pumped into the system. A portion of the feed stream passes across the membrane, yielding a permeate stream (P) with a concentration C_p of the component of interest. Another portion of the feed stream does not cross through the membrane, yielding a retentate stream (S) with a concentration C_s of the component. The performance of the processes can be defined by the following terms.

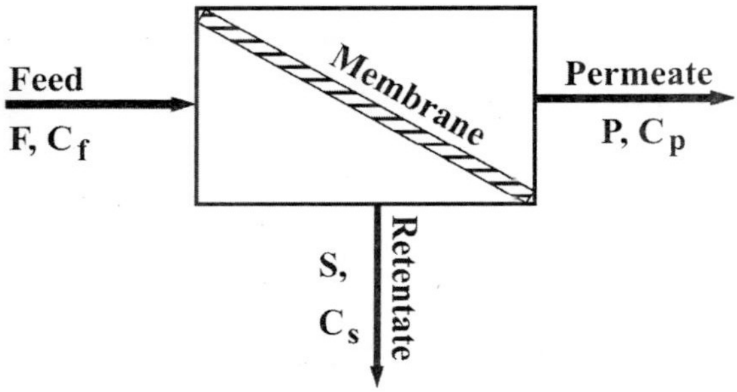

Fig. 16: Single stage of membrane separation

7.1 Rejection coefficient (*R*): It is a measure of the ability of the membrane to separate the component of interest and can be given by-

$$R = \left(1 - \frac{C_p}{C_f}\right)$$

For a perfectly selective membrane which does not allow the component to pass through, $C_p = 0$ and R=100% and for a membrane completely unselective to the component of interest, $C_p = C_f$ and R = 0%

7.2 Concentration factor (F_c): It is the ratio between the concentrations of a particular component in the retentate to the concentrations of that component in feed.

$$F_c = \frac{C_s}{C_f}$$

7.3 Recovery Factor (*Q*): It is the ratio of quantity of permeates obtained to the quantity of feed.

$$Q = \frac{P}{F}$$

7.4 Mass Flux (*J*): It is the amount of a particular component traverse across unit area of membrane per unit time.

7.4.1 *In Reverse osmosis membrane*, mass flux for solvent (J_w) in kg.s^{-1}m^2 is given by-

$$J_w = A\,(\Delta P - \Delta\pi)$$

Similarly, mass flux for solute (J_s) in kg.s^{-1}m^2 is given by-

$$J_s = B\,(C_f - C_p)$$

where, A and B are defined as *permeability of the membrane to the solvent* and *permeability of the membrane to the solute*, respectively; and are expressed as:

$$A = \frac{D_w C_w \bar{V}_w}{RT\delta_m} \text{ and } B = \frac{K_p D_s}{\delta_m}$$

where, ΔP = pressure difference across the membrane (Pa);

$\pi = \frac{n}{V} RT$ = osmotic pressure and $\Delta\pi$ = transmembrane osmotic pressure difference (Pa);

C_f = the concentrations of solute in feed (kg.m^{-3});

C_p = the concentrations of solute in permeate (kg.m^{-3});

D_w = diffusion coefficient of solvent in the membrane (m^2.s^{-1});

D_s = diffusion coefficient of solute in the membrane (m^2.s^{-1});

C_w = average concentration of solvent in the membrane (kg.m^{-3});

$\bar{V}_w$ = partial molar volume of the solvent (m^3.kmol^{-1});

R = the gas constant (8.314 m^3.Pa.K^{-1}.mol^{-1});

T = absolute temperature (°K);

K_p = partition/distribution coefficient (dimensionless); and

δ_m = the membrane thickness (m)

7.4.2 *In Ultrafiltration membrane*, mass flux for solvent (J_w) in kg.s^{-1}m^2 is given by-

$$J_w = \frac{\rho(\Delta P)}{\eta \ R_m}$$

Similarly, mass flux for solute (J_s) in kg.s^{-1}m^2 is given by-

$$J_s = \frac{\rho(\Delta P)}{\eta \ R_m} C_f \sigma$$

R_m is hydraulic resistance of the membrane and is defined as the reciprocal of the hydraulic permeability of the membrane (K_m). K_m is a membrane property and can be expressed by-

$$K_m = \frac{(D_e)^2 \varepsilon}{32\delta_m (\tau^o)^2}$$

σ is a dimensional constant with a value between 0 and 1. The equation to calculate this constant is given by the expressicn:

$\sigma = [2\ (-\chi)^2 - (1-\chi)^4]\ [1-2.104\chi\ \ 2.09\chi^3 - 0.95\chi^5]$

in which, χ is the ratio between the molecular diameter (D_m) of solute and the diameter of the cross section of the membrane channel (D_e); i.e., $\chi = D_m/D_e$.

ε = the porosity of the membrane (%);

D_e = the equivalent diameter of the channel (m),

δ_m = the thickness of the membrane (m);

τ^o = the tortuousness of the channel and is defined as the ration of channel length to membrane thickness ($\tau^o = \delta/\delta_m$) (dimensionless);

ρ = the density of the feed solution (kg.m^{-3});

η = the dynamic viscosity of the fluid (Pa.s);
ΔP= the pressure gradient across the membrane (Pa); and

C_f= the concentration of solute in feed (kg.m^{-3})

Numerical Problems

Problem-1: Calculate the osmotic pressure of a solution containing

(a) 1.0 g mol NaCl/kg of water at 25°C. (The density of the solution is 1036.2 kg.m^3)

(b) 11.7 g NaCl/litre of solution at 25°C.

(c) 10% (w/v) sucrose solution at 25°C.

Solution

(a) Osmotic pressure, $\pi = \frac{n}{V} RT$

$n = 2 \times 1 \times 10^{-3}$ kmol (NaCl gives 2 ions in solution)

$V = 1/1036.5 = 9.65 \times 10\text{-}^4\ m^3$

R= 8.314 × 10^3 m^3 m^3.Pa.K^{-1}.$kmol^{-1}$

T = 273.15+25=298.15 °K

$$\pi = \frac{n}{V} RT = \frac{2\times10^{-3}\times8.314\times10^{3}\times298.15}{9.65\times10^{-4}} = 5.137449\times10^{6}\ \text{Pa}$$

(b) $n = \frac{11.7}{58.5}\times2\times10^{-3} = 4\times10^{-4}$ kmol (Mol. wt. of NaCl is 58.5 and it gives 2 ions in solution)

V = 1*l* = 10^{-3} m^3

R = 8.314 × 10^3 m^3.Pa.K^{-1}.$kmol^{-1}$

T = 273.15+25=298.15 °K

$$\pi = \frac{n}{V} RT = \frac{4\times10^{-4}\times8.314\times10^{3}\times298.15}{10^{-3}} = 9.915\times10^{5}\ \text{Pa}$$

(c) $n = \frac{10}{342}\times1\times10^{-3} = 2.923\times10^{-5}$ kmol (Mol. wt. of sucrose is 342)

V = 100 m*l* = 10^{-4} m^3

R = 8.314 × 10^3 m^3.Pa.K^{-1}.$kmol^{-1}$

T = 273.15+25=298.15 °K

$$\pi = \frac{n}{V} RT = \frac{2.923\times10^{-5}\times8.314\times10^{3}\times298.15}{10^{-4}} = 7.246\times10^{5}\ \text{Pa}$$

Problem-2: In a reverse osmosis experiment conducted at 25°C, a cellulose acetate membrane with an area of 1.44 × 10^{-2} m^2 was used. A salt solution containing 15g NaCl/*l* of solution was passed through the membrane. The permeate solution contained 0.35 g NaCl/*l* (ρ = 997 g/*l*) and the measured flow rate of permeate was 2.98 × 10^{-7} m^3/s. The pressure difference was 5 × 10^6 Pa. Calculate the water and salt permeability constants and salt rejection coefficient.

Solution

Density of the permeate = 997g/1 = 997kg. m^{-3}

$$\text{Mass fraction of solvent in permeate} = \frac{997-0.35}{997} = \frac{996.65}{997}$$

Mass fraction of solvent in permeate $= \dfrac{0.35}{997}$

The mass flow rate of permeate = (2.98×10^{-7} m3.s^{-1}) × (997kg. m^{-3}) = 2.971×10^{-4}

$$\text{Mass flux of solvent}(J_w) = \frac{\text{Mass flow rate} \times \text{Mass fraction of solvent}}{\text{Area of membrane}} = \frac{2.971 \times 10^{-4} \times 996.65}{1.44 \times 10^{-2} \times 997}$$

$$= 2.062 \times 10^{-2} \text{ kg.s}^{-1}\text{.m}^{-2}$$

$$\text{Mass flux of solute}(J_s) = \frac{\text{Mass flow rate} \times \text{Mass fraction of solute}}{\text{Area of membrane}} = \frac{2.971 \times 10^{-4} \times 0.355}{1.44 \times 10^{-2} \times 997}$$

$$= 7.247 \times 10^{-6} \text{ kg.s}^{-1}\text{.m}^{-2}$$

$$\text{Osmotic pressure of feed} = \frac{15 \times 2 \times 10^{-3}}{58.5 \times 10^{-3}} \times 8.314 \times 10^{3} \times 298.15 = 1.271 \times 10^{6}\,\text{Pa}$$

$$\text{Osmotic pressure of permeate} = \frac{0.35 \times 2 \times 10^{-3}}{58.5 \times 10^{-3}} \times 8.314 \times 10^{3} \times 298.15 = 0.030 \times 10^{6}\,\text{Pa}$$

Osmotic pressure of perm eate = 1.271 – 0.030) × 10^6 = 1.241 × 10^6 Pa

$Jw = A\,(\Delta P - \Delta\pi)$

2.062 × 10–2 = A(5 × 10^6 – 1.241 × 10^6) = A × 3.759 × 10^6

$$\therefore A = \frac{2.062 \times 10^{-2}}{3.759 \times 10^{6}} = 5.486 \times 10^{-9}\,\text{kg.s}^{-1}\text{.m}^{-2}\text{.Pa}$$

Concentration of solute in feed (Cf) = 15 g.1^{-1} = 15kg.m^{-3}

Concentration of solute in permeate (Cp) = 0.35 g.1^{-1} = 0.35kg.m^{-3}

$Js = B\,(C_f - C_p)$

7.247×10^{-6} = B (15 – 0.35)

$$\therefore B = \frac{7.247 \times 10^{-6}}{14.65} = 4.947 \times 10^{-7}\,\text{m.s}^{-1}$$

$$\text{Rejection coefficient}(R) = \left(1 - \frac{C_p}{C_f}\right) = \left(1 - \frac{0.35}{15}\right) = 0.977$$

$$\text{Solvent permeability constant}(A) = 5.486\times10^{-9}\,\text{kg.s}^{-1}.\text{m}^{-2}.\text{Pa}^{-1}$$

$$\text{Solute permeability constant}(B) = 4.947\times10^{-7}\,\text{m.s}^{-1}$$

Problem-3: A salt solution containing 8.775 g NaCl/l entering a RO system at a temperature of 20°C at a volumetric flow rate of 12,000 l/h. The system operates at a differential pressure of 4.1 10^6 Pa. The membrane has a rejection coefficient of 0.94 for NaCl. The permeability constant for water is 3.2 10^{-4}. The system operates at a recovery rating of 91% for water. Calculate the flow rate and concentration of NaCl in permeate and retentate stream, flux of permeate and area of osmotic membrane required for the RO system.

Solution

$$\text{Recovery Factor}(Q) = \frac{Permeate}{Feed} = \frac{P}{12000} = 0.91$$

$$\therefore \text{Permeate}(P)\,\text{flow rate} = 12000\times0.91 = 10920\text{l.h}^{-1}$$

$$\therefore \text{Retentate}(S)\,\text{flow rate} = 12000-10920 = 1080\text{l.h}^{-1}$$

$$\text{Concentration of NaCl in feed}(C_f) = 8.775\text{g}/\text{l}$$

$$\text{Rejection coefficient}(R) = \left(1-\frac{C_p}{C_f}\right) = \left(1-\frac{C_p}{8.775}\right) = 0.94$$

$$\therefore C_p = 8.775\times(1-0.94) = 0.5265\text{g}/\text{l}$$

$$F\times C_f = P\times C_p + S\times C_s \text{ (mass balance equation for NaCl)}$$

$$\therefore 12000\times8.775 = 10920\times0.5265 + 1080\times C_s$$

$$\therefore C_s = \frac{12000\times8.775-10920\times0.5265}{1080} = \frac{99550.62}{1080} = 92.1765\text{g}/\text{l}$$

The operation being a continuous one, the average concentration of NaCl in feed side of the membrane will be the weighted average of Feed and retentate stream.

$$\text{Therefore, The average concentration of NaCl in feed side}(\text{C}_\text{c}) = \frac{12000\times8.775+1080\times92.1765}{12000+1080}$$

$$= 15.6614\text{g/l}$$

$$\text{Osmotic pressure in feed} = \frac{15.6614 \times 2 \times 10^{-3}}{58.5 \times 10^{-3}} \times 8.314 \times 10^3 \times 293.15 = 1.305 \times 10^6\,\text{Pa}$$

$$\text{Osmotic pressure in permeate} = \frac{0.5265 \times 2 \times 10^{-3}}{58.5 \times 10^{-3}} \times 8.314 \times 10^3 \times 293.15 = 0.044 \times 10^6\,\text{Pa}$$

Differential osmotic pressure = $(1.305 - 0.044) \times 10^6 = 1.261 \times 10^6$

$$\text{Permeability constant for water} = 3.2 \times 10^{-4}\,\text{kg.s}^{-1}.\text{m}^{-2}.\text{atm}^{-1} = \frac{3.2 \times 10^{-4}}{101325}\,\text{kg.s}^{-1}.\text{m}^{-2}.\text{Pa}^{-1}$$

$$= 3.16 \times 10^{-9}\ \text{kg. s}^{-1}.\ \text{m}^{-2}.\ \text{Pa}^{-1}$$

Flux of water (Jw) = A ($\Delta P - \Delta\pi$) = $3.16 \times 10^{-9}\,(4.1 \times 10^6 - 1.261 \times 10^6)\text{kg.s}^{-1}.\ \text{m}^{-2}$

$$= 8.97 \times 10^{-3}\ \text{kg.s}^{-1}.\ \text{m}^{-2}$$

= $8.97 \times 10^{-3}\ \text{l.s}^{-1}.\ \text{m}^{-2}$(considering density of water to be 1 kg/l)

$$\text{Area of membrane} = \frac{\text{Flow rate}}{\text{Flux}} = \frac{10920\text{l.h}^{-1}}{8.97 \times 10^{-3}\text{l.s}^{-1}.\text{m}^{-2}} = \frac{10920}{3600 \times 8.97 \times 10^{-3}}\,\text{m}^2 = 338.16\text{m}^2$$

References

Baker, RW. (2004). Membranes and modules. In *Membrane Technology and Applications, Second edition*. John Wiley & Sons Ltd, West Sussex, England pp 89-160.

Field, R. (2010). Fundamentals of Fouling. In *Membrane Technology: Membranes for Water Treatment, Volume 4* (eds Peinemann, K.V. and Pereira Nunes, S.), Wiley-VCH Verlag GmbH & Co. KGaA, Weinheim, Germany. doi: 10.1002/9783527631407.ch1, pp1-22.

Kujawski, W. (2000). Application of Pervaporation and Vapor Permeation in Environmental Protection. *Polish Journal of Environmental Studies*. 9(1): 13-26.

Onsekizoglu, P. (2012). Membrane Distillation: Principle, Advances, Limitations and Future Prospects in Food Industry. InTech. Available online at http://www.intechopen.com/ articles/ show/title/membrane-distillation-principle-advances-limitations-and-future-prospects-in-food-industry.

Sablani, SS. (2007). Food Preservation and Processing Using Membranes. In *Handbook of Food Preservation, Second Edition, (ed.* Rahman, MS). CRC press. Pp. 365-385.

Ulbricht, M. (2006). Advanced functional polymer membranes. Polymer. 47(7): 2217-2262.

Wagner, J. (2001). *Membrane Filtration Handbook Practical Tips and Hints, Second Edition.* Osmonics, Inc. p-20.

Wang, H and Zhou, H. (2013). Understand the Basics of Membrane Filtration. *Chemical Engineering Progress.* 109(4):33-40.

19

Food Extrusion

Hridyesh Pandey, Harah P. Sharma, D.C. Joshi and H.N. Mishra

1. Introduction

Extrusion of foods is an emerging technology for the food industries to process and market a large number of products of varying size, shape, texture and taste. Extrusion cooking of cereals is becoming more adaptive due to its various advantages over the traditional food processing techniques (White, 1994). Nowadays the extrusion of food is of major importance in the food-processing sector and it is the only process, which covers a wide range of food products such as ready to eat (RTE) breakfast cereals, snack foods, baby foods etc. Elaborate scientific research, technological development and engineering design, which have been done in this field so far, have led to large-scale industrial production. The main feature of extrusion cooking are that it is versatile process, energy efficient, cost effective and with greater quality and product throughput. Though, the knowledge of plastic extrusion was brought to the realm of food processing. The extrusion cooker is finding application in so many diverse operations that it is being increasingly regarded as a versatile, high temperature short time (HTST) bioreactor.

A simple definition of extrusion is shaping by forcing softened or plasticised material through a die by pressure. The process of extrusion is carried out in extruders, which are based on the principle of the Archimedes screw. The screw is housed in a metal barrel, which can be of variable length and through which cooling water or steam can circulate. Various unit operations like mixing, cooking kneading, shearing, shaping and forming occur in single machine. During extrusion, the cooking temperature could be as high as 180–190°C, but residence time is usually 20–40 seconds (depends on types of extruders and speed of the shaft). For this reason the extrusion process could be called as HTST process.

2. Extrusion Cooking

It is the method of forming substances by forcing them through a small opening known as die to produce product like tubes, rods or any other desired shape.

Food extruder is a device that expedites the shaping process for food ingredients and can be used to cook, form, mix, texturize and shape product under conditions that favor quality retention. Several designs are possible for food extruder. The commonly used extruder contains flighted screws, which pushes the plasticized material forward and creates pressure before discharge end so that it extrudes through die opening. Extruder cooking is accomplished through the application of heat, either directly by steam injection or indirectly through jackets occurring within the dough. Cooking of food ingredients during extrusion results in inactivation of many enzymes, protein denaturation, gelatinization of starch etc (Rossi and Peri, 1980). Other properties such as extrudate expansion and texture of the product are also altered by the interaction of shear heat and moisture in the extruder.

3. Some Terminologies Relevant to Extrusion

Barrel opening (D_b) - This dimension is usually referred to as a barrel opening in which the screw rotates.

Root diameter (D_r) - The diameter of the root of the screw on which the flights are built is the base of the shaft that carries the rotational torque (Figure 1).

Screw clearance (2d) - is the difference between the diameter of the screw and the barrel opening (Figure 1).

Flight height (H) - is the distance between the diameter height of the flight and the diameter of the root of the screw (Figure 1).

Pitch - is the axial distance between the leading edge of the flight at the outside diameter and the leading edge of the same flight in front (Figure 1).

Channel - The helical opening that extends from the feed to the discharge end of the screw. A channel is formed by the inner side of the flight and the top of the screw root diameter and the surface of the barrel opening.

Axial channel width (W)- The channel width is measured from one side of the flight to the next within the channel perpendicular to the angle of the flight (Figure 1).

Axial flight width - is the width of a screw flight in the axial direction. The axial flight width is related to the lead and the axial distance between the flights (Fig. 1).

Flight width - is the width of the screw flight measured perpendicular to the face of the flight, which is sometimes referred to as the thickness of the flight shoulder.

Channel length (Z) - is the length of the screw channel in the Z direction, which can be one or more full turn of the screw helix.

Compression ratio (CR) - This is the ratio obtained by dividing the available feed volume of the screw to the next screw volume developing the pressure needed to process the raw material.

Barrel length to diameter ratio (L/Ds): This is the ratio of the screw diameter to the length of the barrel. If the extruder barrel is in segments, one segment is usually referred as the ratio L/Ds.

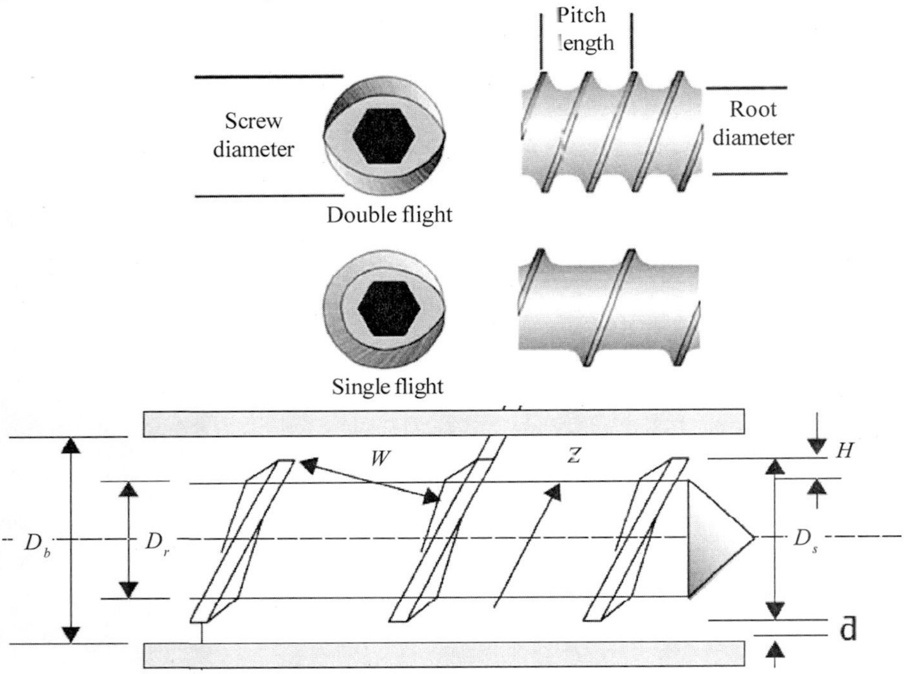

Fig. 1: Screw Terminology (Source: Riaz, 1981)

4. Types of Food Extrusion

There are two main types of extrusion -(i) Hot or High Temperature Extrusion and (ii) Cold Extrusion

4.1 Hot/High temperature extrusion is very commonly used process in the food industry. That is the reason that the extrusion cooking is known as high temperature short time process. In this type of extrusion cooking gelatinization, protein denaturation and enzyme inactivation takes place. All other undesirable flavors are also removed by this process.

4.2 Cold extrusion In this process the product is kept at low temperatures during its passage through the extruder. It could be used to prepare comminuted meat product batters for products such as sausages, frankfurters and other snack products. Using cold extrusion, the traditional mixing and size reduction operations could be performed in a single continuous process rather than as discrete batch operations. Production costs are lesser as compared to hot extrusion process due to the simplicity of the tools needed for it (Kaur et al., 2007).

5. Merits and Demerits of Extrusion Cooking

Various advantages of hot extrusion process are:

a) Development and creation of new food is possible.

b) A wide variety of food with varying texture, size and shape can be produced.

c) It can handle different raw materials of different origin such as plant, animal and microbial source with varying moisture content.

d) Extruder has greater production capacity than the other cooking systems. It can process high amount of finished products per hour (for e.g. as high as over sixteen tons per hour).

e) Extrusion cooking provides high product quality than any other traditional food process because it is having good control over the process parameters.

f) Extrusion provides better mixing of the raw ingredients as well as gives mores flexibility to the product.

g) Extrusion cooking gives more uniformity and consistency to the finished product.

h) Floor space and labor requirements per unit of production are smaller and its energy efficiency results in the low cost of the final product.

i) Extrusion cooking does not offer any process effluents or hazardous cooking waste and provides continuous process facility.

Some of the major disadvantages of extrusion process are

a) Extrusion process is very complex to understand.

b) Sometimes is very difficult to clean and also takes more time for cleaning because of the loose dough consistency.

6. Mechanism of expansion in extrusion

Expansion phenomenon occurs as the transformation of feed stocks such as maize grits, wheat, rice or potato flour, into hot melt fluids inside the extruder barrel, which can be expanded as they emerge from a die. The hot fluid comes out of the machine and its pressure is lowered to atmospheric pressure, this pressure drop causes an extensive flash-off of internal moisture. This high-energy process enables bubbles to be created and expanded within the melt to form continuous foam. The expansion continued until the cell walls reached the limit of their extensibility and ruptured releasing the gas pressure as shown in Fig. 2.

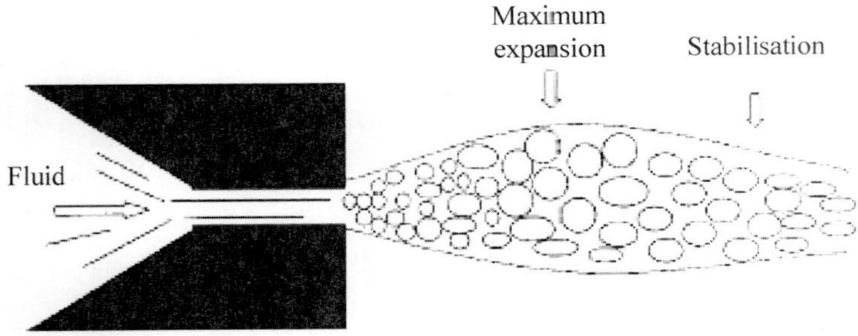

Fig. 2: Mechanism of extrusion of a molten fluid coming out of extruder (Source: Guy, 2001)

Based on visual observations, Fan et al. (1994) suggested that melt expansion at the die consists in an expansion phase followed by a shrinkage phase. The expansion mechanism can be described by the dynamics of bubble growth in a viscous melt (Arhaliass et al., 2009) that assumes the melt as a bubble network surrounded by dough.

7. Components of Food Extruder

A typical extruder is shown in Fig. 3. The common features of an extrusion system are:

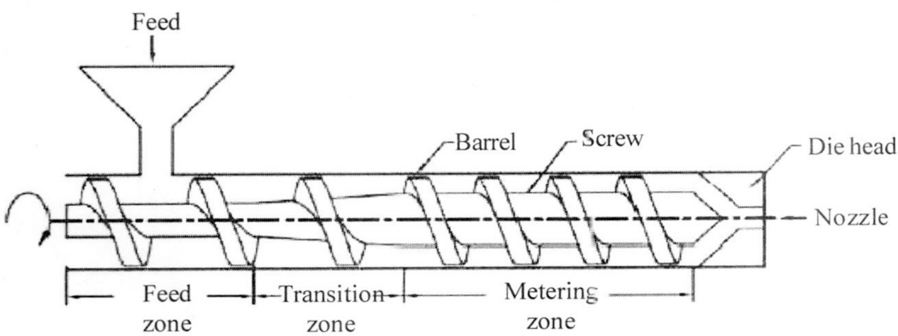

Fig. 3: A single screw extruder (Source: Riyaz, 1981)

7.1 Extruder screw

The function of the screw is to convey the product through the extruder. The geometry of the extruder screw is complex in nature due to its helical shape. The screw profile consists of three zones-

a) Feeding zone: - the feed is compressed in this zone.

b) Transition zone: - the function of this zone is to convey the compressed feed to metering zone.

c) Metering zone: - It receives the compressed feed, homogenizes it and forces it through the die at constant pressure.

7.2 Extruder barrel

It is a pipe like structure in which the extruder screw turns. The clearance between screw channel and barrel should be minimum to reduce the backward leakage flow.

7.3 Die

The main function of the die is to shape the product. It is placed at the downstream of the flow and usually contains small nozzle type openings. The shape and discharge area of the die directly influences the shape and texture of finished product.

7.4 Extruder Feeder

It is one of the pressure feeding device, which contains a compressing screw. These screws deliver dry mix for extrusion from the extruder bin at a variable speed. To achieve this, drying ingredients are usually free flowing and have high quality attributes.

7.5 Temperature control devices

The temperature of the different zones of an extruder and die is usually maintained at different levels. Mechanical energy inputs inside extruder also increases product temperature. Direct or indirect heating by electric heaters helps in controlling temperature. Increasing steam flow rate also increases the temperature.

7.6 Heating arrangements

During extrusion process, direct or indirect heating is usually supplied through electrical heaters mounted on the barrels to feed. In some extruders there is no external heating, so in such cases the heating is due to the mechanical dissipation by viscous forces. Sometimes heating is also carried out by steam injection.

7.7 Driving arrangements

The extruder drive usually consists of an electric motor, a speed reduction system, and a torque transfer system. Some high-speed extruder systems have the main drive motors directly coupled to extruder shaft with some type of torque to protect against high loads.

7.8 Cutter

Assembly that cuts extrudate into pieces of desired length. The cutter assembly is attached to the center of die plate. The cutter shaft is stationary and blades sweep across surface of each die to cut the product into collets.

8. Extruder Classifications

The extruders can be classified on the basis of extent of shear, heating and cooking, and screw type and their arrangement. Some of these classifications are discussed as follows:-

8.1 Classification on the basis of shear

a) **Cold forming extruder-** These are low-shear machines, with smooth barrels, deep flights and low screw speeds generally used for semolina flour, press it through a die with little cooling (Harper, 1981).

b) **High-press forming extruders-** Main function is cooking and forming at temperature greater than 100 °C. Low-shear machines with grooved barrels, compressing screws, used to extrudate pre gelatinized cereal and dough through dies to make pellets for subsequent drying and puffing.

c) **Low-shear cooking extruders-** Main function is cooking and forming at temperature less than 100 °C. Moderate shear machines with high compression screws and grooved barrels to enhance mixing. Heat can be applied to barrel or screw to "cook" the product, denature proteins, inactivate enzymes and starch but puffing is avoided. This machine can make soft-moist foods and meat like snacks such as jerky.

d) **Collet extruders-** These are high shear machines with grooved barrels and screws with multiple shallow flights have been used for making puffed snacks for defatted corn grits. The resulting mass losses moisture and puffs immediately through a die to form a crisp collet (Harper, 1986).

e) **High shear cooking extruder-** These are used as one-step preparation of ready to eat cereals, snack foods, texturization of soy proteins and wafers.

8.2 Classification Based on Heat Generation

a) Polytropic extruders

These types of extruders have provision for alternately adding or removing heat as required by specific process. Example includes cooking extruder with external heating and cooking sections (Rossen and Miller, 1973).

b) Isothermal extruder

They operate at an essentially constant product temperature throughout the entire length of barrel where the water-cooled jackets are also used.

8.3 Classification on the basis of Screw

a) **Single screw extruder** it consists of only one screw which is rotating in the cylindrical barrel. The screw is having deep flights.

b) **Twin screw extruder** was invented to overcome the limitation of the single screw extruder. Although the single screw extruder is most commonly used in the food industry but the twin screw extruder is having more advantages. Twin screw extruders they provide more flexibility and also control the process parameters very easily. A typical food extruder available at College of FPTBE, AAU, Anand is shown in Fig. 4.

Twin-screw extruders have an enhanced ability to convey food materials down the screws length and have several other advantages (Clark, 1978). This is the reason that twin-screw extruders on the commercial scale have replaced most of the single screw extruders.

The twin-screw extruders can be further classified into following on the basis of their screw arrangement:

(i) Counter-rotating twin-screw extruders

These types of extruders are not widely used in the food industry although they are excellent conveyors. They are good in processing for relatively non-viscous materials requiring low speed and long retention times.

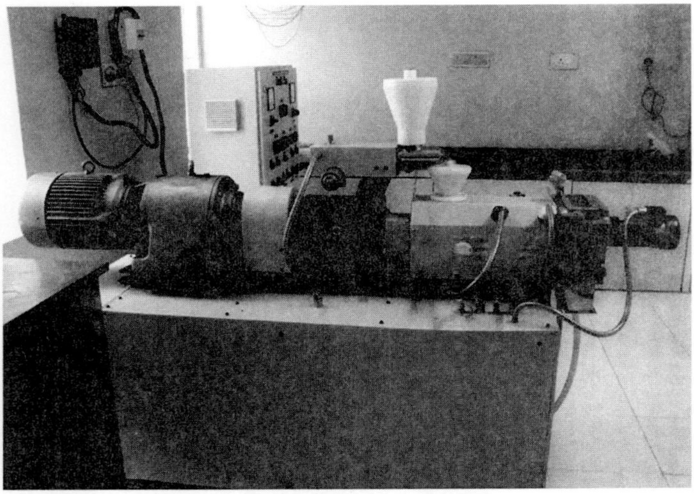

Fig. 4: A typical twin screw extruder

(ii) Co-rotating twin-screw extruders

These are most commonly used in the food and snack food industry and have played a major role in broadening the variety of products that can be made using extrusion technology. High degree of heat transfer can be achieved in these types of extruders (Elsner and Wiedmann, 1985). The system is reported to have better pumping efficiency, good control over residence time distribution, self-cleaning mechanism and uniformity of processing (Schuler, 1986). The screw either rotates in the same or opposite directions. They are further classified on the basis of the positions of the screws and can be given as:

Intermeshing Screw

In these types of extruders, the flight of one screw penetrates the channel of other. Positive pumping action, efficient mixing, and self-cleaning characteristics are taken into consideration. These type of extruders handle viscous, oily, sticky and some products, which will slip in single screw extruder. These extruders have positive pumping action and reduced pulsation at the die.

Non-intermeshing screw

In these extruders, the screws don't engage each other's threads, allowing one screw to twin without interfering with the other. Non-intermeshing screw extruder has two single-screw extruders lying side by side with a small position of barrel (Dziezak, 1989).

The counter-rotating intermeshing screw extruder is best suitable for low viscosity materials and also operates at low screw speeds than co-rotating (Weidmann and Strobel, 1987).

9. Comparison Between Single Screw and Twin Screw Extruder

Following comparison between cold and hot extrusion gives an idea for the selection of single and twin screw extruder to the processor according to their need:

Single screw extruder	Twin screw extruder
a) It contains only one screw	a) It contains two screws either rotating in same direction or in opposite direction.
b) It is used for small samples.	b) It can be used for large samples
c) The product is conveyed by the friction between the barrel and the screw	c) The product is conveyed by the help of double screws.
d) Poor mixing	d) Better mixing
e) Self cleaning is not possible.	e) It is having self wiping features
f) Used for measurement of rheological properties	f) Extrusion of biodegradable products

10. Machine Parameters Affecting the Quality of RTE Extruded Snack

Extruder operators may select parameters such as screw speed, feed moisture, and barrel temperature as primary factors, which in turn determine a secondary set of factors: specific mechanical energy, product or mass temperature and pressure (Meuser and van Lengerich, 1984). These factors influence the viscosity of the food within the extruder barrel, the residence time of the material in the extruder, and the shear applied to the food.

10.1 Effect of Screw Speed and Die Configuration on Product Quality

The screw speed, temperature and moisture content are found to have significant effect on the hardness. β- coefficients show that feed moisture content has a positive effect on extrudate hardness, while screw speed and temperature have negative effects. Feed moisture content is found to have the most significant effect on extrudate hardness, with relative effect strength of 0·6567 (Brinčić et al., 2006). Higher screw speeds may be expected to lower melt viscosity of the mix increasing the elasticity of the dough, resulting in a reduction in the density of the extrudate (Fletcher et al., 1985). Pathania et al. (2013) reported that increase in screw speed and barrel temperature significantly decreased water absorption index (WAI). Higher moisture content in extrusion process can diminish protein denaturation which subsequently lowers water solubility index (WSI) (Badrie and Mellowes, 1991).

Mixtures of maize grits and soybean hull were extruded in a single screw extruder to produce directly expanded extrudates, and the effects of three independent variables, screw speed (100-200 rpm), temperature (100-200°C) and soybean hull content (0-40%) were investigated by Duarte et al. (2009) using a central composite rotatable design and response surface methodology. The response variables used in this study were: specific mechanical energy (SME), sectional expansion index (SEI) and the paste viscosity readings (cold viscosity, peak value, breakdown and setback). The interaction between soybean hull and temperature increased breakdown and reduced setback. SME and SEI decreased as the addition of soybean hull and the temperature increased. No significant effect of increasing screw speed on SME and SEI was observed in this study.

10.2 Effect of Temperature on the End Product

High temperature were required to achieve a significant expansion and reduced density during extrusion cooking in wheat flour (Chiangand and Johnson).The presence of wheat proteins has been shown to have an effect on the expansion of wheat starch. It could be expected that the presence of protein (12.6%) in the wheat flour and fat used in the study would have an influence on the density and expansion observed in the extrudate.

An increase in the barrel temperature will increase the degree of superheating of water in the extruder encouraging bubble formation and also a decrease in melt viscosity (Fletcher et al., 1985) leading to reduced density. The expansion ratio increases significantly with decrease in feed moisture content and increase in screw speed and barrel temperature (Pathania et al., 2013).

10.3 Specific energy consumption.

It is important to understand and interpret the mathematical definition when considering using specific energy consumption (S) in judging quality of product. This value relates to the specific work input from the motor to the material being extruded. It can be obtained using following equation:

$$S = \frac{\text{motor power used}}{\text{output}} \quad (1)$$

$$S = \frac{\text{N x T x}K_w}{\text{output}} \quad (2)$$

where N = screw speed, T = torque, Kw = motor power and Q = output (kg/h).

11 Process Modeling of Extrusion

As the extrusion involves simultaneous mixing, kneading and cooking, it causes a large number of complex changes to a food, including hydration and plastification of starches and proteins, homogenisation, gelation, shearing, denaturation or re-orientation of proteins, and expansion of the food structure. Therefore, the scientific theory of the sequence and nature of these interactions and their effects is also complex. However, computer modelling of fluid flow behaviour and heat transfer inside the extruder barrel has more recently led to a greater understanding of the operation of extruders (Kulshreshtha et al. (1995).

11.1 Modeling Approaches for Single and Twin Screw Extrusion

A typical extruder consists of a barrel inside which one or more helical screws rotate to propel the feed material towards a die opening at the discharge end of the extruder. In food extrusion field, application of model to the processes began less than three decades ago (Frickle et al., 1977). Modeling food extruder depends heavily on developments in plastic extrusion. However, it is important to appreciate the similarities and differences in feed materials. Plastics are more homogeneous in nature that can be characterized chemically and physically. On the other hand, food extrusion uses a variety of food ingredients as feed materials. These materials are biopolymers of starch or protein, which varies in their source, age, prior treatment and thus cause extensive alteration in the chemical and physical nature of food extrudates. Modeling in food extrusion is much more complex process than plastic extrusion.

Most studies concerning modeling in food extrusion can be classified into six as black-box modeling, comprehensive modeling, response surface modeling (RSM), fuzzy modeling, residence time distribution (RTD) curves and finite element approach. The easiest and most widely used approach is done by response surface methodology. This method allows one to establish a mathematical equation in terms of operating conditions, such as moisture content, screw speed, barrel temperature, feed rate, formulation, etc. (Olkku and Vainionpaa, 1980; Antila et al., 1983; Fletcher et al., 1985). Only a limited number of studies have been reported on modeling system parameters from process parameters (Tayeb et al. 1988) or building correlations between system parameters and target parameters (Yacu, 1985; Mueser et al 1987). These results clearly are product- and machine-specific, and the conclusion is limited within the scope of the investigation.

Mueser et al (1987) proposed a system analytical model for the extrusion cooking of starch. This model distinguishes between influencing process and system parameters as well as influenced target parameters. Process parameters are the operating conditions which can be controlled and manipulated directly. System parameters are the properties which are influenced by the process parameters, and subsequently affect product characteristics (target parameters). It is believed that there is an appropriate function to describe the relationship between the process parameters and the system parameters, or between the system parameters and the target parameters.

It is necessary to classify extrusion process variables into manipulated, controlled, and load variables. Since final product attributes such as density, texture, degree of cooking, etc. themselves cannot be continuously monitored on-line, they have to be correlated to measurable process variables such as pressure at the die, temperature at the die and the motor torque developed, which is indicative of the specific mechanical energy (SME) input (Singh and Mulvaney, 1994). A multiple input multiple-output (MIMO) model, based on system analytical model as suggested by Meuser et al. (1987), was developed by Singh and Mulvaney (1994) to control the twin screw cooking process.

Two common approaches have been followed in modelling extrusion operations: dynamic modeling and steady state modeling. Dynamic modeling (Levine et al., 1986) describes the reaction of a process immediately after a perturbation (10–15 sec) and is particularly useful for control and automation, while steady state modeling describes the state of the process after a period long enough for machine stabilization. Between dynamic and steady state models lies the domain of long period (a few minutes) instabilities and metastable states (Roberts and Guy 1987) that probably can be explained by qualitative models.

The steady state model presented by Kulshreshtha et al (1991) and its dynamic version, Kulshreshtha and Zaror (1992), represent one of the recent approaches in one-dimensional mechanistic modelling for twin-screw cooking extrusion (Elsey et al, 1997). Both the steady state and the dynamic model are based on elemental heat and energy balances. Given the screw speed, feed rate, moisture content, feed temperature and barrel temperature profile this model calculates the temperature and pressure profiles along the barrel and the overall shaft-power required.

Another ideal model, which has been used by several authors (with different degree of success) to scale up both single and double screw food extruders, considers the flow as isothermic and Newtonian among other assumptions (Linko et al., 1981). In order to obtain the same product characteristics, some of the so-called primary criteria (such as residence time, mixing intensity and shear, product temperature and die pressure) have to be kept constant in the scaling up process Gonzalez et al (2001).

Though regression techniques are commonly used, difficulties arise when dealing with the complex characteristics of some systems. Regression is usually limited to linear and static systems, and conventional nonlinear regression algorithms are clumsy when handling systems like the extrusion process with multiple inputs and outputs. One limitation to traditional mathematical modeling is that the mathematical relationships describing each process of the system must be closely approximated to obtain good results. One of the newer approaches in modelling of food extruder is application of neural network for a multiple input and multiple output (MIMO) system. Dynamic changes in torque, specific mechanical energy, and pressure were modelled using Neural network (NN) and models were developed for prediction of individual product properties, which performed better than the regression models. Multiple input and multiple output (MIMO) models were developed to simultaneously predict five product properties or three product properties from three input parameters (Ganjyal et al., 2006).

A more recent approach to research in modeling of food extrusion is the 3-dimensional flow modeling using finite element computer package. Works on either 2 or 3-dimensional finite element analysis for food extrusion in literature are limited. Adekola (1998) carried out on 2-dimensional flow simulation in twin-screw food extruder die using finite element analysis.

11.2 Fluid flow in extrusion

The flow regime is envisaged as being composed of three distinct types of flow: drag flow, pressure flow, and net flow. The drag flow between the barrel and screw for Newtonian flow can be obtained using following equation:

$$v(y) = Vz\frac{y}{H} \tag{3}$$

$$Q_P = \frac{-WH^3\Delta P}{12\mu\ L} \equiv \frac{-\beta\Delta P}{\mu} \tag{4}$$

Where, W is the channel width, H is the channel depth and V_z is the flow velocity. The drag flow effectively pumps the food through the extruder.

Extruders usually have some flow restriction (like a die) at the end of the extruder. This creates a pressure gradient along the screw that works against the flow through the screw, which is expressed as:

$$Q_D = \frac{WV_Z}{H}\int_0^H y\,dy \equiv \frac{WV_Z}{H}\frac{H^2}{2} \equiv \frac{WV_Z H}{2} \tag{5}$$

The net volumetric flow rate through the extruder is sum of pressure flow and drag flow-

$$Q = Q_p + Q_D \tag{6}$$

The literature survey reveals that food extrudates are highly non-Newtonian, exhibiting shear-thinning behavior, where the apparent viscosity of a material decreases with increasing shear. The shear viscosity η can be a function of local shear rate, moisture contents, temperature and pressure. For the fluid under investigation, a power law model can be employed to describe the cereals rheology on the basis of available experimental data (Harper, 1978):

$$\eta = K\gamma^{m-1} \exp\frac{K_n}{T}\ exp(-K_m C) \tag{7}$$

In the above equation, K is the consistency, m is the power law index, T is the temperature, while C is the content of moisture. The above model neglects extruder leakage flows. Similar to pressure flow, these flows tend to decrease extruder output.

11.3 Mixing in Extruder

For mixing in a twin-screw extruder, the power number is

$$P_0 = \frac{E_V}{\mu N^3 D^5} \tag{8}$$

In the above equation, Po is the power number, ρ is the density of fluid, D is the diameter of the extruder, N is the speed of rotation and E is the energy supplied. The above equation is valid for no-pressure flow. In extrusion, the pressure flow is typically very small relative to the rotational flow.

11.4 Heat Transfer Modeling in Extruder

Knowledge of the heat transfer encountered in food extrusion is essential for the scale-up and design of food extruder and temperature control systems. Heat transfer coefficient is essential for food extrusion modeling, scale up and design of food extruder. Various models for heat transfer in extruder had already been reviewed by Van Zuilichem et al. (1990). Some of the important out of these include-

11.4.1. Penetration Theory Based Model

The theory of heat penetration was developed by Jepson (1953). This theory considered that after the flight of the rotating screw has wiped a certain area on the inner barrel surface a fresh layer of melts becomes attached to same region and remains there for one revolution in twin-screw extruder. The amount of heat penetrating into this layer during that time by pure conductive heat transfer is then removed with the polymer layer and is thought to be homogeneously distributed throughout the bulk of the polymer in the screw channel. This process ofheat penetration during extrusion can often be considered as a non-stationary penetration in a semi-infinite medium.

11.4.2 Yacu Model

The model as described by Yacu (1985) is a simulation of a twin-screw co-rotating food extruder. The extruder is divided into three main sections: (a) solid conveying zone; (b) melt pumping zone; (c) melt shearing zone (achieved by reverse screw elements). In these sections, the temperature and pressure profiles are predicted from knowledge of feed and extruder characteristics. The following assumptions were made in trying to simulate these three sections- (i) the food melt rheology is described by a non-newtonian, nonisothermal viscosity model, taking into account the effect of moisture and fat content; (ii) steady-state behaviour and uniform conditions exist; (iii) the melt flow is highly viscous and in the laminar flow regime; (iv) gravity effects are negligible and ignored; (v) heat losses to the screw shafts are neglected; (vi) only the net flow in the axial direction is considered. Detailed mathematical equations for different section had been presented in Yacu (1985) and Van Zuilichem et al (1990). Although the most thorough description of non-newtonian behaviour had been given by Yacu (1985) but the value of heat transfer coefficient found by it was much lower than predicted by the penetration theory.

12. Applications of Extrusion Technology

12.1 Breakfast Cereals

They are also known as ready to eat cereals. Breakfast cereals are most commonly consumed in the form of snacks or sweet snacks. They are nutritionally very sound. Direct expanded, light density snacks, starch; breading and breakfast cereals are examples of high-shear products. In the production of breakfast cereals mainly single screw extruders are used.

Two types of extrusion cooking processes are very common. One is direct expanded extrusion cooked breakfast cereals and another is pellet to flacks extrusion cooked breakfast cereals. In the direct expanded breakfast cereals the flour is generally mixed with water below 20% and the process used may be either single screw or twin screw. In the pellet to flakes extrusion cooked breakfast cereals, the flour is mixed with water upto 22-26%. For this process only twin screw extruders are used.

Breakfast cereals are manufactured to satisfy the major population i.e. children, nutritional and health conscious consumers and fitness oriented consumers due to its nutritional advantage so regarded as fortified expanded products. Nowadays breakfast cereals are available in the market in the form of flaked cereals, puffed cereals and cereal mixes. Commonly used raw materials for ready to eat breakfast cereals are: yellow corn flour meals, wheat starch, sugar, colors, salt and barley malt extract. Extrusion cooking of breakfast cereals is known as the thermo mechanical cooking because it requires temperature in the range of 140-180^{0}C, time in the range of 20-60s and moisture content below 26%. By varying these conditions new product can be developed.

12.2 Snacks

A traditional technology grew up in the Americas where maize or Indian corn was a major cereal crop. The grains of maize were cooked in water until they softened and could be separated from their hulls and milled into dough. This dough was rolled into thin layers, cut into pieces and baked or fried to form tortilla chips (Sharma, 1998). These traditional products were dry and crispy and pleasant to eat. For the modern industry there are several forms of snack products, which include formed dough products from maize derivatives, half-product or pellet snacks, biscuits and breadsticks, directly expanded extruded snacks, popcorn and puffed rice, wheat etc.. Generally, breakfast food and snacks should have an expanded structure and a crispy texture. Extrusion-cooking and puffing are used to produce direct expanded products such as snacks, breakfast cereals and pet food.

Snacks are generally consumed as light meals. Many common snack foods such as baked collets are corn based and are produced on simple single screw, high pressure cookers. the extruded snacks can be classified on the basis of their functional changes taking place in the raw material e.g. raw cut vegetable snacks, formed dough products from potato derivatives, half products or pellet snacks, directly expanded extruded snacks, popcorn and puffed wheat etc.. It is important to have moisture control of expanded products between 8-10% on wet basis. The product may require an additional drying for the required crispness. The product may be shaped/flavored and filled snacks ready for baking. These snack foods come in variety of shapes such as tubes, wheel finger, and hats mushroom sticks. Shape of the snack directly directly depends upon the die. Many types of proteins and protein enrichments may be added to snack recipes such as meats, dairy products (cheese, yogurt) and legume proteins (soy pea, bean). These type of ingredients have very useful effects on texture-quality and flavor of final products. Among these ingredients, salt is very useful in assisting with uniform moisture migration throughout pellet after drying.

Emulsifiers may be useful to reduce stickiness during cutting and other processing steps. Some co extruded products are also available in the market e.g. co extrusion of biscuits with soft filling.

12.3 Baby Foods

For the growth of the child most of the nutrients are required at that stage. Extrusion cooking is the best process to transform all the nutrients present in the raw material into product. The commonly used raw materials for baby foods are rice, corn, legumes and fruits etc. In the market baby foods are available as baby cereals and baby foods. Baby cereals are highly soluble and are extruded with high shear rates. The raw material for baby cereals are fruit juices, cane sugar/fructose and also fortified with vitamins and minerals. Baby foods are highly soluble and they are fully gelatinized. In the production of baby foods the main thing is to gelatinize the starch otherwise the child will not be able to digest that food. They are prepared by adding natural sweeteners and other nutritional substances. The common thing between both the products is that they are always served by adding liquid. It is used as supplement and is always in the powdered form. Extrusion cooking of baby foods provides absence of birefringence in the starch granule, rapid and high rehydration capacity and better digestibility. The other important thing is the size of the product it should not be so large.

12.4 Confectionary Products

Confectionary products including caramels and toffees, clear hard candies, wine gums and liquors are the best suitable examples of extrusion cooking process (Best, 1994). In this process, the material to be extruded is fed to the orifice by means of multiple rollers or screws. The cross-sectional design of orifice determines the shape of final units. Many extruders have a different type of orifices producing ropes that can be cut into either bars or small pieces. Chocolates can also be manufactured using the extrusion process, it helps in the development of desirable flavors and removal of undesirable flavors. In case of chewing gum the flavor is lost during manufacturing, but the extrusion helps in adequate mixing of the flavor with the base gum.

12.5 Texturized Vegetable Products

They are also known as "fabricated foods" or " texturized plant proteins (TVP)". Meat extenders represent the largest position of texturized vegetable food protein. These products are hydrated to 50-65% moisture and blended with meat or meat emulsions to extend these products. "Breaded chicken patties with 30% of meat are one of the meat analogues. Meat analogue piece can be flavored and formed into sheets, disks, patties, strips and other shapes. Meat free hot dogs, hamburgers, sausages, meat snacks are also included as extruded products (Bazua, 1979). Proteins are effectively denatured during the moist thermal process of extrusion. Denaturation of protein, lower solubility, destroys the biological activity of enzymes and toxic proteins (Smith, 1975). As mechanical and thermal energy are applied during extrusion, the macromolecules in the protein portion of raw materials loose their original structure and results in plastic like mass. Some of the volatile flavors are also flashed off.

12.6 Enzyme Inactivation and Sterilization

Extrusion cooking is the high temperature short time process although most of the nutrients are heat sensitive but in the extrusion cooking the treatment is given only for a very short period so it is not so much harmful to the nutrients. Most of the enzymes are deteriorative in nature. Extrusion cooking helps to inactivate those enzymes. In case of food plants there are some substances which reduces the absorption of nutrients, extrusion is the process which helps to remove those antinutritional substances. Protease inhibitors are inactivated in extrusion cooking. Extrusion cooking is also used for sterilization of food materials as well as pharmaceutical materials.

12.7 Pet Foods

These are medium-shear stress products. The first requirement in expander, prepared pet foods as dog foods is to gelatinize the starch. Some dog foods are extruded through shaped dies and are clearly cut into shaped particles like bones. The product is dried and coated with fat (Rokey, 1994). Dunsford et al. (2002) showed that canines seemed to favour more thermally cooked product. It is also finding application in preparation floating food and sinking foods for fish and other aquatic animals.

12.8 Pasta Products

These are low shear stress products. Semolina, purified middling of durum wheat, is preferred as raw materials for production of pasta products. The uniform size of semolina make easier to mix with water and amber color is of desirable quality (Harper, 1981a). These are also called fortified-precooked cereal products as semi-product or macaroni products.

12.9 Animal Feed

The extruder cooker can be used to precook animal feeds to improve digestibility and palatability. Feeds and fish foods are manufactured using pre-gelatinized starches; generally they are combined with soybean meal (Smith and Singh, 1996).

Extruder is also finding its application for the beverage powder, restructured muscle foods, blended products, instant soup and gravy bases, conventional bakery products, sugar decrystallization and transformation, modified starches and starch degradation products etc.

13. Case Study on RTE Snack Product for Diabetics

Wild et al. (2004) reported that the worldwide number of people with diabetes is expected to double in 30 years, increasing from 171 million in 2000 to 366 million in 2030. However, within a span of six year, Shaw et al. (2010) reported that the world prevalence of diabetes among adults (aged 20-79 years) will be 6.4 % affecting 285 million adults in 2010, and will increase to 7.7 % (439 million adults) by 2030. Between 2010 and 2030, there will be a 69 % increase in numbers of adults with diabetes in developing countries and a 20 % increase in developed countries. India is currently experiencing an epidemic of type 2 diabetes mellitus and has the largest number of diabetic patients (Raghuram, 1991; Misra et al., 2011). It is reported that, in india, there is an increase in diabetes prevalence among rural population at a rate of 2.02 per 1000 population per year (Misra et al., 2011).

The recommendations of American Diabetes Association (2004) on nutritional requirements for diabetes are similar to that of European Association for Study of Diabetes (EASD) which advises an intake of 60-70 o/o of daily energy from carbohydrates and l0 o/o energy from fats with intake of saturated fats limited to 7 %. The remaining l5-20 % of energy requirement is satisfied by protein; dietary intake of fiber is encouraged and salt intake is limited to less than 6 g/day (Choudhary, 2004).

The food materials, which have shown to improve the conditions of the diabetic subjects, include, Bengalgram flour, parboiled rice flour, uncooked corn starch, Indian gooseberry, fenugreek seeds, soy polysaccharides (Pandey and Mishra 2004b; Pandey, 2007).

Different type of food products have been developed for diabetics viz. bars, beverages, biscuits, breads, confectionary, chocolate, dessert, jam, pasta products. Traditional products like pongal, dhokla, laddu and uppuma (Pandey and Mishra, 2007). The functional features of anti-diabetic food products include resistant starch, added fiber, fortification with vitamins and minerals, incorporation of bioactive components such as flavones, saponins, antioxidants, fatty acids etc. (Pandey and Mishra, 2004a).

Extrusion cooking is a multi-step, multi-function thermal/mechanical process. Application of extrusion cooking in the food industry covers a wide variety of extruded products like breakfast cereals, texturized vegetable proteins, extruded snacks, pet foods, confectionary products etc (Kaur et al., 2007). Extrusion is largely affected by the processing parameters such as feed moisture, temperature of the barrel, screw speed and feed composition (Peri et al., 1983). All these parameters affect the quality of extruded product in direct or indirect way. Shirani and Ganesharanee (2009) investigated the effects of fenugreek flour inclusion on the physical characteristics and glycaemic index (GI) of chickpea-rice based extruded products. The GI of extruded products based on 100 % chick pea, 70:30 chick pea-rice blends (control) and 15 % fenugreek polysaccharide were 56± 1.15, 68±2.00 and 43±1.15, respectively.

India is having a snack food market with a value of US$3 billion and is growing due to rising incomes and a booming economy. In view of the imparting health benefits to Indian diabetic population of 50.8 million (Misra et al., 2011) a study on ready-to-eat (RTE) snack food products utilizing extrusion technology was undertaken with the following specific objectives.

(i) Standardization of processing aids for hot extrusion process and organoleptic evaluation of the developed product

(ii) Evaluation of the antidiabetic potential of the developed RTE snack food.

(iii) Storage studies and evaluation of shelf life of the developed product.

13.1 Standardization of Processing Aids and Sensory Evaluation

The addition of non-starch polysaccharides (NSP) guar gum (Ellis et al., 1981) and carboxy methyl cellulose enhances the dietary fiber content of the product and has been used in extruded product (Parada, 2011). The processing aids guar gum (0.9-2.1%) and carboxy methyl cellulose (0.5-2.5 %), were standardized to develop RTE food product using hot extrusion technology.

The antioxidant activity in term of DPPH scavenging capacity was maximum (644.68 pmole/g at 10 min) for processing aids consisting of guar gurn and carboxy methyl cellulose respectively in 0.78 %o and 0.25 % on flour weight basis. The minimum antioxidant value of 617.56 pmole/g at 10 min was obtained for guar gum and carboxy methyl cellulose respectively in 0.50 % and 0.18 % on flour weight basis. The maximum value of total phenol (TP) as gallic acid equivalent (GAE) was 978.44 mg/100 g obtained for processing aids consisting of guar gum and carboxy methyl cellulose respectively in 0.78 % and 0.25 % on flour weight basis. The minimum value of TP was 934.44 (GAE) mg/l00 g, obtained for a processing aid combination of guar gum and carboxy methyl cellulose respectively in 0 .22 oh and 0 .25 % on flour weight basis. Following equations were developed to predict total polyphenol (TP) in terms of GAE and antioxidant activity (AA) in hot extruded product with processing aids (guar gum (X_1) and carboxy methyl cellulose (X_2)).

$$TP=956.43 +15.65*X_1 - 2.08*X_2 + 0.25* X_2^2 + 1.63x X_1* X_2 (r^2 = 0.9996)$$

$$AA= 628.16 + 5.41*X_1 + 3.40* X_2 + 4.13 *X_1^2 -2.54*_{X2}^2 + 3.28*X_1* X_2^2 ((r^2 = 0.9820)$$

The optimum values of processing aids to be added in extruded product were guar gum 0.3% and carboxy methyl cellulose 0.2%. The experimental values for total polyphenols and antioxidant activity in hot extruded snack prepared by addition of optimized processing aids were 941.12 mg GAE/100 g and 624.32 µmole/g at l0 min. The predicted values for total polyphenols and antioxidants were 943.77 mg GAE/100 g, and 625.2.9 µmole/g at 10 rnin. Typical response surface showing effect of processing variable on expansion ratio is shown in Fig. 5.

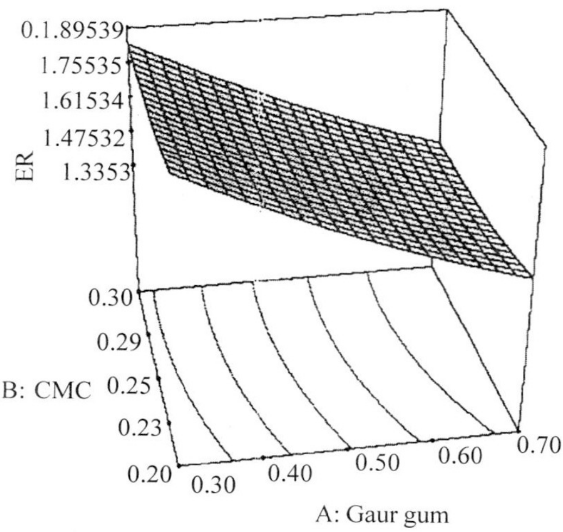

Fig. 5: Effect of processing aids on expansion ratio of extruded snacks

13.2 Assessing antidiabetic potential of the developed product using animal studies

Fifteen Wistar male rats were housed according to the local guidelines on animal welfare in three different groups of 5 (of approximately same initial weight) in clean individual cages with feeders and diabetes was induced in Wistar rats by the injection of alloxan (150 mg/kg, i.p.).They were kept under a 12/12 h day/ night cycle, with free access to drinking tap water and fed with distinctive standard diets A and B. The standard diet 'A' consisted of standard balanced pellet diet composed of: corn starch 80 %, com oil 10 %, non nutritive cellulose 5 %, minerals 4 % (Ca, Cu, Fe, K, Na, Se, Li, Zn, Mg, Mn), vitamin mixture 1 % (vitamin A and B complex). Diet 'B' consisted of developed RTE extruded snack product (0.03 g/kg body weight per day). Before entering the experiment, all the animals were fed with the standard diet for laboratory rats. The weight throughout the study period was measured. Blood glucose was measured once weekly after an overnight fast.

There was a reduction of 34.69 % in blood glucose level in alloxan treated rats fed on prepared RTE extruded product (0.03 g/kg body weight per day) after 3 weeks, however in alloxan treated rats fed on normal diet (0.03 g/kg body weight per day) the change in blood glucose level was 0.025 % only, over the same period.

13.3 Storage studies and evaluation of shelf life

For storage studies, the prepared RTE product samples (10 g) were separately filled in metalized polyester (MP) and high density polyethylene (HDPE) pouches (100 mm x 100 mm) and the pouches were heat sealed. The sealed pouches were placed in desiccators maintained at 90% RH and 40°C temperature in thermostatically controlled incubator. One packet each was taken out from the desiccators at regular interval of time for a period of 90 days and analyzed for moisture content, water activity, total polyphenols and antioxidant capacity. Shelf life of RTE food product was calculated using the equation reported by Labuza (1984). The shelf life was evaluated on the basis of loss of crispiness (>16, +ve peaks), which was attained when moisture content of 7.1 % db (Pardeshi, 2008). The shelf life of developed RTE extruded product was72 days in MP pouches and 48 days under accelerated storage (90% RH and 40 °C).

14. Conclusion

Food extrusion is still an art though in recent years several researchers have aimed at understanding extrusion phenomena in the light of the effect of extruder dimensions like geometry of screw and die, length to diameter ratio of extruder, barrel grooves. The extrusion process variables viz. temperature of extrusion, rate of shear, protein moisture and fat content of feed, particle size of feed material and pressure profile impart significant effect on finished product characteristics. Extrusion cooking technology will be successful in Indian industries with the removal of obstacles such as development of high precision food extruder, curb on import technology and upgrdadation of indigenous technology and optimization of the process variables of extrusion to achieve desirable characteristics of the products, suitable to Indian taste.

References

American Diabetes Association 2004. Nutrition principles and recommendations in diabetes. Diabetes Care, 27: 536-546.

Antila J, Seiler K, Seibel W, Linko, P 1983. Production of flat bread by extrusion cooking using different wheat/rye ratios, protein enrichment and grain with poor baking ability. J Food Engng 2:189-210

Arhaliass A, Legrand J, Vauchel P, Fodil-Pacha F, Lamer T, Bouvier, JM. 2009. The effect of wheat and maize flours properties on the expansion mechanism during extrusion cooking. Food Bioprocess Technol 2:186–193

Badrie N, Mellowes, WA. 1991. Texture and microstructure of cassava flour extrudate. J Food Sci, 56: 1319–1322.

Bazua, CD. 1979. Extruded corn flour as an alternative to lime-heated corn flour. Food Science, 44: 940

Best, ET. 1994. Confectionary extrusion. In: "The Technology of Extrusion Cooking". N. D. Frame, ed. Blackie, London.

Bhatnagar, S. 1994. Extrusion processing conditions of amylose-lipid complexing. Cereal Chemistry, 71: 587-593

Brinèiæ M, Tripalo B, Ježek,Drvar, N, Ukrainczyk M. 2006. Effect of twin-screw extrusion parameters on mechanical hardness of direct-expanded extrudates. Sadhana Vol. 31(5): 527–536.

Chandrasekhar PR, Chattopadhyay PK 1989. Heat transfer during fluidized bed puffing of rice grains. J Food Process Engng 11:147-157.

Chiangand BY, Johnson, JA. 1977. Gelatinization of starch in extruded products. Cereal Chemistry (54): 436-443.

Choudhary P. 2004. Review of dietary recommendations for diabetes mellitus. Diabetes Research and Clinical Practice, 65S: S9-S15.

Clark, JP. 1978. Texturization by extrusion. Journal of Texture Studies. 9: 109.

Clark, JP. 1986. Texturization processes in the cereal foods industry. Food Technol 40(3): 91-93.

Duarte G, Carvalho CWP, Ascheri JLR. 2009. Effect of soybean hull, screw speed and temperature on expanded maize extrudates. Braz. J. Food Technol 12(3):205-212.

Dunsford B, Plattner, B, Greenbury, D, Rokey, G. 2002. The influence of extrusion processing on pet food palatability. Proceedings of Pet Forum Chicago, Illinois, USA. Watt Publishing Inc. Illinois, USA.

Dziezak, JD. 1989. Single, twin-screw extruders in food processing. Food Technology, 43(4): 163-174.

Ellis PR, Apling, EC, Leeds, AR, Bolster NR. 1981. Guar bread: acceptability and efficacy combined. Studies on blood glucose, serum insulin and satiety in normal subjects. British J Nutrition, 46:267-276.

Elsey J, Riepenhausen J, McKay B, Barton GW, Willis M. 1997. Modeling and control of a food extrusion process. Computers Chem Engng 21:S361-S366

Elsner G, Wiedmann, W. 1985. Cooker extruder for the production of gums and jelly articles. Impulse Food Suppl. 2.

Fan J, Mitchell JR, Blanshard JMV. 1994. A computer simulation of the dynamics of bubble growth and shrinkage during extrudate expansion. J Food Engng 23:337–356.

Fletcher SI, Richmond P, Smith AC 1985. An experimental study of extrusion cooking of maize grits. J Food Engng 4:291-312.

Fletcher, SI Richmond, P, Smith A. 1985. An experimental study of twin-screw extrusion cooking of maize grits. J Food Engng 4: 291-312.

Frickle A L, Clarke JP, Mason, TF 1977. Cooking and drying of fortified cereal foods: extruder design. Chem Engng Prog Symp Ser 73(163):134-145

Ganjyal G, Hanna MA, Supprung P, Noomhorm A, Jones D. 2006. Modeling selected properties of extruded rice flour and rice starch by neural networks and statistics. Cereal Chem 83(3):223–227

Gonzalez RJ, Torres RL, Greef DM. 2001. Application of an ideal model to the scaling up of a laboratory extruder. J Food Engng 48:45-51

Harper, JM. 1981. Macaroni extrusion of foods. CRC Press Inc., Florida, Vol. I: 20-29.

Harper, JM. 1986. Extrusion texturization of foods. Food Technology, 40(3): 70-76.

Hoseney, RC. 1986. Breakfast cereals. Principles of cereal science and technology. Ch. 13. Amer Assoc Cereal Chem, St Paul, Minnesota, U.S.A.

Jepson OH 1953. Future extrusion studies. Ind Chem Engng 45:992–1000

Kaur M, Pandey H, Mishra, HN. 2007. Recent advances in food extrusion. Beverage Food World. 34(4):46-50.

Kulshreshtha MK, Zaror CA, Jukes DJ, Pyle, DL. 1991. A generalised steady state model for twin screw extruders. Trans IChemE Part C Food and Bioproducts Proc 69(C4):189-199

Kulshreshtha MK, Zaror CA, Jukes DJ. 1995. Simulating the performance of a control system for food extruders using model-based set-point adjustment. Food Control 6(3): 135–141.

Kulshreshtha MK, Zaror CA. 1992. An unsteady state model for twin screw extruders. Trans IChemE Part C Food and Bioproducts Proc 70(C4):21-28

Labuza TP. 1984. Moisture Sorption: Practical aspects of isotherm measurement and use. American Association of Cereal Chemists, St. Paul, Minnesota.

Levine L, Rockwood, J 1986. A correlation for heat transfer coefficients in food extruders. Biotechnol Prog 2:105-108

Linko P, Colonna P, Mercier C 1981. High-temperature short-time extrusion cooking. In: Advances in Cereal Science and Technology. Vol. 4, Y. Pomeranz (ed). American Association of Cereal Chemists Inc., St. Paul, Minnesota, p145-235

Misra P, Upadhyay, RP, Misra A, Anand, K. 2011. A review of the epidemiology of diabetes in rural India. Diabetes Research and Clinical Practice 92:303-311.

Mueser F, Pfaller W, van Lengerich B 1987. Technological aspects regarding specific changes to the characteristic properties of extrudates by HTST-extrusion cooking. In: Extrusion Technology for the Food Industry, Connor CO (ed). Elsevier Applied Science, New York, pp. 35-53

Murugesan G, Bhattacharya KR. 1986. Studies on puffed rice. I. Effect of processing conditions. J Food Sci Technol 23(4): 97-202.

Olkku J, Vainionppaa, J 1980. Response surface analysis of HTST extrusion texturized starch-protein-sugar paste. In: Food Process Engineering Vol. 1. Food Processing Systems. Linko P, Malkki Y, Olkku J, Larinkari J (eds). Applied Science, London.

Pandey H, Mishra HN. 2007. Development of cereal pulse based extruded products for diabetics. ICFOST-2007 Souvenir, 78.

Pandey H, Mishra, HN. 2004a. Formulation processing and quality evaluation of preventive/ curative health food for diabetics. Food and bioprocess engineering, Proceedings of International Conference on emerging technologies in Agricultural and Food Engineering Department, IIT, Kharagpur, ANANMAYA Publishers, New Delhi, India, pp. 244-249.

Pandey H, Mishra, HN. 2004b. Diet Management for diabetes. In "Project file column of Education Times supplement of Times of India" National Daily News Paper, March, 29, 2004.

Pandey, H. 2007. Food Product Development for Type II Diabetes. In "Cereal and Food World" AACC International Annual Meeting, October 7-10, San Antonio, Texas, USA. 456.

Pardeshi IL. 2008. Development of Cereal Based Ready-to-Eat Snack Foods. Unpublished Ph.D. thesis, Post Harvest Technology Centre, IIT, Kharagpur (W.8.), India.

Paruda J, Aguilera JM Brennan, C 2011. Effect of guar gum content on physical and nutritional properties of extruded products. J Food Engng 103: 324-332.

Pathania S, Singh B, Sharma S, Sharma V, Singla S. 2013. International J Engng Research and Applications. 3(3): 1040-1049

Payne FA, Taraba JL, Saputra D. 1989. Review of puffing processes for expansion of biological products. J Food Engng 10:183-197.

Peri C, Barberi R, Casiraghi EM. 1983. Physical, chemical and nutritional quality of extruded corn germ flour and milk protein blends. J Food Technol 18:43.

Raghuram TC, Swaran P, Sharma, RD. 1991. Diet and Diabetes. National Institute of Nutrition, Indian Council of Medical Research, Hyderabad.

Riaz, MN. 1981. Extruders in food applications. Technomic Publishing Company Incorporation, Pennsylvania.

Riaz, MN. 2000. Introduction to extruder and their principles. In: Riaz, MN editor Extruders in food application. Boca Raton (FL): CRC Press. p 1-23.

Roberts SA, Guy RCE 1987. Metastable states in a food extrusion cooker. J Food Engng 6:103-112

Rokey, G. 1994. Better extrusion. Pet Food Industry, 36 (4): 10.

Rossen JL, Miller RC. 1973. Food extrusion. Food Technology, 27(8): 46-53.

Rossi, M, Peri, C. 1980. Effects of extrusion cooking on structural and functional characteristics of sunflower protein. In "Food Engineering and Processing Applications" P. Linko, Y. Malkki, J. Olkko, and Larinari. Journal of Applied Science Publishers, London.

Schuler, EW. 1986. Twin-screw extrusion cooking for food processing, Cereal Foods World, 31: 413–416.

Shirani G, Ganesharanee, R. 2009. Extruded products with Fenugreek (Trigonella fbenum-graecium) chickpea and rice: Physical properties, sensory acceptability and glycaemic index. J Food Engng 90:44- 52

Singh B, Mulvaney SJ 1994. Modeling and process control of twin-screw cooking food extruders. J Food Engng 23:403-428

Smith, OB. 1975. Extrusion cooking. In "New Protein Foods", ed. N. M Altschul, Academic Press, New York.

Smith, RC, Singh, N. (1996). New applications of extrusion cooking technology. Indian Food Industry. 15: 14-23.

Sullivan, JF, Craig JC Jr. 1984. The development of explosion puffing. Food Technol 38(2): 52-5,131.

Tayeb J, Vergnes B, Della Valle G 1988. A basic model for a twin screw extruder. J Food Sci 53:1047-1056

Van Zuilichem DJ, Van der-Laan E, Kuiper E 1990. The development of a heat transfer model for twin screw extruders. J Food Engng 11:187-207.

Varnalis AI, Brennan JG, MacDougall DB. 2001. A proposed mechanism of high temperature puffing of potato. Part I. The Influence of blanching and drying conditions on the volume of puffed cubes. J Food Engng 48:361-367.

Weidmann W, Strobel E. 1987. Processing and economic advantages of extrusion cooking in comparison with conventional processes in the food industry. In: "Extrusion Technology for the Food Industry" C. O. 'Connor, ed., Elsevier, London.

White, G. 1994. Defining the true meaning of snacks. Food Technology International Europe, 2:115-117.

WHO. 1999. Definition, diagnosis and classification of diabetes mellitus and its complications. Report of a WHO-consultation, Part 1: Diagnosis and classification of diabetes mellitus. World Health organization, Geneva.

Wild S, Roglic G, Green A. 2004. Prevalence of Diabetes: Estimates for the Year 2000 and Projection for 2030. Diabetes Care 27:1047-1053

Yacu WA 1985. Modeling a twin screw co-rotating extruder. J Food Eng 8:1-21

20

Microwave Assisted Processing of Food

S.S. Shirkole and Prag Prakash Sutar

1. Introduction

Thermal processing usedextensively in thefood industry for preservation and shelf life extension of the products (Arnoldi 2001). Heat treatments remain most important in food processing unit operations, not only to achieve desirable changes in eating quality but also for preservative effect by inactivation of microorganisms, destruction of enzymes and insects.Heat is applied in various unit operations through adifferent mode of heat transfer such as conduction, convection, and radiations. Conduction is heat transfer by microscopic collisions of particles and movement of electrons within thebody. Convection is the heat transfer by the movement of groups of molecules in a fluid. Radiation is the heat transfer by electromagnetic waves from one body to another. There are different merits and demerits of each mode of heat transfers. There arevarious demerits and drawbacks associated with conventional heating methods used for thermal processing of food, but microwave heating could be a potential alternative (Fig.1). It is because of the microwave can heat more rapidly and uniformly with direct interaction between microwave energy and products (Datta and Nelson 2000). Microwaves are anelectromagnetic wave withfrequency band from 300 MHz to 300 GHz. There are different applications of microwave energy in the food industry such as drying, thawing, sterilization, pasteurization, heating and reheating (Ayappa, Davis et al. 1991). Drying is theremoval of moisture content from food material up to predetermined level to increase the shelf life of food products and its preservation (Mujumdar 2000). Hot air drying is widely used method for drying of agricultural food materials, but it leads to considerable changes in quality characteristics of food products such as color, taste, nutritional content, and density. Also shifting of interior solutes from center to surface of product due to extended drying time and high temperature (Maskan 2001). Application of microwaves in the drying operation becomes popular because of minimum quality deterioration, rapid drying and effective distribution of heat (Dýaz, Martýneż-Monzo et al. 2003).

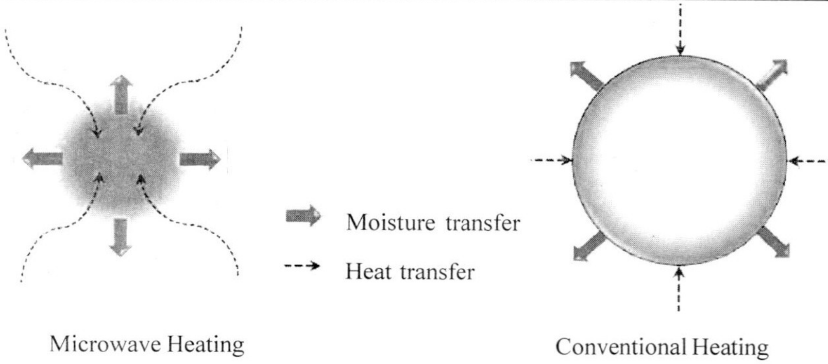

Microwave Heating

- Microwaves penetrate the materials and create rapidly changing fields
- The heat is uniformly distributed throughout the material

Conventional Heating

- Heat transferredmolecule by molecule from the outside
- Overheating can occur on the outside

Fig. 1: Microwave Vs Conventional Heating

Blanching of vegetables and fruits is carried out to inactivate enzymes and softening vegetable tissues. Microwave assisted blanching leads to effluent free process as compared to hot water bath process (Lanqingzi 2005). Thawing is the process to change the frozen form to liquid or semiliquid state. The distinctive property of microwaves to penetrate and generate the heat within theinterior of product sample make them potential in accelerating thawing process (Li and Sun 2002). Pasteurization and sterilization operations are used to inactivate the microorganisms and enzymes in food products. Microwave assisted thermal processing of liquid food mainly focuses on thedestruction of microorganisms (Al-Hilphy and Ali 2013), enzyme inactivation (Salazar-González, San Martín-González et al. 2012) and physicochemical properties (Iuliana, Rodica et al. 2015).

Microwave (MW) assisted thermal processing is an appropriate way as it is assumed to be a fast, it results in volumetric heating of product, clean, safe and easy to use technology. For thermal unit operations application of microwaves is the topic of research forsome years ago.The purpose of this chapter is to discuss the application of microwave in theprocessing of food and to focus latest developments and its effect on microorganisms, enzymes, and physicochemical properties of food products with providing recent research findings (Constantin and Csatlos 2010), (Vadivambal and Jayas 2007).

2. Microwave heating mechanism and principle

During microwave assisted processing, heating of the food product is caused due to absorption of microwaveradiationby the productand results into interior heating of product. Some important facts that need to keep in mind that dielectrics and microwaves are not a form of heat but heat is generated through its interaction with materials. Many mechanisms are involved in this energy conversion such as ionic conduction, dipole rotation, dipole stretching, interface polarization, piezoelectricity, ferroelectric hysteresis, nuclear magnetic resonance and ferromagnetic resonance (Mujumdar 2014). Microwaves penetrate into food products and interacts with solid particles and water available in aproduct which results in volumetric heating due to dipole action of solvent and movement of dissolved ions. The dipole action (Fig.2) results because of deviations in magnetic and electric fieldalso, water is the key source for microwave interaction due to its polarity.Heating of the product might also take placebecause of oscillatory movement of ions which generates heat because of high-frequency oscillating electric field (Datta and Davidson 2000)

There are several factor affects microwave assisted heating, uniformity and distribution of heat but, dielectric properties of the food material to be heated and penetration depth of microwaves are the most important of them.

2.1 Dielectric properties of food material

The dielectric properties arevery essentialto design and development of both microwave assisted heating process and equipment. Methods used for measurement of dielectric properties of food materials depends on thedegree of accuracy and frequency required also, food being tested. The process of measurement involves generation of a microwave signal at the frequency of interest, directing it into theproduct to be tested, measurement of changes in the signal due to product and then estimation of dielectric properties of product based on changes occurred.The capability of material to transform microwave energy into heat can be implicit by dielectric properties of material to be heated. The most important property which governs microwave heating is the dielectric permittivity (ε^*), expressed in equation 1. The real part of thedielectric property is known as dielectric constant, which indicates thecapability of material to store electric energy and imaginary part of thedielectric property is known as dielectric loss factor, which indicates the capability of material to transformelectric energy into heat. The microwave dielectric properties of liquid water in listed in Table 1 for various frequencies at temperatures of 20 and 50°C from data given by (Hasted 1973).

$$\varepsilon^* = \varepsilon' - j\,\varepsilon'' \qquad (1)$$

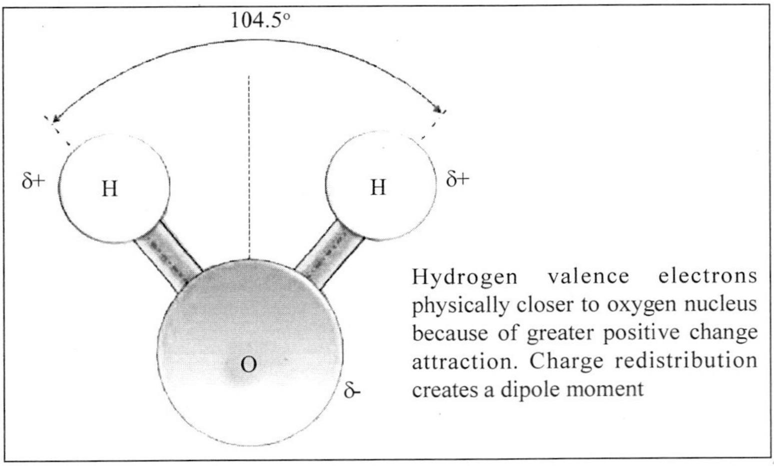

Fig. 2: Water molecule dipole moment

where, ε* is dielectric permittivity,

ε' is dielectric constant,

ε" is dielectric loss factor, and

j =√" (-1).

The ratio of dielectric loss to dielectric constant is given by loss tangent and expressed as,

$$\tan \tan \delta = \frac{K''}{K'} = \frac{\varepsilon''}{\varepsilon'} \tag{2}$$

where, k' – relative dielectric constant, k' = $\varepsilon'/\varepsilon_o$

k" – relative dielectric loss, k" = $\varepsilon''/\varepsilon_o$

here, ε_o is the permittivity of free space (ε_o= 8.854 × 10^{-12} F/m).

The dielectric properties of the product mainly influenced by thefrequency of microwaves and operating temperature used. On the basis of microwave absorption, products are categorized into three categories:

i. High dielectric loss materials or absorber, which absorbs microwaves very strongly.

ii. Low dielectric loss materials or transparent, microwaves are passes through the materials with little attenuation.

iii. Opaque, which reflects the microwaves.

Table 1: Microwave dielectric properties of water at indicated Temperature and frequency adapted from (Hasted 1973)

Frequency, *GHz*	Dielectric constant (ε') Temperature		Dielectric loss factor (ε") Temperature	
	20°C	50°C	20°C	50°C
0.6	80.3	69.9	2.75	1.25
1.7	79.2	69.7	7.90	3.60
3.0	77.4	68.4	13.0	5.80
4.6	74.0	68.5	18.8	9.40
7.7	67.4	67.2	28.2	14.5
9.1	63.0	65.5	31.5	16.5
12.5	53.6	61.5	35.5	21.4
17.4	42.0	56.3	37.1	27.2
26.8	26.5	44.2	33.9	32.0
36.4	17.6	34.3	28.8	32.6

Hence, knowledge of dielectric properties is essential to categorize the materials on the basis of microwaves absorption also, needed for designing of dielectric drying equipment and process design.

The microwave power penetration depth (Dp) can be defined as the distance at which the power density drops to a value of 1/e from its value at the surface and expressed as (Metaxas and Meredith 1983).

$$D_P = \frac{C}{\sqrt{2\pi f}\left[K'\left\{\sqrt{1+\left(K''/K'\right)^2}-1\right\}\right]^{1/2}} \tag{3}$$

where, c is the velocity of light, $c = (\mu_o \varepsilon_o)^{-1/2}$,

μ_o is the permeability of free space ($\mu_o = 4\pi \times 10\text{-}^7$ H/m)

Above equation (eqn. 3) is applicable for food materials which are non-magnetic in nature.The microwave power which varies with the square of the electric field is given as,

$$q = \frac{1}{2}\omega\varepsilon_o K''|E|^2 \tag{4}$$

where E is the intensity of electric field, ω is the angular frequency. Apart from penetration depth and dielectric properties of materials other factors also affect microwave assisted food processing such as thedesign of microwave cavity (geometry and size of cavity), microwave frequency, product moisture content, composition and density of the product, size and shape of the product (Nelson and Datta 2001). The moisture content of the product or amount of water available in theproduct is the key factor in determining its dielectric properties as water is a good absorber of microwaves.

2.2 Factors affecting microwave heating

Various physico-thermal, dielectric and electrical properties of the food products decide the absorption of microwave and heating behavior of the product. Such factors discussed shortly:

2.2.1 Moisture content and temperature dependency

Free moisture present in the product is one of the key factor governs the heating of product in amicrowave oven (Nelson and Kraszewski 1990). When product moisture content is high, it willresult efficient heating of the product as it has high dielectric loss factor (Schiffmann 1986). Also, thechemical composition of food product and dipole movement associated with free moisture present in the product affects dielectric properties of the product. The temperature dependency of dielectric constant is a bitdifficult, and it may vary with thetemperature of the product.

2.2.2 Effect on composition of food product

The dielectric properties of food productsalso influenced by thechemical composition of the product which complicates the investigation of dielectric properties of whole product on the basis of single ingredient or components. The organic components of thefood product may be considered as transparent to energy as they are dielectrically inert ($\varepsilon' < 3$ and $\varepsilon'' < 0.1$) (Mudgett 1985). Also, salt content is influencing the dielectric properties of theproduct(Ohlsson and Bengtsson 1975). The loss factor increases slightly with thetemperature at lower fat contents and higher moisture (To, Mudgett et al. 1974).

2.2.3 Effect of microwave frequency

The penetration depth increases with adecrease in microwave frequency and water content (Coronel, Simunovic et al. 2003). Equation expresses the relation between penetration depth and wavelength:

$$D = \lambda_0 (K')^{1/2} \Big/ 2\pi K'' \qquad (5)$$

where

D = Penetration depth (at which 63% of incident energy is absorbed),

λ_0 = wavelength in free space (cm),

K' = relative dielectric constant, $\pi = 3.1415$ and

K" = relative dielectric loss.

Penetration depth increases with wavelength and decrease in frequency of microwave.The dielectric loss factor is highly affected by frequency, for aqueous solutions a low frequency generates a smaller dielectric loss.

2.2.4 Product parameters

Product parameters such as mass, density, and its geometry also affect the microwave assisted heating profile of the food products. Larger objects absorb a large amount of microwave energy and take alonger time to heat. The air present in the product makes the product good insulator (Schiffmann 1990). The shape and size of aproductdecide how theproduct will heat during microwave heating. Food products with corners and edges shown tendency localized heating because of multi-directional distribution of microwaves (Campanone and Zaritzky 2005).

2.3 Non-uniformity in temperature distribution

Even though microwave assisted heating of product is avolumetric phenomenon and is more uniform as compared to many other conventional heating methods but non-uniformity in temperature distribution is one of themostimportant problem associated with microwave heating. One of the major drawbacks of microwave assisted heating is the presence of hot spots in numerous zones depending upon thegeometry of the food product (Campanone and Zaritzky 2005).Due to lack of uniform temperature distribution few regions of food product heated quickly and larger extent, whereas the remaining part gets heated to a lesser extent (Vadivambal and Jayas 2010). Few proposed solutions to minimize non-uniformity in temperature distribution are (i) combination of microwave and conventional heating methods (ii) appropriate microwave oven design (iii) controlling thegeometry of food products (iv) controlling heat cycle and (v) heating of product at low microwave power overan extended period. Also, the thickness of the product samples can be limited up to 25 mm (Ohlsson and Thorsell 1984).

3. Microwave assisted cooking

During last two decades' domestic microwave ovenbecomes an essential in most kitchens. Different parts of domestic microwave oven are shown in (Fig.3) such as a transformer for high voltage power source, a high voltage capacitor connected to magnetron and transformer via a diode to the chassis, magnetron cavity transforms high voltage electric energy into microwave radiation, control panel, waveguide to connect or guide microwave power from the magnetron cavity to oven cavity, turntable and base plate arrangement for homogeneity in distribution of heat, metal waveguide stirrer, limit switch for safety purpose to turn off magnetron system if door is opened while operation.Microwave assisted cooking of food product is one of the major application of microwave energy. During cooking microwaves affect quality, color retention, taste and texture of foodstuffs and are reviewed in this section. Quality alterations during conventional and microwave baking of breadwere studied (Içöz, Sumnu et al. 2004).They noticed in bakingoperations; it is important to obtain good texture and browning at fixed level of moisture. In microwave-assisted baking, enough surface browning and crust formation were not possible because of the air surrounding product sample was cold, and moisture removing from product sample gets condensed in contact with cold air, which results in alack of crispness. They had placedsusceptors at the bottom of the sample for crust formation and surface browning. Susceptors are made up of material which absorbs microwaves and converts microwave energy into heat to supply the heat to food products by conduction and convection.

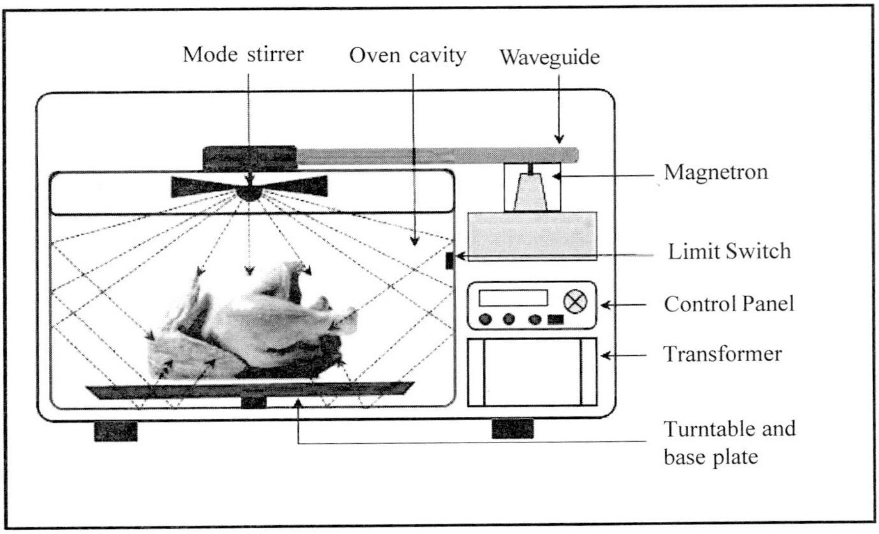

Fig. 3: Domestic Microwave Oven

Cooking efficiency and energy usage of domestic appliances like LPG pressure cooker and electric rice cooker (ERC) and microwave was compared (Lakshmi, Chakkaravarthi et al. 2007). Microwave assisted cooking of rice was in theshorterperiod as compared other cooking methods but energy efficiency of microwave assisted cooking has to be improved so that it can be competitive with other cooking methods.

4. Microwave assisted drying

Drying is the process of removal of moisture content from food products to predetermined level by creating vapor pressure difference between product and surrounding using different methods. The main goal of drying is to enhance shelf life and preservation over along duration without altering quality characteristics. There are different drying methods used such as sun drying, solar drying, hot air drying, fluidized bed drying, spray drying, recirculatory air drying, freeze drying and osmotic drying (Mujumdar 2000). Drying methods are applied according to product form, operational and industrial need, resources available, climatic condition and targeted outputs and so on. Also, there are different technical issues associated with conventional drying operation such as extended drying time, quality alterations in theproduct, energy efficiency, temperature uniformity within product and textural changes. Now days microwave assisted drying of food commodities gaining popularity because of many advantages such as quick heating, volumetric heating, less quality alteration of product and improvement of thequalityof some food products and very effective during falling rate period. Volumetric heating leads to vapor transmission from inside of product to surface as pressure gradient is developed. Thus, shrinkage of product sample can be prevented in microwave assisted drying. The combination of microwave and other drying methods can enhance product quality and efficiency of drying which far superior to that practicable by only conventional methods or microwave alone (Zhang, Tang et al. 2006).

5. Microwave assisted blanching

Blanching is pretreatment process applied to fruit and vegetables at temperature enough to inactivate enzyme present in thetissue. Blanching inhibits the enzyme activity, sets the color of theproduct and shortens thetime required for drying. Some food industries using microwave blanching to blanch fruits and vegetables as it is effluent free technology. Microwave assisted blanching permits efficient heat transfer with little or no water requirement, which decreases nutrients lost to leaching as compared to immersed water blanching (Lin and Brewer 2005). Microwave blanching preserve nutrients better as compare to conventional water or steam blanching because of adecrease in leaching losses during microwave blanching (Ramesh, Wolf et al. 2002). Color retention and quality of frozen peas were superior in microwave blanching as compared to steam and water blanching (Lin and Brewer 2005).

6. Microwave assisted sterilization

Sterilization or pasteurization of food products refers to elimination or destruction of all forms or targeted or pathogenic microorganisms in food products to improve storage life and food safety (Nott and Hall 1999). There are different methods applied tomicroorganisms such as irradiation, ethylene dioxide fumigation, high-pressure processing and steam sterilization, etc.,and its application varies according to product to product. To ensure destruction of pathogenic microorganisms, food materials are maintained at particular time and temperature. Pasteurization also includes inactivation of unwanted enzymes which causes cloud loss in certain juices. Microbes or enzymes inactivated by microwaves at sub-lethal temperature was described by one or more of the following theories such as selective heating, rupture of thecell membrane, magnetic field coupling and electroporation (Kozempel, Annous et al. 1998). The selective heating proposes that microwaves heat the microorganisms selectively and reach the temperature greater than of the surrounding fluid which results in thedestruction of microorganisms more quickly. The electroporation theory suggests, the electrical potential across the cell membrane affects pores, which results in leakage of cellular materials. According to cell membrane rupture theory, cell membrane gets ruptured because of thevoltage across the cell membrane. As per magnetic coupling theory, thecoupling of critical molecules like DNA or protein with electromagnetic energy disruptures the internal components of the cell. With the proper conditions for applying microwaves and based on knowledge of dielectric properties of food products desired process lethality can be achieved (Salazar-González, San Martín-González et al. 2012).

7. Microwave assisted extraction

There are different methods used for extraction of the various components from food matrix such as Soxhlet extraction, steam distillation, hydrodistillation and simultaneous distillation-extraction. Recently there has been extensive interest in application of microwave heating for extraction of different components such as essential oil from herbs (Lucchesi, Chemat et al. 2004), aromas(Chee, Wong et al. 1996), phenols (Proestos and Komaitis 2008), pesticides (Onuska and Terry 1993) and dioxins (Kodba and Marsel 1999) have been extracted efficiently from different matrices such as food, plant materials, animal tissues, soils and sediments. The main benefits of microwave assisted extraction are aminimum requirement of solvent and energy, reduction of extraction time(Lucchesi, Chemat et al. 2004). All reported research findings showed that microwave-assisted solvent extraction (MAE) is a viable alternative to conventional techniques.

8. Drawbacks in microwave assisted processing of food

Microwave drying of food products resultsinpoor quality if not correctly applied (Adu and Otten, 1996). Maintaining heating uniformity is essential to confirm thehigh quality of product and microbiological safety (Błaszczak, Gralik et al. 2002). The difficulty with microwave assisted processing is that heating behavior of product is affected by thickness, its dielectric properties, and geometry. Non-uniformity in microwave heating is due to theuneven spatial distribution of electromagnetic field inside the cavity (Manickavasagan, Jayas et al. 2006). The factors affect thedistribution of electromagnetic field can be categorized into two groups: cavity effects and interaction of product. Cavity effects are due to thedesign of cavity, theshape of thecavity, thelocation of inlet point for themicrowave. Product interaction includes dielectric loss factor, product thickness, penetration depth, size and shape of the product (Kelen, Ress et al. 2006).

9. Concluding remarks

The mechanism of microwave heating, factors affecting microwave heating, different microwave assisted unit operations and drawbacks associated with theprocessing of food products are discussed. Microwave assisted heating has several advantages as compared to conventional heating methods but microwave heating is still not applied widely for commercial purposes because of some technical as well as cost factors. It needs noteworthyresearch targeted at enhancements in certain areas. Microwave assisted processing of food products required to conduct at large scale or pilot scale level rather than laboratory level so that outcomes might be applicable for industry level. Further, more investigation on intricate nature of microwave-product interaction needs to be done for better understanding of microwave-assisted processes. Thus, microwave assisted processing of food products to appear to have great potential in the near future.

Reference

Adu, B. and L. Otten (1996). "Microwave heating and mass transfer characteristics of white beans." Journal of Agricultural Engineering Research 64(1): 71-78.

Al-Hilphy, A. R. S. and H. I. Ali (2013). "Milk Flash Pasteurization by the Microwave and Study its Chemical, Microbiological and Thermo Physical Characteristics." Journal of Food Processing & Technology 2013.

Arnoldi, A. (2001). "Thermal processing and food quality: analysis and control." Thermal technologies in food processing: 138-159.

Ayappa, K., et al. (1991). "Analysis of microwave heating of materials with temperature dependent properties." AIChE Journal 37(3): 313-322.

B³aszczak, W., et al. (2002). "Effect of ã radiation and microwave heating on endosperm microstructure in relation to some technological properties of wheat grain." Food/Nahrung 46(2): 122-129.

Campanone, L. and N. Zaritzky (2005). "Mathematical analysis of microwave heating process." Journal of food engineering 69(3): 359-368.

Chee, K. K., et al. (1996). "Optimization of microwave-assisted solvent extraction of polycyclic aromatic hydrocarbons in marine sediments using a microwave extraction system with high-performance liquid chromatography-fluorescence detection and gas chromatography-mass spectrometry." Journal of Chromatography A 723(2): 259-271.

Constantin, A. and C. Csatlos (2010). "Research on the influence of microwave treatment on milk composition." Bulletin of the Transilvania University of Bra°ov 3: 52.

Coronel, P., et al. (2003). "Temperature Profiles Within Milk after Heating in a Continuous flow Tubular Microwave System Operating at 915 MHz." Journal of Food Science 68(6): 1976-1981.

Datta, A. and S. Nelson (2000). Fundamental physical aspects of microwave absorption and heating in handbook of microwave technology for food applications, CHIPS Publications, USA.

Datta, A. K. and P. M. Davidson (2000). "Microwave and radio frequency processing." Journal of Food Science 65(s8): 32-41.

Dýaz, G. R., et al. (2003). "Modelling of dehydration-rehydration of orange slices in combined microwave/air drying." Innovative Food Science & Emerging Technologies 4(2): 203-209.

Hasted, J. B. (1973). "Aqueous dielectrics."

Içöz, D., et al. (2004). "Color and texture development during microwave and conventional baking of breads." International Journal of Food Properties 7(2): 201-213.

Iuliana, C., et al. (2015). "Impact of microwaves on the physico-chemical characteristics of cow milk." Romanian Reports in Physics 67(2): 423-430.

Kelen, A., et al. (2006). "Mapping of temperature distribution in pharmaceutical microwave vacuum drying." Powder Technology 162(2): 133-137.

Kodba, Z. C. and J. Marsel (1999). "Microwave assisted extraction and sonication of polychlorobiphenils from river sediments and risk assesment by toxic equivalency factors." Chromatographia 49(1-2): 21-27.

Kozempel, M. F., et al. (1998). "Inactivation of microorganisms with microwaves at reduced temperatures." Journal of Food Protection® 61(5): 582-585.

Lakshmi, S., et al. (2007). "Energy consumption in microwave cooking of rice and its comparison with other domestic appliances." Journal of food engineering 78(2): 715-722.

Lanqingzi, D. Y. F. J. L. (2005). "Application of microwave on plant fiber blanching [J]." Journal of Radiation Research and Radiation Processing 6: 007.

Li, B. and D.-W. Sun (2002). "Novel methods for rapid freezing and thawing of foods–a review." Journal of food engineering 54(3): 175-182.

Lin, S. and M. Brewer (2005). "Effects of blanching method on the quality characteristics of frozen peas." Journal of food quality 28(4): 350-360.

Lucchesi, M. E., et al. (2004). "Solvent-free microwave extraction of essential oil from aromatic herbs: comparison with conventional hydro-distillation." Journal of Chromatography A 1043(2): 323-327.

Manickavasagan, A., et al. (2006). "Non-uniformity of surface temperatures of grain after microwave treatment in an industrial microwave dryer." Drying Technology 24(12): 1559-1567.

Maskan, M. (2001). "Drying, shrinkage and rehydration characteristics of kiwifruits during hot air and microwave drying." Journal of food engineering 48(2): 177-182.

Metaxas, A. and R. Meredith (1983). "Industry Microwave Heating." London: Peter Peregrinus Ltd.

Mudgett, R. (1985). "Dielectric properties of foods." Microwaves in the food processing industry: 15-37.

Mujumdar, A. S. (2000). Drying technology in agriculture and food sciences, Science Publishers, Inc.

Mujumdar, A. S. (2014). Handbook of industrial drying, CRC Press.

Nelson, S. and A. Kraszewski (1990). "Grain moisture content determination by microwave measurements." Transactions of the ASAE 33(4): 1303-1305.

Nelson, S. O. and A. K. Datta (2001). "Dielectric properties of food materials and electric field interactions." FOOD SCIENCE AND TECHNOLOGY-NEW YORK-MARCEL DEKKER-: 69-114.

Nott, K. P. and L. D. Hall (1999). "Advances in temperature validation of foods." Trends in Food Science & Technology 10(11): 366-374.

Ohlsson, T. and N. Bengtsson (1975). "Dielectric food data for microwave sterilization processing." Journal of Microwave Power 10(1): 93-108.

Ohlsson, T. and U. Thorsell (1984). "Problems in microwave reheating of chilled foods." Foodservice Research International 3(1): 9-16.

Onuska, F. and K. Terry (1993). "Extraction of pesticides from sediments using a microwave technique." Chromatographia 36(1): 191-194.

Proestos, C. and M. Komaitis (2008). "Application of microwave-assisted extraction to the fast extraction of plant phenolic compounds." LWT-food science and technology 41(4): 652-659.

Ramesh, M., et al. (2002). "Microwave blanching of vegetables." Journal of Food Science 67(1): 390-398.

Salazar-González, C., et al. (2012). "Recent studies related to microwave processing of fluid foods." Food and Bioprocess Technology 5(1): 31-46.

Schiffmann, R. (1986). "Food product development for microwave processing." Food technology (USA).

Schiffmann, R. (1990). "Microwave foods: basic design considerations." Tappi journal 73(3): 209-212.

To, E., et al. (1974). "Dielectric properties of food materials." Journal of Microwave Power 9(4): 303-315.

Vadivambal, R. and D. Jayas (2007). "Changes in quality of microwave-treated agricultural products—a review." Biosystems Engineering 98(1): 1-16.

Vadivambal, R. and D. Jayas (2010). "Non-uniform temperature distribution during microwave heating of food materials—A review." Food and Bioprocess Technology 3(2): 161-171.

Zhang, M., et al. (2006). "Trends in microwave-related drying of fruits and vegetables." Trends in Food Science & Technology 17(10): 524-534.

21

Ohmic Heating and its Applications in Food Processing

Hilal A. Makroo, Sajad A. Sofi and Brijesh Srivastava

1. Introduction

Heat transfer from a hot surface is a general method or technique of heating liquids and slurry type of food materials. In such cases the heat can be generated either directly by a heating element or indirectly from a hot medium like steam or hot water via a heat exchanger (e.g. shell and tube, plate). Therefore for the heat transfer during heating by these techniques a temperature gradient will be required, which usually causes the fouling on the surface of the heating element or heat exchanger. Afterwards the fouling not only reduces the heat transfer rate but also results in quality deterioration within the product which is being processed. Moreover heat transfer is an issue when heating a high viscous material or liquid foods with particulates due to non uniform heating. Therefore to overcome the limitations of conventional method of thermal processing novel techniques such as microwave or ohmic heating etchas been developed.

The concept of ohmic heating of foods is not new. In the 19th century, numerous processes were patented that used electrical current for heating flowable food materials (Ruan et al. 2002). Ohmic Heating, as the name indicates is a term related to Ohm or electricity or electrical resistance. According to J.P. Joule (1841) when electricity is allowed to flow through a conductor, some heat gets generated within the conductor which is commonly known as joules effect, the phenomenon of heat generation is also known as ohmic heating or joules resistance heating (Zareifard et al. 2014).Ohmic heating is a novel electro-heating method accepted by the food industry for processing of a broad range of food products. Industry trials have demonstrated excellent product quality by the application of this technology (Ruan et al. 2002a). During ohmic heating, the electrical energy given to the system is ideally used only for heat generation. Ohmic heating appears to offer an attractive alternative to the conventional thermal processing; because it heats material by internal heat generation i.e. heat is generated directly with in the food rapidly and evenly (Knirsch, et al. 2010). Conventional heat exchangers work indirectly, by passing heat from a surface into the fluid, which gives rise to fouling and corrosion(Samaranayake et al. 2005). Ohmic heating has got other advantages over conventional methods of thermal processing such as less damage to rheological, textural, organoleptic and nutritional properties, suitability for particulate foods, besides being energy efficient. Therefore in the present chapter you will get familiar with the principle and concept, applications, andadvantages & limitations of ohmic heating in food processing.

2. Ohmic Heating Principle

Electric current conducting material may contain different kind of charged particles like electrons in metals and ions or molecules (proteins) in most of the food material. The charged particles begin to flow when an electric potential is applied to these conductors,this phenomenon is commonly known as electric current. Once the electric currentis permitted to flow through these materials, the motion of the charged particles leads to the vibration of molecules resulting in the raise in the temperature due to generation of heat known as ohmic heating(Sastry et al. 2014) (Eq.1). Particularly in food material the electric current flows due to the motion of the charged particles towards the electrodes of opposite polarity. Both alternating (AC) and direct (DC) current can be responsible for the ohmic heating phenomenon (Fig.1).

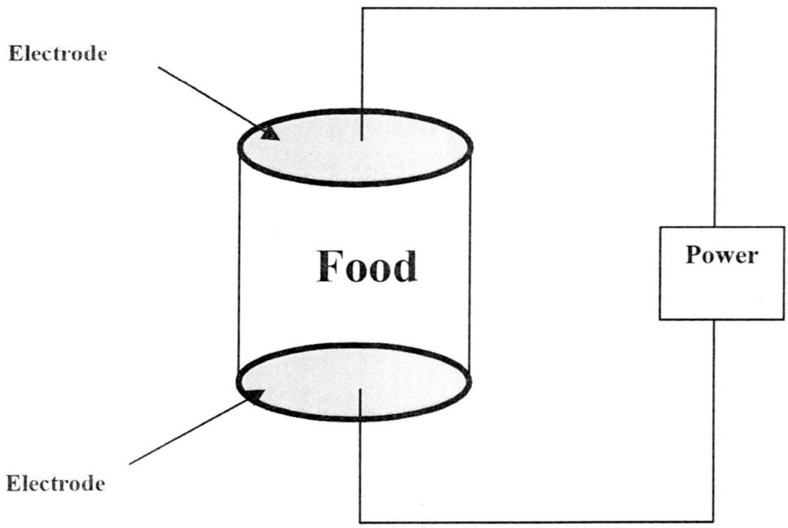

Fig. 1: Ohmic heating principle

The heat generation during ohmic heating may be characterized by the following equation (Sastry et al. 2014)

$$Q = E^2 \times \sigma \tag{1}$$

Where:

Q = rate of heat generated per unit volume, w/m^3

E = electric field strength (volt per unit length), V/m

σ = electrical conductivity, S/m

$$\sigma = \left(\frac{1}{R}\right)\left(\frac{L}{A}\right) \text{(Zareifard et al. 2014)} \tag{2}$$

R = resistance, ohm

L = distance between electrodes, m

A = area of the electrodes, m^2

The rate of energy generation during ohmic heating is dependent on voltage gradient and electrical conductivity of the material (Ruan, et al. 2002). The electrical conductivity is the most critical property effecting energy generation, for most of the solid food material undergoing conventional heating (Knirsch, et al. 2010), electrical conductivity increase sharply with temperature at around 60°C as a result of break-down of cell wall materials. To generate the heat within a food material by the ohmic effect its electrical conductivity must not be zero. Therefore completely non polar materials like oils alcohol or sugar which do not conduct electricity cannot be heated by the method of ohmic heating. Hence the applicability of ohmic heating is dependent on product electrical conductivity, however most food material contains good amount of water and charged ions viz acids, salts or proteins, due to which it have significant electrical conductivity which render the flow of electric current. The drag on ionic movement goes on reducing due to decrease in the viscosity of the material as the temperature starts to increase hence the electric conductivity also goes on changing during the ohmic effect.

When a constant voltage control is used heating becomes more efficient at higher temperatures because of increase in conductivity. A material at room temperature exhibit lower conductivity than fluid carrier, yet become more conductive at higher temperatures, with obviously strong implication on heating rates. Since electrical conductivity is related to the concentration of ionic constituents, it is possible to alter porous solid property by infusion of salts. For liquids, σ-T relationship is linear regardless of the mode of heating used and electrical conductivity decreases with increasing pulp content, as a result of non-polar constituents in the pulp.In a fluid containing particulate material, if the electrical conductivity of the two phases is similar, heat could be generated at the same or equivalent rate in both phases during ohmic heating process due to the flow of current in both the phases of the materials (Fig.2). In other conditions, heat can also be generated faster in the particulate than in the continuous phase (Ruan, et al. 2002a).

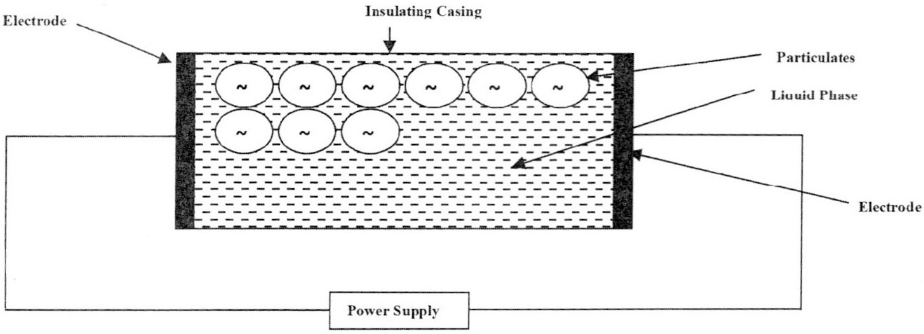

Fig. 2: Simultaneous flow of current in particulates and liquid of a suspension

Ohmic methods thus could be used as an alternative method for processing particulate food at the rate of HTST (high temperature short time) processes(Ruan, et al. 2002a), but without the limitation of non uniform heating in particulate food mixtures as it occurs in conventional HTST. At a specific temperature, electrical conductivity depends upon the resistance of the material and the geometry of the ohmic heating chamber (Eq.2). However σ can be changing linearly with temperature (in most of the liquid materials), or non-linearly (solid materials), the expression for the linear change in electric conductivity with the temperature can be modeled as (Sastry et al. 2014)

$$\sigma_T = \sigma_o [1+mT] \quad (3a)$$

Or

$$\sigma_T + \sigma_r [1 + m(T-T_r)] \quad (3b)$$

Where:

σ_T = electrical conductivity at temperature, T

σ_o = electrical conductivity of material at 0°C

σ_r = electrical conductivity at any reference temperature, T_f

m = temperature coefficient of electrical conductivity, °C^{-1}

T = Temperature at which σ is to be calculated, °C

As the product heats during ohmic heating it does not experience a large temperature gradient within itself, the liquid and the particulate are heated virtually simultaneously, which results in less heat damage to the liquid phase and prevents overcooking of the outside of the particulate. Another major advantage is that there are no heat transfer surfaces; this reduces the possibility of deposit formation (fouling) and transfer of burnt particles to the finished product.

3. Factors Affecting Ohmic Heating Process

Following are the major factors which do affect the ohmic heating process:

3.1 *Electrical conductivity*

The heat generation rate during ohmic heating critically dependents on electrical conductivity (Ruan, et al. 2002a). The rate of ohmic heating is directly proportional to the square of theelectric field strength and the electrical conductivity;hence the design of the effective ohmic heaters depends on the electrical conductivity of foods.

3.2 *Electric field strength*

Higher the electric field intensity, higher is the electrical conductivity and rapid is theheating rate. The field strength applicationresults in increasing fluid motion through the capillaries, which is directly proportional toelectrical conductivity (Kaur, & Singh 2015). Higher strength fields will results in higher rate of microbial inactivation, enhanced rate of extraction

3.3 *Particle Size and concentration of material*

The rate of heating in ohmic heating is affected by particle size, as the particle size increases the electric conductivity decreases thus the rate of heating goes on decreasing (Kaur, & Singh 2015). Therefore the heating efficiency reduces in liquids during ohmic heating as the particle size increases.Particle concentration is animportant factor for the heating rateof the two phases during ohmic heating. Brix value, concentration or the percentage of the suspended particles in a liquid phase during ohmic heating significantly affects the rate of heating.

3.4 *Ionic Concentration*

The structural changes in the tissue like cell wall proto-pectin breakdown, expulsion of non conductive gas bubbles, softening, and lowering in aqueous phase viscosity causes the increase in the electrical conductivity during ohmic heating of biological tissue which leads to the increase in the ionic mobility (Kaur, & Singh 2015; Makroo et al. 2019). Therefore as the ionic concentration goes on increasing, rapid will be the rate of heating.Hence a uniform ionic concentration will be essentially required to heat a material effectively by the ohmic heating process.

3.5 *Particle Location*

Particle orientation and position also play important role during ohmic heating.The ohmic heating behavior of a mass of particles is influenced by their location with respect tothe electrodes (Kaur, & Singh 2015). It has been found that for the parallel condition, the liquid phase heated faster than the solid phase whilein the series condition, the reverse was observed.

3.6 *Electrodes*

Electrodes of different material (Stainless steel, titanium, aluminum etc) and shapes (dimensions) heats with different rate during ohmic heating and the electrolysis may lead to the transfer of ions of electrode from electrode to the material during ohmic heating, therefore the selection of appropriatematerial and dimensions of the electrode are the important factor to be considered (Samaranayake et al. 2005). Also the electrodes are a potential cause of heat loss during ohmic heating, heat losses from the periphery of the food material depends significantly on the type of product containment cell and electrodes, which results in unacceptably high temperature gradients within the product (Kaur, & Singh 2015).

4. Prominent and Possible Applications of Ohmic Heating in Food Processing

Processing of the food is being continued since centuries, initially people started to preserve the food material by sun drying, pickling and or icing. With the advancement in science people became conscious about the quality of food and hence as the time passes modern techniques of food processing are being developed to overcome the drawbacks of old or conventional methods. The processes are developed while considering the quality of food, efficiency, convenience and safety of the technique. In this context ohmic heating has been found superior as compared to conventional as well as to the various modern techniques of the food processing (Makroo et al. 2020). Below is the major food processing applications of ohmic heating:

4.1 *Cooking*

Cooking is a process where heat is applied to the food material with or without other ingredients to bring certain desired changes in organoleptic characteristics like flavor, color, texture etc. Because of the quick and uniform heating capabilities, ohmic heating can be used as a new technique for the cooking of the food. The ohmic heating assisted cooking has been found successful in reducing the cooking time and energy consumption and producing safer food products (Farid et al. 2014). Cooking time and rate of heating may vary from material to material, therefore cooking rate or time can be regulated by adjusting the applied voltage during ohmic heating cooking process. Generally after cooking of the food, consumer demand depends mainly upon its taste andthe overall quality,the quality tests has showed that in ohmically and conventionally cooked meat products the overall quality in terms of fat, moisture and textural properties were similar.

4.2 *Sterilization and pasteurization*

Conventional, batch method of sterilization has been nowadays swapped with the modern, continuous method for the efficiency, convenience and safety. However, the shifting from a batch to a continuous processing in order to achieve a high-temperature short-time (HTST) treatment of solid–liquid mixtures is difficult and mostly it is a practical challenge (Ruan et al 2002). Ohmic heating is a promising way for aseptically processing fluids containing large particles which is impossible to achieve using other heating techniques (Makroo et al. 2019). Hence ohmic heating can be applied for the sterilization of wide range of food products.Most of the liquid food materials can be pasteurized efficiently and more precisely by the application of ohmic heating, with greater control and regulations.

4.3 *Blanching*

The main objective of the blanching process in food processing is to inactivate the quality deteriorating enzymes in fruits and vegetables by the application of heat. In conventional method of blanching heat transfer occurs by the conduction into the solid or particulate structure of fruits and vegetables, this is a time consuming process whereas during ohmic blanching heat is generated rapidly within the particulate matter itself (Icier & Bozkurt, 2014). Ohmic blanching of multi phase food material can be made more efficient by making the electrical conductivity of all the phases similar. When the electrical conductivity of food material

4.4 *Thawing*

Thawing or the opposite process of freezing, because of the distinction in thermodynamic (higher heat capacity and lower thermal conductivity of water than ice) properties of water and ice, thawing is more time consuming as compared to freezing to the same gradient of temperature. During ohmic thawing heat is generated due to the resistance of frozen food to the flow of electric current hence the product gets thawed rapidly and volumetrically. It is a novel method in significantly shorter time than that of conventional method. The lesser time of thawing helps in maintaining quality of the product with less chances of microbial attack (Seyhun et al. 2014). Ohmic thawing is having some limitations such as good contact of the food with electrodes, electrical properties, shape and size also plays an important role during ohmic thawing of frozen foods

4.5 *Extraction*

The application of ohmic heating for the extraction increases the efficiency and the rate of extraction such as extraction of solutes, dye, oil, juices from the vegetable tissue (Kumar et al. 2014). The rate of extraction can be controlled by varying the process parameters such as time, temperature, voltage per unit length (V/cm), frequency of electric current etc. such as experimentally it has been proved that lowering the frequency of alternating current significantly increased the efficiency of extraction of oil which may be due to the electroporation of the plant tissue during ohmic heating.

5. Effects of Ohmic Heating on Various Aspects

5.1 Food quality

The ohmic heating requires less time comparatively to process a food material to the same degree as by conventional methods. Therefore overall quality of food material can be maintained to a superior level by reducing length of processing time by ohmic heating (Ruan et al. 2002a). In many studies it has been found that food quality of ohmically processed food material is either superior or similar as yielded by conventional methods (Knirsch et al. 2010; Kaur, & Singh 2015; Saxena et al. 2016). Moreover research is still required to understand the effect that ohmic heating may have on the nutritive, organoleptic, and functional properties of various foods. Essential research must be conducted regarding possible effects of ohmic heating on the bio-molecules, thus contributing to improve and optimize food processing. A detailed understanding of current applications must be obtained and further research efforts is essential to contribute to the validation of Ohmic heating in novel applications. Ohmic heating in food industry are used for preheating, cooking, blanching, processing, extraction of valuable compounds without compromising the nutritional and organoleptic properties, reduce processing time and better yield (Icier & Ilicali, 2005). The nutritional, sensory and processing parameters are interdependent and ohmic heating without compromising quality parameters of food products.Ohmic heating as new cooking method applied to meat patties reduced the cooking time without affecting the quality parameters such as moisture and fat content (Ozkan et al., 2004). Ohmic as semi cooking treatment in meat balls retained moisture and fat, and higher yield with good sensory acceptability (Sengun et al., 2014). The pasteurization by ohmic heating improved thephysiochemical and sensory qualities of fermented red pepper paste as compared to conventional method (Cho et al., 2016). Texture is one of the quality parameter of food products determine its acceptability by the end use consumers. Gahavain et al (2019) reviewed the effect of ohmic heating on texture of fruits (apple, pineapple) and vegetables (tomato, cauliflower, potato, carrot, turnip and radish) showed improved texture as compared to conventional method of heating. The quality parameters of tomato like moisture, pH, lycopene andacidity are important in terms of sensory and consumer acceptability. Uses of ohmic heating as processing method to tomato concentrate preserve the quality parameters due to rapid uniform heating method (Fadavi et al., 2018; Makroo et al. 2016). The ohmic heating have been used as alternate for steam for heating processing in shrimps for higher yield, and maintaining color and nutritional value (Lascorz et al., 2016).The pasteurization of citrus juice by ohmic heating retains most of carotenoids with good organoleptic and nutritional quality (Achir et al., 2016). The ohmic heating also has application in extraction of pigments from different sources with improve in quality. Ohmic heating assisted extraction produce colorant pigment from black rice bran with improve in physico-chemical properties, colorant yield, anthocyanin pigments and bioactive compounds with good quality as compared to conventional method (Loypimai et al., 2015). Thus ohmic heating is a minimally processing technique with wide applications in retains the nutrient loss, good quality and acceptability of food products to end use consumers.

5.2 Enzymes

The food quality deteriorating enzymes such as lipoxygenase, polyphenoloxidase, peroxidase, etc may be present either inherently in food or can be produced by microorganism while growing on it (Saxena et al. 2017). Pectin methylesterase is an enzyme that has been found in essentially every plant tissue, several fungi and bacteria. Therefore the inactivation of these enzymes is an important concern while processing most of the food materials (Kumar et al. 2014). Electrical fields, applied during ohmic heating caused faster inactivation of lipoxygenase, pectin esterase, peroxidase and polyphenoloxidase as compared to conventional heating (Kaur, & Singh 2015; Saxena et al. 2018). However the trend of the inactivation has been reported similar in conventional and ohmic heating methods of thermal treatment of food materials such as the inactivation of exogenous pectin methylesterase in apple juice and cloudberry jams has been found to follow first-order kinetic by conventional as well as ohmic heating methods. The shelf life of fruit and vegetable can be increased by controlling their enzyme system and extended their shelf life for some extended period of time. The ohmic heating also has applications in controlling the enzymes of fruits and vegetables and increases their shelf life. The enzymatic browning in fruits and vegetable during cutting and peeling produce melanin and benzoquinone from phenolsdue to polyphenol oxidase enzyme and reduce their quality. The blanching treatment with main aim of reduce phenol oxidase through ohmic heating were reported by Pham et al. (2014) in pine apple, Delfiya &Thangavel (2016) in coconut water and Abedelmaksoud et al. (2018) inapple juice. Ohmic heating reduces the activity of various enzymes such as Polyphenoloxidase in sugarcane juice (Saxena et al. 2018) and watermelon juice (Makroo et al. 2017), pectin esterase enzyme in apple juice (Jacob et al., 2010), orange juice (Demirdoven & Baysal, 2015) and tomato paste (Makroo et al., 2016); peroxidase enzyme in pea puree (Icier et al. 2006), Jackob et al., 2010) in potatoes and carrot and in sugarcane juice (Brochier et al., 2016); Lipoxygenase in soya bean by (Castro et al., (2004) and alkaline phosphatase in bovine milk by Jackob et al., (2010) and in raw milk (Castro et al., 2004) .

5.3 Microorganisms

The primary mechanisms of microbial inactivation in ohmic heating are thermal in nature (Kaur, & Singh 2015). It has been also proved by many researchers that ohmic heating may have mild non thermal effects on microbial activity as well (Vicente, et al. 2014; Makroo et al. 2020). The study on few microorganisms in some food materials has revealed the non-thermal effect may be due to the presence of electric field leading to the cellular damages of the microorganisms (Ruan et al. 2002b). Ohmic heating at lower frequencies (50-60Hz), the electric field causes the charge build up hence form pores known as electroporation of cell wall of the microbes which ultimately gives the lower D value as compared to conventional thermal techniques applied for same time and temperature combinations (Knirsch, et al. 2010). However more research is required to entirely understand all the effects produced by ohmic heating. The effects of the applied electric field, the incident electric current and the applied electric frequency during ohmic heating for different microorganisms and foods (at molecular and cellular level) still need to be studied more intensely. It would be significantly economic for industries if confirmed, the electroporation effect or any other temporary permeabilization effect on cellular membranes that occurs during ohmic heating (Ruan, et al. 2002b). The ohmic heating maintains the safety and shelf life of food products by decreasing microbial load and destruction of pathogenic microorganisms. The ohmic heating replaces the conventional methods of thermal processing such as sterilization and pasteurization in various food products for microbial destruction without affecting consumer acceptability. Ohmic heating combined with other processing treatments increase the efficiency of ohmic heating in microbial inactivation in treated foods. Ohmic heated orange juice showed inactivation of microbial spores with better shelf life and good sensory acceptability (Leizerson & Shimoni, 2005a). Apple cubes treated with ohmic heating combined with osmatic dehydration reduce their total plate and yeast count (Moreno et al. 2013). The pathogenic microorganism such as *E coli, Salmonella typhimurium; Listeria monocytogenes* reduced them below the detection limit in orange juice, tomato juice, apple juice, and skim milk during ohmic heating (Lee et al. 2012; Park and Kang 2013; Park et al. 2017). Milk and milk products have limited period of shelf life and easily vulnerable to microorganism and thermal degradation of nutrients. Ohmic heating can be used a replacement for conventional methods of processing for microbial degradation with retention of most of nutrients. Milk treated with ohmic heating reduced the spores of *Streptococcus thermophiles* as compared to conventional method (Sun et al., 2008). Therefore understanding, characterizing and modeling this phenomenon is required in order to optimize and possibly exploit its effects.

6. Advantages and Limitations of Ohmic Heating Process

6.1 Advantages

- Continuous generation of heat without heat transfer surfaces
- No upper limit for the desired temperature
- Easy to monitor the process
- Electric energy directly converted into heat hence much efficient
- Rapid and uniform heating of materials having higher viscosity or containing particulates
- Ideal process for shear-sensitive products because of low flow velocity;
- Minimum fouling on the surfaces and less cleaning requirements
- Simpler process control and lower maintenance cost
- Environmentally friendly system
- Maintenance of the color and nutritional value of food

6.2 Limitations

- More installation cost of ohmic food processing systems to that of conventional retorting, freezing and heating in a conventional tubular heat
- Not applicable to the food material which are rich in oil or fat
- Not suitable for dry powder or food material with minimum water content
- In case pathogens are present in a fat globule like structure within the food material, it can feel less heat therefore will not get exposed to the desired temperature or may survive through the whole treatment process

Example 1

A voltage of 442 V was applied for 1.5 minutes to different food materials in a cylindrical ohmic cell of volume 250 cm^3 and 8.84 cm long. The electrical conductivity observed is mentioned in table below, assume that no change took place in the respective electrical conductivity during the experiment find the total amount of heat generated in each sample.

Material	Tomato juice	Orange Juice	Yoghurt	Milk	Mixed fruit juice
Conductivity (S/m)	0.863	0.567	0.609	0.390	0.191

Solution

$$EFS = \frac{Voltage}{Length}$$

$$= \frac{442}{8.84} \times 10^2 \, \mathrm{V/m}$$

$$= 50 \times 10^2 \mathrm{V/m}$$

Volume of the cell = 250 cm^3 = 250 $\times 10^{-6}$ m^3

Time = 1.5 min = 90 sec

Therefore by applying the Eq.1

$Q = E^2 \times \sigma$

The total amount of heat generated will be:

$Q = [E^2 \times \sigma] \times$ volume $\times$ time (joules)

Hence for:

I. Tomato juice

$=(50\times10^2)^2\times0.863\times250\times10^{-6}\times90$

$=485437$ joules or 485.4kJ

II. Orange juice

$=(50\times10^2)^2\times0.567\times250\times10^{-6}\times90$

$= 318937$ joules or 318.9kJ

III. Yoghurt

$=(50\times10^2)^2\times0.609\times250\times10^{-6}\times90$

$=342562$ joules or 342.5kJ

IV. Milk

$=(50\times10^2)^2\times0.390\times250\times10^{-6}\times90$

$=21937$ joules or 219.3kJ

V. Mixed fruit juice

$=(50\times10^2)^2\times0.191\times250\times10^{-6}\times90$

$=107437$ joules or 107.4kJ

Example 2

Homogenous fruit juice having an electrical resistance of 0.23Ω at 0°C and temperature coefficient of 0.15/°C was allowed to heat in a cylindrical ohmic cell 13.26 cm long with circular stainless steel electrodes 0.2 cm thick and 3.6 cm radius on the both ends. Calculate the electrical conductivity of the material at 56°C

Solution

Area of the electrode $= \pi r^2$

$= 3.14\times3.6\times3.6 = 40.69\text{cm}^2 = 40.69\times10^{-4}\text{m}$

Length $= 13.6\text{cm} = 13.6\times10^{-2}$

$R = 0.23\Omega$

Applying equation (2)

$$\sigma_0 = \left(\frac{1}{R}\right)\left(\frac{L}{A}\right)$$

$$= \left(\frac{1}{0.23}\right)\left(\frac{13.6\times10-2}{40.69\times10-4}\right)$$

$= 1.45$ S/m

Hence σ_{56} can be calculated by the equation (3a)

$\sigma_{56} = \sigma_0\ [1 + mT]$

$= 1.45[1+0.15\times56]$

$= 13.63$ S/m

Example 3

Tomato juice was thermally treated by ohmic heating (24V/cm) at 90°C to inactivate the pectin methyl esterase (PME) enzyme. The enzyme activity was reduced to a residual activity of 10.2% in 45sec; consider that the enzyme inactivation followed the first order reaction. Calculate the value of reaction constant "k".

Solution:

First order equation:

$$\frac{\text{Final activity}}{\text{Initial activity}} = \exp(-kt)$$

$$\frac{0.102}{1} = \exp(-k\times45)$$

$$\frac{\text{Ln}(0.102)}{45} = -k$$

$k = 0.0507 s^{-1}$

Example 4

Ohmic heating process was used to process vegetable juice sample with initial count of 2.3×10^2 of an aerobic bacteria. After the ohmic (32V/cm) heat treatment of 2.5min at 80°C and at 90°C the bacterial count was reduced to 1.6×10^{-1} and 3.8×10^{-4} respectively. Find the D_{80}, D_{90} value of the pathogenic bacteria during the ohmic processing

Solution:

We Know that:

$$log\left[\frac{N_f}{N_0}\right]=-\left[\frac{t}{D}\right]$$

Therefore $D_{80} = -\left[\frac{2.5}{\log\left(1.6\times10^{-1}/2.3\times10^{2}\right)}\right]$

$= 0.791$ min

Similarly $D_{90} = -\left[\frac{2.5}{\log\left(3.8\times10^{-4}/2.3\times10^{2}\right)}\right]$

$= 0.432$ min

References

Abedelmaksoud, T. G., Mohsen, S. M., Duedahl-Olesen, L., Elnikeety, M. M., &Feyissa, A. H. (2018). Optimization of ohmic heating parameters for polyphenoloxidase inactivation in not-from-concentrate elstar apple juice using RSM. Journal of Food Science & Technology, 55(7), 2420–2428.

Achir, N., Dhuique-Mayer, C., Hadjal, T., Madani, K., Pain, J. P& Manuel Dornier, M. (2016).Pasteurization of citrus juices with ohmic heating to preserve thecarotenoid profile. Innovative Food Science and Emerging Technologies 33, 397–404.

Brochier, B., Mercali, G. D., & Marczak, L. D. F. (2016). Influence of moderate electric field on inactivation kinetics of peroxidase and polyphenol oxidase and on phenolic compounds of sugarcane juice treated by ohmic heating. LWT- Food Science and Technology, 74, 396–403.

Castro, I., Macedo, B., Teixeira, J. A., & Vincente, A. A. (2004). The effect of electric field on important food-processing Enzymes: Comparison of inactivation kinetics under conventional and ohmic heating. Journal of Food Science, 69(9), C696–C701.

Cho, W. I., Yi, J. Y., & Chung, M. S. (2016). Pasteurization of fermented red pepper paste by ohmic heating. Innovative Food Science & Emerging Technologies 34, 180–186.

Delfiya, D. S. A., &Thangavel, K. (2016). Effect of ohmic heating on polyphenol oxidase activity, electrical and physicochemical properties of fresh tender coconut water. International Journal of Food Engineering, 12(7), 691–700.

Demirdoven, A. H. &Baysal, T. (2015). Effects of electrical pre-treatment and alternative heat treatment applications on orange juice production and storage. Food and Bioproducts Processing, 94, 443–452.

Fadavi, A., Youseû, S., Darvishic, H., &Mirsaeedghazi, H.(2018). Comparative study of ohmic vacuum, ohmic, and conventional-vacuum heating methods on the quality of tomato concentrate. Innovative Food Science and Emerging Technologies 47, 225–230.

Farid, M. M., Kong, W. J. and Ozkan, N. (2014). Ohmic Cooking of Food. In: H. S. Ramaswamy, M Marcotte, S. Sastry and K. Abdelrahim. Ohmic Heating in Food Processing. CRC Press, Taylor & Francis Group, ISBN 13: 978-1-4200-7109-2 (eBook - PDF)

Gavahiana, M.,.Tiwari,B. K., Chua, Y., Ting, Y., &Farahnaky, A. Food texture as aûected by ohmic heating: Mechanisms involved, recent ûndings, beneûts, and limitations Trends in Food Science & Technology 86 (2019) 328–339.

Icier, F. & Bozkurt, H. (2014). Ohmic Blanching. In: H. S. Ramaswamy, M Marcotte, S. Sastry and K. Abdelrahim. Ohmic Heating in Food Processing. CRC Press, Taylor & Francis Group, ISBN 13: 978-1-4200-7109-2 (eBook - PDF)

Icier, F., &Ilicali, C. (2005). Temperature dependent electrical conductivities of fruit purees during ohmic heating. Food Research International 38(10), 1135–1142.

Icier, F., Yildiz, H., &Baysal, T. (2006). Peroxidase inactivation and colour changes during ohmic blanching of pea puree. Journal of Food Engineering 74, 424-429

Jakob, A., Bryjak, J., Wójtowicz, H., Illeova, V., Annus, J., &Polakovic, M. (2010). Inactivation kinetics of food enzymes during ohmic heating. Food Chemistry, 123,369–376.

Kaur, N & Singh, A K. (2015). Ohmic Heating: Concept and Applications - A Review. Critical Reviews in Food Science and Nutrition, DOI: 10.1080/10408398.2013.835303.

Knirsch, et al. (2010). Ohmic heating: a review. Trends in Food Science & Technology 21, 436-441.

Kumar, J. P., Ramanathan, M. and Ranganatha, T. V. (2014). Ohmic Heating Technology in Food Processing - A Review. International Journal of Engineering Research & Technology 3 (2): 1236-1241.

Lascorz, D., Torella, E., Lyng, J. G., & Arroyo, C. (2016). The potential of ohmic heating as an alternative to steam for heat processing shrimps. Innovative Food Science & Emerging Technologies, 37, 329–335

Lee, S., Sagong, H., Ryu, S., & Kang, D. (2012). Effect of continuous ohmic heating to inactivate Escherichia coli O157:H7, *Salmonella Typhimurium* and *Listeria monocytogenes* in orange juice and tomato juice. Journal of Applied Microbiology, 723–731.

Leizerson, S., &Shimoni, E. (2005a). Effect of ultrahigh-temperature continuous ohmic heating treatment on fresh orange juice. Journal of Agricultural and Food Chemistry, 53(9), 3519–3524.

Loypimai, P., Moongngarm, A., Chottanom, P., &Moontree, T. (20150. Ohmic heating-assisted extraction of anthocyanins from black rice bran to prepare a natural food colourant. Innovative Food Science and Emerging Technologies 27, 102–110.

Makroo, H. A., Prabhakar, P. K., Rastogi, N. K., & Srivastava, B. (2019). Characterization of mango puree based on total soluble solids and acid content: Effect on physico-chemical, rheological, thermal and ohmic heating behavior. *LWT*, *103*, 316-324.

Makroo, H. A., Rastogi, N. K., & Srivastava, B. (2016). Enzyme inactivation of tomato juice by ohmic heating and its effects on physico-chemical characteristics of concentrated tomato paste. Journal of Food Process Engineering, 40(30), 12464.

Makroo, H. A., Rastogi, N. K., & Srivastava, B. (2020). Ohmic heating assisted inactivation of enzymes and microorganisms in foods: A review. *Trends in Food Science & Technology*.

Makroo, H. A., Saxena, J., Rastogi, N. K., & Srivastava, B. (2017). Ohmic heating assisted polyphenol oxidase inactivation of watermelon juice: Effects of the treatment on pH, lycopene, total phenolic content, and color of the juice. *Journal of Food Processing and Preservation*, *41*(6), e13271.

Moreno, J., Simpson, R., Pizarro, N., Pavez, C., Dorvil, F., Petzold, G., et al. (2013). Influence of ohmic heating/osmotic dehydration treatments on polyphenoloxidaseinactivation , physical properties and microbial stability of apples (cv .Granny Smith). Innovative Food Science & Emerging Technologies, 20, 198–207.

Ozkan, N., Ho, I., & Farid, M. (2004). Combined ohmic and plate heating of hamburger patties: Quality of cooked patties. Journal of Food Engineering, 63, 141-145.

Park, I., & Kang, D. (2013). Effect of electropermeabilization by ohmic heating for inactivation of Escherichia coli O157:H7, Salmonella enterica serovar typhimurium, and Listeria monocytogenes in buffered peptone water and apple juice. Applied and Environmental Microbiology 79(23), 7122–7129.

Park, I., Ha, J., & Kang, D. (2017). Investigation of optimum ohmic heating conditions for inactivation of Escherichia coli O157:H7, Salmonella enterica serovar Typhimurium, and Listeria monocytogenes in apple juice. BMC Microbiology, 17(117), 1–8.

Pham, H., Jittanit, W., &Sajjaanantakul, T. (2014). Effect of indirect ohmic heating on quality of ready-to-eat pineapple packed in plastic pouch. Songklanakarin Journal of Science and Technology, 36(3), 317–324.

Ruan, et al. (2002). Ohmic Heating. In: P. Richardson. Thermal Technologies in Food Processing. Woodhead Publishing Limited. ISBN 1 85573 558

Ruan, et al. (2002a). Ohmic Heating. In: K. J. Henry and C. Chapman. The Nutrition Hand Book for Food Processors. Woodhead Publishing Limited. ISBN 1 85573 665 9 (e book)

Samaranayake, C. P., Sastri, S.K. and Zhang, H. (2005). Pulsed Ohmic Heating - A Novel Technique for Minimization of Electrochemical Reactions During Processing. Journal of Food Science, 70 (8): E460-E465.

Sastry, S. K., Heskitt, B. F., Sarang, S. S., Somavat, R. and Ayotte. K. (2014). Why Ohmic Heating? Advantages, Applications, Technology, and Limitations. In: H. S. Ramaswamy, M Marcotte, S. Sastry and K. Abdelrahim. Ohmic Heating in Food Processing. CRC Press, Taylor & Francis Group, ISBN 13: 978-1-4200-7109-2 (eBook - PDF)

Saxena, J., Ahmad Makroo, H., & Srivastava, B. (2017). Effect of ohmic heating on Polyphenol Oxidase (PPO) inactivation and color change in sugarcane juice. *Journal of Food Process Engineering*, *40*(3), e12485.

Saxena, J., Makroo, H. A., & Srivastava, B. (2016). Optimization of time-electric field combination for PPO inactivation in sugarcane juice by ohmic heating and its shelf life assessment. *LWT-Food Science and Technology*, *71*, 329-338.

Saxena, J., Makroo, H. A., Bhattacharya, S., & Srivastava, B. (2018). Kinetics of the inactivation of polyphenol oxidase and formation of reducing sugars in sugarcane juice during Ohmic and conventional heating. *Journal of Food Process Engineering*, *41*(3), e12671.

Sengun, I. Y., Kendirci, P., & Icier, F. (2013). Effect of ohmic treatment on quality characteristic of meat : A review. Meat Science, 93(3), 441–448.

Seyhun, N., Sumnu, S.G and Ramaswamy, H. S. (2014). Ohmic Heating as Thawing and Tempering Technology. In: H. S. Ramaswamy, M Marcotte, S. Sastry and K. Abdelrahim. Ohmic Heating In Food Processing. CRC Press, Taylor & Francis Group, ISBN 13: 978-1-4200-7109-2 (eBook - PDF)

Sun, H., Kawamura, S., Himoto, J., Itoh, K., Wada, T., & Kimura, T. (2008). Effects of ohmic heating on microbial counts and denaturation of proteins in milk. Food Science and Technology Research, 14(2), 117–123.

Vicente, et al. Vicente, A. A., Pereira, R. N., Penna, T. C. V.and Knirsch, M. (2014). Electricity Effects on Microorganisms and Enzymes. In: H. S. Ramaswamy, M Marcotte, S. Sastry and K. Abdelrahim. Ohmic Heating In Food Processing. CRC Press, Taylor & Francis Group, ISBN 13: 978-1-4200-7109-2 (eBook - PDF)

Zareifard, M. R., Marcotte, M., Ramaswamy, H. S. and Karimi, Y. (2014). Electrical Conductivity: Importance and Methods of Measurement. In: H. S. Ramaswamy, M Marcotte, S. Sastry and K. Abdelrahim. Ohmic Heating In Food Processing CRC Press, Taylor & Francis Group, ISBN 13: 978-1-4200-7109-2 (eBook - PDF).

22

Pulsed Electric Field Processing of Foods

Akhil K.G., Madaraboina Venkateswara Rao, Akash Pare and Sunil C.K.

1. Introduction

Thermal processing treatments are the conventional methods that have been broadly employed in the food industry to guarantee the microbiological safety of food materials at a minimal cost. However, these thermal treatments cause undesirable changes in flavor, nutritive value, texture, or color of the finished product that impact food quality (Pan et al., 2017). Growing the modern consumer's demand towards the minimally processed and "fresh-like" taste foods has challenged the food industry to develop novel non-thermal preservation techniques as an alternative to thermal processing. The application of novelnon-thermal techniques has been extensively investigated and applied, including ultrasound treatment, gamma irradiation, light processing, high-pressure processing, high-intensity magnetic field, and high intensity pulsed electric field (PEF). Among them, PEF has been recognized as the mostuseful and promising technology because of its reduced heating effects and short treatment times concerning other non-thermal techniques. PEF technology has been widely usedfor microbial decontamination and inactivation, enzyme modification, extraction, drying and processing of egg, meat and seafood, milk, and milk products (Altunakar et al., 2002).

PEF is a novel non-thermal preservation technology with exceptional ability to extend the shelf life of food products by inhibiting microbial growth during the processing of foods and has the advantages of food safety and minimal influence onflavor, sensory and nutritional characteristics (Lado & Yousef, 2002). This chapter presents the outline of the principal, mechanism, equipment, and implementation of PEF for the processingof food products and its effects on quality parameters.

2. Principle of PEF

The principle of PEF involvesthe discharge of high intensity short electrical pulses (typically 20 – 80 kV/cm) into the food products such as liquid or semi-solid foods, which is placed between a set of electrodes for a few microseconds to milliseconds. The high electric field pulses in the range of 20 – 50 kV/cm strength are applied for pasteurization to inactivate microorganisms. The application of high voltage electric fields to the foods causeslocal structural changes and a rapid breakdown of the cell membrane resulting in the increment in electrical conductivity and permeability of cellular material. The changes may be due to the appearance of electric charge carriers in food materials such as a large concentration ofions, referred to as "*electroporation*"(Agarwal & Raghav, 2019). When the electric field is applied to a cell membrane, thesehigh voltage electrical pulses ensures the temporary perforation and destabilization of proteins and lipid bilayers. As electric fields are detached from cells, the pores can realign themselves if the induced poresremain smaller than themembrane area. Suppose the intensity and strength of the electric field are too high. In that case, there is an irreversible pore formation in the microorganism that willconsequently start leaking small compounds or even undergoes cell lysis or cell death (Ortega-Rivas, 2012).

The magnitude of voltage needed for the critical breakdown of the cell membrane depends on the biological cell type, shape, size, diameter, growth phase of the microorganism, and the medium in which the microorganism is present. The average diameters of biological tissue and microorganism cells vary from 10 μm to 1 mm and 10 nm to 1 μm. Normally, the voltage of trans-membrane (u_m)induced on the cell membrane possess to an external electric field is given as (Ortega-Rivas, 2012)

$$u_m = \alpha d_c E\cos\theta \tag{1}$$

where, α depends on cell shape (α = 0.75 for spherical cell; α = 1 for rectangular cell);

d_c = cell diameter in cm;

E = electric field strength in kV/cm;

θ = angle between the direction of the electric field and a point on the membrane surface.

3. PEF Treatment System

The typical PEF processing system isa simple electrical system represented in Fig.1. It is mainly consisting of a high-voltagepulse generating system, a treatment chamber, a suitable product handling system, and a set of monitoring and controlling devices (Jayathunge et al., 2019). The temperature, high voltage, high current, and high-intensity pulses can be monitored with thermocouples, current monitors, high voltage probes, and oscilloscopes. The high-voltage pulse generating system is the most crucial part of the PEF system since it produces the high electric pulses from the fast discharge of the electrical energy within a short duration (Alirezalu et al., 2020).

These high voltage electric pulses are applied to the electrodes' set and then conduct those pulses to the foodswhile passed through the treatment chamber. The resultant conduction of food material mainly depends on its conductivity, which varies as a function of ions present in the food material (Jongyingcharoen & Ahmad, 2014). The system is designed so that the first step will be a pre-heating of the food with heat exchangers. After the PEF treatment, it requires a cooling system such that the food product may be out of the ohmic heating effects and cooled back to the ambient temperature for the packaging. The packaging is done with aseptic packaging to prevent contamination of the food material (Toepfl et al., 2014).

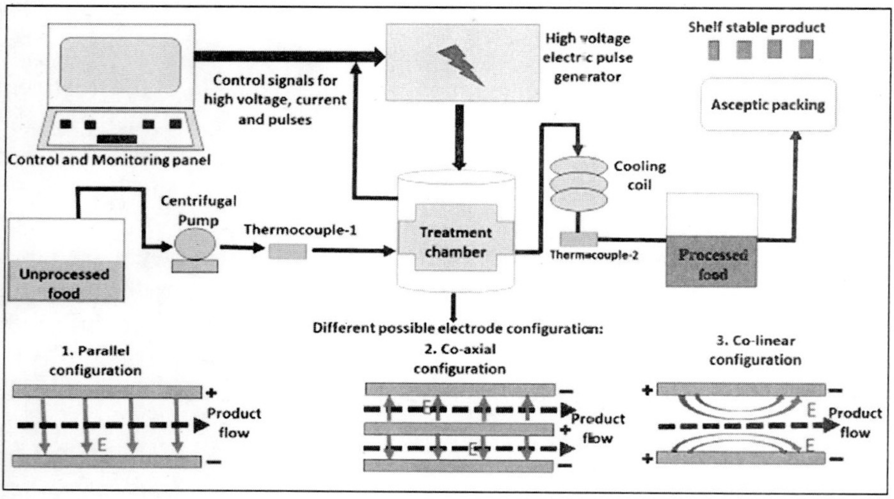

Fig.1: PEF system and treatment chamber configurations.
1. Parallel chamber; 2. Coaxial chamber; and 3. Co-linear chamber

3.1. Treatment Chamber-Design and Development

The treatment chamber is the central part of the PEF system. Food is being exposed to the high voltage electric pulses, which consists of 2 electrodes, i.e., one on a high voltage and other on-ground potential encased in the insulating medium. An insulating spacer is used to alter the chamber volume. The treatment chamber can be static or continuous, whereas continuous chambers are most suitable for food preservation purposes and static chambers for dehydration and extraction (Yogesh, 2016). The main aspects of designing a continuous PEF treatment chamber with higher capacities are to allow the high flow rate of foods (liquid or semi-liquid) through the treatment chamber that being pumped or conveyed in a continuous mode.

In contrast, the high voltage pulses are being applied to the foods at ambient, elevated, or refrigeration temperatures. The continuous operability with short treatment time is the key benefit that PEF processing can easily be incorporated into the existing industrial processing lines (Álvarez et al., 2006). Materials used to develop the treatment chamber have to be food grade, autoclavable, and chemically inert concerning foods, and consequently, their electrochemical properties should consider. The insulating material separates the electrodes in the treatment chamber in different geometric configurations. Stainless steel has been extensively utilized for the electrodes, and the other inert materials are gold, carbon, metal oxides, and platinum. The insulating materials used are nylon, polypropylene, polythene, plexiglass, polysulphone, and polyvinyl chloride (Toepfl et al., 2014).

The main factors influencing the treatment were the geometry of the treatment chamber, processing parameters, and the pulse modulators' design. The electrodes' arrangement in the treatment chamber is represented in three geometric configurations (Fig.1): 1. Parallel plate, 2. Co-axial, and 3. Co-linear (Gómez et al., 2019). Parallel plate systems are generally used for batch mode and provide the most uniform electric field in a large area. However, in boundary regions, the treatment intensity gets reduced between the electrodes while the other configurations are used for continuous flow operations. The distribution of the electric field intensity inside the treatment chamber should be homogeneously distributed to avoid the arcing areas caused by the high electric fields (Ortega-Rivas, 2012). Distribution of the electric field intensity in a treatment chamber mainly depends on the treatment chamber's characteristics, basically the shape of electrodes, a gap of the chamber, gas impurities in electric materials, etc. The optimization of the PEF treatment chamber is of great significance to apply the electric fields properly into the food materials and keep away the unpredicted phenomena during the PEF treatment, such as deposition of particles, electric field perturbations, heat accumulation, gas-bubble formation, and arching.

However, the treatment chamber design should be based on mathematical simulations to evaluate the electric field strength(Huang & Wang, 2009).

3.2. High Voltage Pulse Generating System

The pulse generators will convert the low-intensity voltage into high-intensity electric fields and provide the latter to the PEF chambers. It is mainly comprised of a charging unit and a discharging unit. PEF generation requires a slow charging and rapid discharge of electrical energy within a short duration of time, as the pulse width is short compared to the pulse duration. The charging voltage needed to generate the pulses of sufficient electric field strength is majorly dependent on the electrode distance (Elez-Martínez et al., 2012; Toepfl et al., 2014).

The pulse forming networks (PFN) is an integral part of the pulse generator, generating and forming a different kind of pulse configurations. The high voltage pulses are supplied to the system through a high voltage pulse generator by maintaining the required intensity, shape, and duration. The high-frequency power supply is mainly provided from a high voltage direct current (DC) generator, which converts the simple voltage (110 or 220 V) of alternating current (AC) to the high voltage DC (Altunakar, 2002). This generated electric power is stored in a capacitor bank connected in a parallel or series, and discharged into the PEF treatment chamber across the protective resistor and high voltage switch within a few microseconds when the circuit allows. The stored energy (W) of the capacitor is given by the equation;

$$W(J) = CV^2/2 \quad (2)$$

Where C is capacitance (F), and V is charging voltage (V)(Yogesh, 2016). Several instruments can be used as the discharge switch, including thyratrons, semiconductor switches, or high power transistors, gas, or vacuum spark gaps. A type of discharge switch plays a crucial role in determining the highest repetition rate and the maximum rate of voltage and current it can withstand and also influence the efficiency of the PEF system (Toepfl et al., 2014).

3.3. Types and Characteristics of Pulses

In PEF processing, the high-intensity electric pulses in the range of 10–80 kV/cm are applied as the short duration pulses of a few micro to milliseconds using a proper pulse forming networks (PFN). The impact of PEF on food material is a function of pulse duration and pulse energy. The square or rectangular wave, oscillating wave, bipolar, or exponential decay pulses are the most commonly used for the PEF processing of food products (Fig.2) (Nowosad et al., 2020).Among them, oscillating wave pulses have a low-efficiency level. It inhibits the microbial cells in liquid foods from being exposed to high-voltage electrical strengths continuously. The exponential decay or bipolar pulses has a very high voltage surge but with a prolonged decay rate, causing the long tail section to generate heating. However, it's not enough to inactivate the microbes in the food material. The square or rectangular wave pulses are much better than another two kinds of pulses. It can generate a stable peak voltage for the prolonged time interval, which will have a practical impact on the microbes in food materials (Toepfl et al., 2014).

4. Mechanisms of Microbial Inactivation

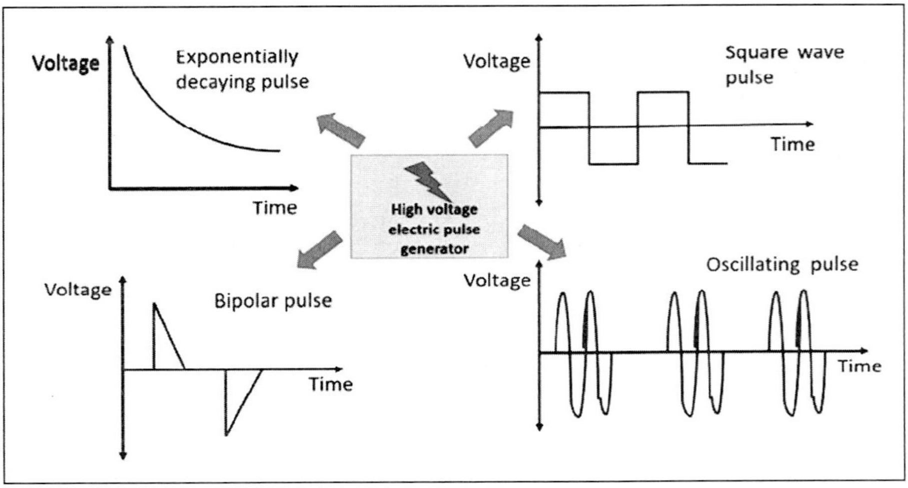

Fig. 2: Schematic representation of types of pulses

Pulse electric field has been found to have enough lethal effects on various vegetative bacteria, mold, and yeast. The electric pulse-field alone has sufficient inactivation power. In contrast, the PEF, along with the combination of heat or low temperature as other hurdles, has been a research activity in the current scenario (Agarwal & Raghav, 2019). The microorganism's inactivation is mainly associated with cell breakdown and its electromechanical instability. Simultaneously, the application does is adjusted using electric field intensity, treatment time, and the number of pulses. The microbial inactivation mechanism of PEF treated food samples has not been thoroughly clarified. Two principal theories have been proposed for microbial inactivation of foods, namely;

1. Theory of dielectric breakdown
2. Theory of Electroporation

4.1Theory of dielectric breakdown: When an electric field is applied to the cell suspension, it produces the potential difference across the cell membranes (~trans-membrane potential) and results in a sharp improvement in permeability and conductivity. When this induced trans-membrane potential goes beyond the critical value of the electric field (E_c), it causes a rupture in many cellular membranesdestroys cell membranes (Fig.3). Correspondingly, the application of an external electric field of nearly 10 kV/cm for *Escherichia coli* inactivation, which is generally accepted standard for pasteurization of foods. The theory of dielectric breakdown is referred to as electro-plasmolysis, and this can also be explained by the theory of *electroporation* (Ortega-Rivas, 2012).

4.2 Theory of electroporation: The electroporation theory is mainly based

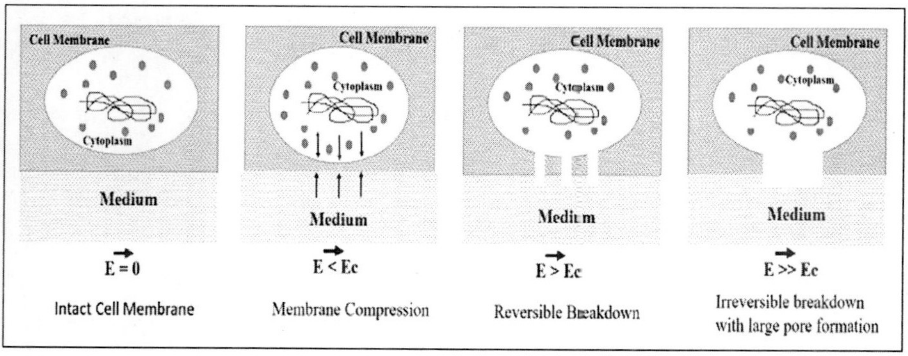

Fig. 3: Schematic representation of the compression and possible breakdown by pore formation of a biological cell subjected to a pulsed electric field treatment (E). Ec is the critical electric field (Gómez et al., 2019)

on the formation of the pores in the protein or lipid domains of the cellular membrane. Due to the permeability of lipid bilayer to ions and the electric charges of lipid molecules, the lipid bilayers in cell membranes are susceptible to the applied electric field. These electric charges will cause the lipid molecules to reorient themselves under a high electric field, forming the hydrophilic pores and impairing the lipid bilayer against ions (Fig.4). However, the electroosmotic effect and physiological impact may also impact electroplasmolysis efficiency (Ortega-Rivas, 2012; Vorobiev & Lebovka, 2019).

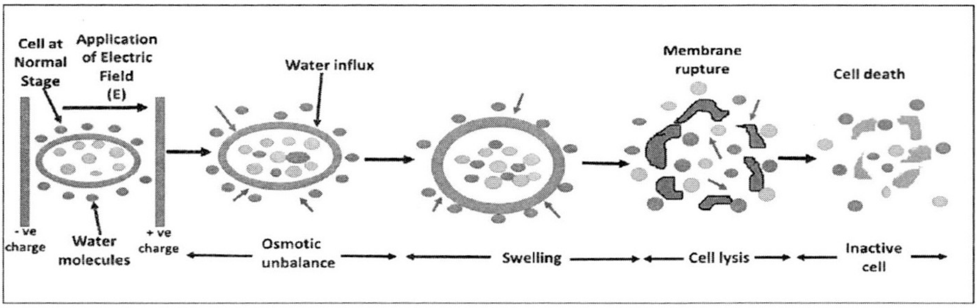

Fig. 4: Different stages of electroporation in a cell membrane induced by the PEF process through osmosis. The red arrow indicates the electric field intensity, green dots are pores, and blue dots are water molecules.

5. Factors Determining the Inactivation Efficiency of PEF Processing

The critical factors determining the inactivation efficiency of PEF processing can be classified as:

1. Processing factors
2. Treatment medium factors
3. Biological factors

5.1. Processing Factors

5.1.1 Electric Field Strength (E): Electric field strength or Electric field intensity is defined as the ratio of voltage across the food product (V) to the electrodes' distance. It represented as

$$E = V/D \quad (3)$$

Inactivation of microorganisms improves with an increase in the E over a critical electric field strength (E_c). E_c is defined as the minimum threshold field intensity required for the formation of pores in a cell membrane, and it is mainly dependent on cell size, shape, and orientation in the electric field. Equation 3 clearly shows that E is inversely proportional to D, i.e., with improvement in D, the higher voltage is needed to obtain the desired E_c (Álvarez et al., 2006).

5.1.2 Pulse shape and Pulse width: The inactivation of microbes by PEF has done with the high-intensity short pulses. These pulses may be of monopolar or bipolar while considering the polarity of the waves. The bipolar pulses have one positive and one negative pulse. The application of either exponential or square waveform bipolar pulses also resulted in 50-60% more inactivation of various microorganisms than monopolar pulses. Moreover, the lethal efficiency of PEF under different forms of pulses (monopolar or bipolar) mainly depends on the electric field strengths of the pulses (Álvarez et al., 2006).

The effect of pulse width is still controversy concerning the effect on the microbial lethality. The previous studies have found that microbial inactivation is independent of the pulse width and also some studies have also found that higher inactivation occurs when treated with longer pulses. These treatments lead to temperature development to about 40°C. Even this small rise in the temperature could improve the inactivation of microorganisms. At the same time, studies have also found that shorter pulses with the same specific energy and applied at higher field strengths were more active on the inactivation of several microorganisms than the more full pulses applied at lower field strength. These shorter pulses offer a joule heating effect and thus reduces the migration of compounds from electrodes to the treatment medium (Álvarez et al., 2006).

5.1.3 Treatment time: Total time or treatment duration is determined by using equation 4.

$$t = n\tau_p \tag{4}$$

Where 'n' is the number of pulses applied and 'τ_p' is the pulse width, the treatment time is depended on the number of pulses and pulse width. There will be an improvement in the treatment time with an increase in a number of pulses and pulse width, which improves the system's energy consumption, and consequently, there is a rise in the product temperature (Álvarez et al., 2006).

5.1.4 Frequency: Most of the frequencies applied for microbial inactivation range from 1 to 500Hz. If the applied number of pulses is the same, microbial inactivation is independent of the pulse frequency. Still, if the pulse frequency gets increased, then the treatment time decreases, leading to an increase in treatment temperature (Álvarez et al., 2006).

5.1.5 Specific Energy: Specific energy is the energy applied per unit mass. The microbial inactivation improves with the increase in the specific energy. It is a vital processing parameter that influences the cost of operation(Álvarez et al., 2006).

5.1.6 Temperature: The fundamental need of the PEF technology is to inactivate microorganisms at nonlethal temperatures to avoid the food matrix's thermal effects. Whereas the lethal effects of PEF treatment increase with an increase in the medium temperature. The inactivation has been found with both lethal and non-lethal temperatures. The inactivation of the microorganisms is increased with an increase in the temperature, as the temperature increment results in the diminution of the minimum transmembrane potential needed to induce a dielectric breakdown of the cell membrane. The diminution is said to be due to the phospholipids' temperature-dependent phase transition from gel to liquid crystalline phase (Álvarez et al., 2006).

5.2. Treatment Medium Factors

The food that has been treated in PEF is generally a complex matrix with different proportions, which consists of different proportions of components in it. Notably, the proteins, lipids, carbohydrates, and other food constituents can confer the protective effect against the PEF. The food composition is also a significant factor, as the substrate's composition, like charge ions, may lead to changes in the effect of PEF on the microorganisms in it (Altunakar, 2002).

Several studies have concluded that pH, water activity, ionic strength, and conductivity significantly influence the inactivation efficiency of PEF. Inactivation of *Salmonella* was observed to be highest in neutral and above neutral pH values. The *L.monocytogenes,* which are developed in high acid foods such as citrus juices, are found to have higher inactivation with PEF. A medium with higher water activity has higher microbial inactivation with PEF technology (Syed et al., 2017). The electrical conductivity has an inverse relation with PEF treatment, which indicates that fluid medium having the lower electrical conductivity and ionic strength are most effective while treating the PEF. This is because of the significant difference in an ionic concentration between the suspension and cell cytoplasm. This large ionic gradient makes it easy to increase the ionic substance across the cell membrane, weakening its structure and making it more susceptible to the PEF. High electrical conductivity foods are difficult to process under as they produce low peak electric fields across the treatment chambers due to high current (Álvarez et al., 2006).

5.3. Biological Factors

The effect of PEF for the inactivation of microorganisms is highly dependent on the morphological and physiological characteristics of the microorganisms treated under PEF. The effect varies with the change in cell type, cell size and shape, cell density, the cell arrangement in suspension, and dielectric breakdown (Toepfl et al., 2014). The gram-positive is said to have a rigid teichoic acid in its cell, making it more resistant to PEF than the gram-negative cells. The PEF inactivation varies with different species and different growth phase characteristics of the species. The treatment with PEF technology does not inactivate bacterial spores and viruses. The yeasts are easily killed by the PEF technology compared with other cells with smaller diameters due to a lower breakdown of trans-membrane potential because of its larger diameter. The effectiveness of PEF treatment is inversely related to the population density of the microorganisms (Syed et al., 2017).

6. Applications of PEF in Food Processing Operations

The principal advantages of using the PEF method in food processing operations like food preservation, drying, freezing, and extraction against conventional thermal treatments are:

1. Improved extraction yield
2. Increased mass transfer
3. Reduced processing time
4. Decreased intensity of the conventional extraction parameters (i.e., extraction temperature, solvent concentration)
5. Facilitation of purified extract (i.e., reducing grinding)
6. Inactivation of pathogenic bacteria with minimizing the degradation of heat liable compounds (e.g., flavors, proteins), and
7. Reduction of environmental impact and energy costs (Barba et al., 2015).

6.1. Food Preservation

As compared with conventional thermal treatment, the PEF system not only inactivates the pathogenic bacteria and also the quality deteriorating enzymes with minimizing the losses of the quality, sensory properties, and other heat liable compounds of food. Inadequate milk pasteurization causes the growth of pathogenic bacteria such as *Escherichia coli*, *Listeria* spp. and *Pseudomonas*. PEF treatment induces the 5-6 log reduction of gram-positive and gram-negative bacteria in whole milk at 22-28 kV/cm for 17-101 μs at 50°C (Syed et al., 2017). Additionally, Sharma, Oey, Bremer, & Everett, (2018) exhibited that PEF treatment inactivated the alkaline phosphatase and also reduced the plasmin oxidase (7%) and xanthine activity (30%) while bovine milk is subjected to the 25.7 kV/cm for 34 μs after heating to 55°C and it held for 24s. Similarly, the application of PEF combination with manothermosonication caused the mean reduction of 5.6 log cfu/ml of *Listeria innocua* in a milk-based smoothie. The combination of PEF with 60 ppm peracetic acid also resulted in a reduction of *E. coli* and *Listeria innocua*in blueberries without changing its color and appearance. PEF treatment improved the phenolic compounds and anthocyanins of blueberries by 25% and 10% (Nowosad et al., 2020).

6.2. Drying

Water removal from food products such as fruits and vegetables for increasing preservation has been a significant concern in energy and cost concerns. Several studies reported that the application of PEF for foods decrease its resistance to the diffusion by destroying its cell structure and improve the heat and mass transfer between the cells during drying; for example, pre-treatment of potato cubes with PEF treatment with a field strength of 1-2 kV/cm leads to decrease in the drying time for fluidized bed drying. Similarly, the drying time of PEF (0.9 kV/cm, 50 Hz, 20μs, after 1000 pulses) pre-treated parsnip and carrot slices are reduced the drying time to 28% at 70°C and to 21% at 60°C as compared to the untreated samples (Nowosad et al., 2020). However, the application of reversible permeabilization of PEF (65 pulses of 650 V/cm, 150 μs pulse width, 760 μs between pulses) to the basil leaves resulted in the reduction in drying time by 57% for air drying, 33% for vacuum drying and 25% for freeze-drying. Among them, PEFpre-treatment with vacuum dried basil samples were closest to the fresh leaves regarding smell and color determined by the sensory panel members (Telfser & Gómez Galindo, 2019).

6.3. Freezing

Formation of ice crystals during the freezing of foods is a significant drawback, can destroy the cell tissue so that after thawing the food products, mainly leafy vegetables, soft fruits will lose their shape and become sodden. Mostly, PEF treatment was applied in combination with vacuum impregnation in the presence of cryoprotectants (sucrose, glucose, trehalose, and fructose), which caused the leaf cells to remain viable and retain turgor after the process of freezing and thawing cycle. Carrot discs treated with PEF after soaking in different texturizing and cryoprotectant agents had greater ûrmness than untreated samples after thawing (Nowosad et al., 2020). The PEF system induces multicellular tissues' electroporation, leading to better connections between the intra and extracellular tissues, further allowing the increased probability of ice nucleation and faster ice propagation after freezing, consequently reducing the freezing and thawing time. For example, (Wiktor, Schulz, Voigt, Witrowa-Rajchert, & Knorr, 2015) reported that PEF treatment reduced the total freezing and thawing time of apple tissue by 3.5-17.2% and 71.5%, respectively.

6.4. Extraction

Extraction is one of the principal operations in the food industry used for obtaining valuable compounds, and generally, it involves chemical and thermal treatment of food. Several studies have reported that application of PEF can enhance its extraction efûciency, decrease in extraction time and minimize damages to the extracted nutrients; for example, citrus fruits (orange, lemon, and pomelos) treated with 3 kV/cm of PEF enhance the juice yield by 59% for lemons, 37% for pomelos, and 25% for oranges; however, peels with 10 kV/cm of PEF improved the extraction of polyphenols to 50%; and it also enhances the extracts

Table 1: Application of PEF in food processing

Process	Food products	PEF parameters	PEF effects	Reference
Preservation	Milk	25.7 kV/cm for 34 µs time using bipolar square wave pulses; pulse width(20 µs); frequency (20 Hz); after pre-heated to 55 °C with holding time of 24s;	• Inactivated the alkaline phosphatase and microbiologically stable for 21 days at 4°C. • Reduced the plasmin oxidase and xanthine activity by 7 % and 30 %	(Sharma et al., 2018)
	Fresh blueberries	Treated to 2 kV/cm for 2, 4, and 6 min using bipolar square wave pulses at pulse width (1 µs) and frequency (100 pulses/sec)+60 ppm PAA (peracetic acid)	• 3 log reduction of *E. coli* and *Listeria* and 2 log/g reduction of native microbiota. • The color and appearance did not change but softened the texture of berries. • Conc. Of phenolic compounds and anthocyanin's improved by 25% and 10%.	(Jin et al., 2017)
	Low-alcohol red wine	20–50 kV/cm with square-wave pulse; Pulse width of 1 µs; Frequency of 500–1500 Hz	• Decontamination effect increased as the ûeld strength and frequency increases	(Puligundla et al., 2018)
	Apple juice	No. of cycles: 4,6, and 8 (each cycle consists of 50 pulses);Electric field strength: 30 kV/cm	• Inactivated the food spoilage microorganisms without changing its vitamin C and total polyphenols content under 72 h of refrigeration storage	(Dziadek et al., 2019)
	Stevia and ginger	20 kV/cm; 270 µs powder	• Highest inhibition zone diameter of 15.11±0.11mm was obtained without affecting its functional properties	(Pina-Pérez et al., 2018)

(Contd.)

Drying	Basil leaves	65 pulses of650 V/cm, 150 μs pulse width, 760 μs between pulses	• Reduced the dryingtimes by 57 % for air drying at 40°C, 33 % for vacuum drying and 25 % for freeze drying	(Telfser et al.,2019)
	Parsnip and carrot	20 μs pulse width, 50 Hz frequency, 0.9 kV/cm, after 1000 pulse	• Drying time reduced to 28% at 70°C and to 21% at 60°C, compared to the untreated samples	(Alam et al.,2018)
	Onions	0.36 – 1.07 kV/cm using exponential monopolar pulses; Specific energy: 0.2 – 20 kJ/kg;	• Reduction of drying time by 30.3 %, 21.4 % and 31 6.4 % at 45 °C, 60 °C and 75 °C to reach a residual moisture of 7 %.	(Ostermeier et al., 2018)
Freezing	Apple tissue	Electric field intensity(E) :1.85,3, and 5 kV/cm;Pulse numbers: 10, 50, and 100; Pulse width: 15, 20, and 24 μs. Specific energy: 5, 63, 8 and 80 kJ/kg.	• Reduced the total freezing and tha wing time by 3.5-17.2% and 71.5%. • Maximum (8.9%) and minimum (1.6%) mass loss was observed at 50 pulses at 5kV/cm, and 10 pulses at 3 kV/cm during freeze-thaw cycle.	(Wiktor et al., 2015)
	Beef muscle	E: 1.4 kV/cm;Pulse width: 20 μs; Frequency: 50 Hz; Total specific energy: 250 kJ/kg.	• Changed the micro-structure of meat tissue without affecting its free fatty acid profiles. • Enhanced the tenderness and purge loss but not cooking loss.	(Faridnia et al., 2015)
	Baby spinach leaves	Two trains of bipolar, rectangular pulses with amplitude of 350V, with 10 s interval between trains.Each train consisted of 500 pulses of 200 μs pulse width and 1600 μs of space between the pulses (frequency 500 Hz)	• Improved freezing tolerance by applying vacuum impregnation and PEF in the presence of cryoprotectants such as fructose, sucrose, glucose and trehalose. These combinational methods caused that leaf cells remained viable and the leaves retained turgor after the freeze-thaw cycle	(Demir et al., 2018)

(Contd.)

Extraction	Citrus fruits and its peels	3 kV/cm field strength for citrus fruits (orange, lemon and pomelo); 10 kV/cm field strength for its peels;	• Enhanced the juice yield by 25% for oranges, 59% for lemon, and 37% for pomelos, improved the extraction of polyphenols up to 50%	(El Kantar et al., 2018)
	Blueberry fruits	1,3, and 5 kV/cm of electric field strengths with 10 kJ/kg of specific energy	• Enhanced the juice yield by 28%; Extracted juice had higher anthocyanin content (+60%), total phenolic content (+ 43%), and antioxidant activity (+ 31%).	(Bobinaitė et al., 2015)
	Blue berry	1-10 kJ/kg at 3 kV/cm field strength	• Enhanced the juice yield by 36%. Largest increment of anthocyanin content (55%) and antioxidant capacity (41%)	(Pataro et al., 2017)

of various intercellular compounds, like sugar from sugar beet and phytosterols from maize germs(Nowosad et al., 2020). Additionally, Bobinaitė et al., (2015) also resulted that PEF (1,3 and 5 kV/cm; 10 kJ/kg) pre-treatment improved the blueberry juice yield (+ 28%) and also enhanced its total phenolic content (+ 43%), total anthocyanin content (+60%) and antioxidant activity (+ 31%). Similarly, (Pataro et al., 2017) also observed the increments in blueberry juice yield (+32%), and also its anthocyanin (+ 55%) and antioxidant activity (+ 36%) while treated with PEF at 3 kV/cm with an energy input of 10 kJ/kg. PEF processing could be a promising approach for the efficient extraction of fruit juices and its antioxidants.

7.0 Conclusion

PEF processing, a non-thermal methodhas received considerable attention in research as well as industrial applications due to its potential for maintaining safety and minimal or without affecting the quality attributes (sensory and nutrition properties) of food. PEF is found to be a substitute and as well an alternative technique for all the conventional thermal treatments for preservation of foods. PEF processing has also been applied for drying, freezing and extraction enhancement, and enzyme inactivation. However, the commercial exploitation of the PEF in the food processing sector is comparatively low compared to its use in the lab and pilot-scale models. To increase the commercial scale use, the PEF models should be made in a cost-effective method that could be affordable to the industries. The PEF can be used as a continuous method by packing the products hygienically and giving them a low-temperature storage environment. Overall PEF processing seems to be a promising non-thermal technology for microbial inactivation to maintain better food quality and safety.

References

Agarwal, N., & Raghav, P. K. (2019). Pulsed electric field processing of foods - a review. International Journal of Engineering Research and Modern Education, 1(1): 111-1118.

Alam, M. R., Lyng, J. G., Frontuto, D., Marra, F., & Cinquanta, L. (2018). Effect of Pulsed Electric Field Pretreatment on Drying Kinetics, Color, and Texture of Parsnip and Carrot. Journal of Food Science, 83(8): 2159–2166.

Alirezalu, K., Munekata, P. E. S., Parniakov, O., Barba, F. J., Witt, J., Toepfl, S., Lorenzo, J. M. (2020). Pulsed electric field and mild heating for milk processing: a review on recent advances. Journal of the Science of Food and Agriculture, 100(1): 16–24.

Alvarez, I., Condon, S., & Raso, J. (2006). Microbial inactivation by pulsed electric fields. In: Raso-Pueyo, J. & Heinz, V. (Eds.), *Pulsed electric fields technology for the food industry* (pp. 97-129). Springer, Boston, MA.

Barbosa-Cánovas, G. V., & Altunakar, B. (2006). Pulsed electric fields processing of foods: an overview. In: Raso-Pueyo, J. & Heinz, V. (Eds.), *Pulsed electric fields technology for the food industry* (pp. 3-26). Springer, Boston, MA.

Barba, F. J., Parniakov, O., Pereira, S. A., Wiktor, A., Grimi, N., Boussetta, N., Vorobiev, E. (2015). Current applications and new opportunities for the use of pulsed electric fields in food science and industry. Food Research International, 77:773–798.

Bobinaitë, R., Pataro, G., Lamanauskas, N., Đatkauskas, S., Viðkelis, P., & Ferrari, G. (2015). Application of pulsed electric field in the production of juice and extraction of bioactive compounds from blueberry fruits and their by-products. Journal of Food Science and Technology, 52(9): 5898–5905.

Demir, E., Dymek, K., & Galindo, F. G. (2018). Technology Allowing Baby Spinach Leaves to Acquire Freezing Tolerance. Food and Bioprocess Technology, 11(4): 809–817.

Dziadek, K., Kopeæ, A., Dró¿d¿, T., Kie³basa, P., Ostafin, M., Bulski, K., & Oziemb³owski, M. (2019). Effect of pulsed electric field treatment on shelf life and nutritional value of apple juice. Journal of Food Science and Technology, 56(3): 1184–1191.

Elez-Martínez, P., Sobrino-López, Á., Soliva-Fortuny, R., & Martín-Belloso, O. (2012). Pulsed electric field processing of fluid foods. In:P. J. Cullen, B. K. Tiwari, V.Valdramidis, (Eds.), *Novel thermal and non-thermal technologies for fluid foods* (pp. 63-108). Academic Press.

El Kantar, S., Boussetta, N., Lebovka, N., Foucart, F., Rajha, H. N., Maroun, R. G., … Vorobiev, E. (2018). Pulsed electric field treatment of citrus fruits: Improvement of juice and polyphenols extraction. Innovative Food Science and Emerging Technologies, 46: 153–161.

Faridnia, F., Ma, Q. L., Bremer, P. J., Burritt, D. J., Hamid, N., & Oey, I. (2015). Effect of freezing as pre-treatment prior to pulsed electric field processing on quality traits of beef muscles. Innovative Food Science and Emerging Technologies, 29: 31–40.

Gómez, B., Munekata, P. E. S., Gavahian, M., Barba, F. J., Martí-Quijal, F. J., Bolumar, T., Lorenzo, J. M. (2019). Application of pulsed electric fields in meat and fish processing industries: An overview. Food Research International, 123: 95–105.

Huang, K., & Wang, J. (2009). Designs of pulsed electric fields treatment chambers for liquid foods pasteurization process: A review. Journal of Food Engineering, 95(2): 227–239.

Jayathunge, K. G. L. R., Stratakos, A. C., Delgado-Pando, G., & Koidis, A. (2019). Thermal and non-thermal processing technologies on intrinsic and extrinsic quality factors of tomato products: A review. Journal of Food Processing and Preservation, 43(3): 1–15.

Jin, T. Z., Yu, Y., & Gurtler, J. B. (2017). Effects of pulsed electric field processing on microbial survival, quality change and nutritional characteristics of blueberries. LWT - Food Science and Technology, 77: 517–524.

Jongyingcharoen, J. S., & Ahmad, I. (2014). Processing Effects on the Functional Components during Product Development. In: A. Noomhorm, I. Ahmad, A.K. Anal, (Eds.).*Functional Foods and Dietary Supplements: Processing Effects and Health Benefits*, 297–324.John Wiley & Sons

Lado, B. H., & Yousef, A. E. (2002). Alternative food-preservation technologies: Efficacy and mechanisms. Microbes and Infection, 4(4): 433–440.

Nowosad, K., Sujka, M., Pankiewicz, U., & Kowalski, R. (2020). The application of PEF technology in food processing and human nutrition. Journal of Food Science and Technology.

Ortega-Rivas, E. (2012). High-Voltage Pulsed Electric Fields. In: E. Ortega-Rivas,(Eds.). *Non-thermal Food Operations*. Springer US.

Ostermeier, R., Giersemehl, P., Siemer, C., Töpfl, S., & Jäger, H. (2018). Influence of pulsed electric field (PEF) pre-treatment on the convective drying kinetics of onions. Journal of Food Engineering, 237: 110–117.

Pan, Y., Sun, D. W., & Han, Z. (2017). Applications of electromagnetic fields for non-thermal inactivation of microorganisms in foods: An overview. Trends in Food Science and Technology, 64: 13–22.

Pataro, G., Bobinaitë, R., Bobinas, È., Đatkauskas, S., Raudonis, R., Visockis, M., … Viðkelis, P. (2017). Improving the Extraction of Juice and Anthocyanins from Blueberry Fruits and Their By-products by Application of Pulsed Electric Fields. Food and Bioprocess Technology, 10(9): 1595–1605.

Pina-Pérez, M. C., Rivas, A., Martínez, A., & Rodrigo, D. (2018). Effect of thermal treatment, microwave, and pulsed electric field processing on the antimicrobial potential of açaí (Euterpe oleracea), stevia (Stevia rebaudiana Bertoni), and ginseng (Panax quinquefolius L.) extracts. Food Control, 90: 98–104.

Puligundla, P., Pyun, Y. R., & Mok, C. (2018). Pulsed electric field (PEF) technology for microbial inactivation in low-alcohol red wine. Food Science and Biotechnology, 27(6): 1691–1696.

Sharma, P., Oey, I., Bremer, P., & Everett, D. W. (2018). Microbiological and enzymatic activity of bovine whole milk treated by pulsed electric fields. International Journal of Dairy Technology, 71(1): 10–19.

Syed, Q.A., Ishaq, A., Rahaman, U. U., Aslam, S., &Shuka, R. (2017). Pulsed electric field technology in food preservation: a review. Journal of Nutritional Health and Food engineering, 6(5):168-172.

Telfser, A., & Gómez Galindo, F. (2019). Effect of reversible permeabilization in combination with different drying methods on the structure and sensorial quality of dried basil (Ocimum basilicum L.) leaves. Lwt, 99: 148–155.

Toepfl, S., Siemer, C., Saldaña-Navarro, G., & Heinz, V. (2014). Overview of Pulsed Electric Fields Processing for Food. In: D. W. Sun (Eds.), *Emerging Technologies for Food Processing*(pp. 93-114). Academic Press.

Vorobiev, E., & Lebovka, N. (2019). Pulsed electric field in green processing and preservation of food products. In: F. Chemat, E. Vorobiev (Eds.), *Green Food Processing Techniques*(pp. 403-430). Academic Press

Wiktor, A., Schulz, M., Voigt, E., Witrowa-Rajchert, D., & Knorr, D. (2015). The effect of pulsed electric field treatment on immersion freezing, thawing and selected properties of apple tissue. Journal of Food Engineering, 146: 8–16.

Yogesh, K. (2016). Pulsed electric field processing of egg products: a review. Journal of Food Science and Technology, 53(2): 934–945.

23

Atmospheric Cold Plasma and its Applications in Food Processing

Dipsikha Kalita, Pramod K. Prabhakar and Brijesh Srivastava

1. Introduction

In recent years, consumers have been demanding convenient, healthy and safe foods which have 'fresh-like' characteristics with a long shelf-life. These demands include important food quality such as appearance, texture, taste and nutritional content which are strongly impacted by the way it is processed. Another major concern among the consumers is food safety and quality. The shelf-life of the food products are mostly influenced due to the action of various harmful microorganisms and enzymes. The conventional food processing technologies, such as thermal treatments (pasteurisation and high temperature sterilisation), evaporation, drying, frying and baking can assure the microbiological safety and/or stability of processed food products, but affects the food quality and nutrients (heat sensitive vitamins, flavour and other bioactive compounds) (Zhang et al., 2019)

With the increasing demands for food safety and quality, the non-thermal emerging technologies viz. high hydrostatic pressure, irradiation, ultrasound, pulse electric field, pulse light, oscillating magnetic field, micro and ultrafiltration introduced to the food processing industry with a concept of LTST (Low Temperatures Short Time) which employ minimal or no changes on essential nutrients and acceptable sensory properties (aroma and flavours).

Cold plasma (CP) technology is one of the important emerging and novel (non-thermal) food processing technologies, which has captivated the attention of many readers, researchers and processors across the globe and this because of its food and biodecontamination potential. Plasma was discovered in the year 1879 by a British physicist named Sir William Crookes, however, the term "Plasma" was applied by Irving Langmuir, an American chemist in 1929. According to current research trends, cold plasma is a powerful and profitable technology for the food processing industry. The application of non-thermal technology is highly favourable for microbial inactivation of food products involving spores, pathogenic organisms which is attributed to the presence of reactive species and charged particles encompassed in the quasi-neutral plasma gas (Chizoba Ekezie et al., 2017).

2. Fundamentals of Plasma technology

Plasma is basically considered as the fourth state of matter and constitutes 99% of the universe. It comprises different species such as electrons, positive and negative ions, atoms, molecules, radicals, and photons (Tendero et al., 2006). Plasma occurs in either excited or ground state, having a net neutral charge. It is induced and influenced under variable temperatures and pressures by energizing or exciting a neutral gas and that is why it can be categorised into thermal and non-thermal plasma. Thermodynamic and electrochemical feasibility is required to be explored for better understanding of plasma science.

2.1 Thermodynamics

Plasma is a fourth state of matter energy which indicates higher order of energy from solid to liquid to gas and finally an ionised state in the form of plasma. The gas can be ionized by providing required energy from either by applying heat or exposing under a simultaneously acting electrical and magnetic field (energy intense light or electric field). The plasma is electrically conductive, internally interactive, and strongly responsive to electromagnetic fields due to presence of the free electric charges, viz., ions and electrons (Misra et al., 2018).

As per the fundamental laws of collision mechanics, when two bodies with different masses (m_1 and m_2) collide, the ratio of their masses (m_1/m_2) will allow the extent of energy exchange among them. The exchange of energy in context of gas plasma can be approximated/ predicted under certain conditions (Misra et al., 2018) (i) Energy sharing is nearly complete when ions and neutral particles are approximated to have a similar order of mass (m_1/m_2 H ≈ 1). (ii) Minor fraction of total energy is exchanged provided electrons are much lighter than ions and neutral ($m_1/m_2 << 1$). (iii) The electrons' temperature (T_e) approaches to be much higher than that of the global gas, ions, and neutrals (T_g) temperature (Te>>Tg) which is a consequence of application of electric field to a gas. However, reactive radicals are generated during further collisions among electrons. These non-equilibrium thermodynamic conditions are quite exciting for generating the chemical reactive species which are sustained under ambient gas temperature and atmospheric pressure. This electrical gas discharge at low temperature is referred to as cold plasma (Misra et al., 2018).

2.2 Electrochemistry

In non-equilibrium plasma, the simultaneous generation of several reactive chemical species places it as a potential antimicrobial tool and an efficacious chemical method to functionalise the chemical and associated properties of food and biomaterials (Misra et al., 2018). As stated by Mishra et al. (2018), a series of reactions takes place in plasma discharge which results in the formation of reactive metastables and radicals mentioned below:

(i) the electron impact reactions (vibration, excitation, dissociation, attachment, and ionization),

(ii) ion–ion recombination to form neutrals,

(iii) interactions between ions and molecule,

(iv) penning ionization,

(v) quenching processes,

(vi) recombination of neutrals via three body reaction schemes, and

(vii) photoemission, photo absorption, and photoionization

The concentration of plasma chemical species depends on their generation and rate of loss. The rate formation of these reactive species is imputed to the temperature and electron density, whereas the rate of loss is determined by material being treated and discharge configuration (Misra et al., 2018)

3. Classification of plasma and its sources

Plasma is mainly classified into two categories based on the conditions in which they are created viz. thermal and non-thermal plasma (Tendero et al., 2006). The detailed classification of plasma is presented in Fig. 1.

3.1 High temperature plasma or Thermal plasma

It implies that all species (electrons, ions and neutral species) are in a thermal equilibrium state. Examples involve Laser fusion plasma. This category of plasma is not used in food applications.

3.2 Low temperature plasma or Non-thermal plasma

Non-thermal plasma also known as cold plasma and is present in non-local thermodynamic equilibrium (non-LTE) state. In cold plasma the temperature of the heavy particles remains lower than the electronic ones. The inelastic collision between electrons and heavy particles are excitatory or ionizing and therefore the temperature of the heavy particles does not rise. Non-thermal plasma can be generated by an electric discharge in a gas at low pressure or using microwaves.

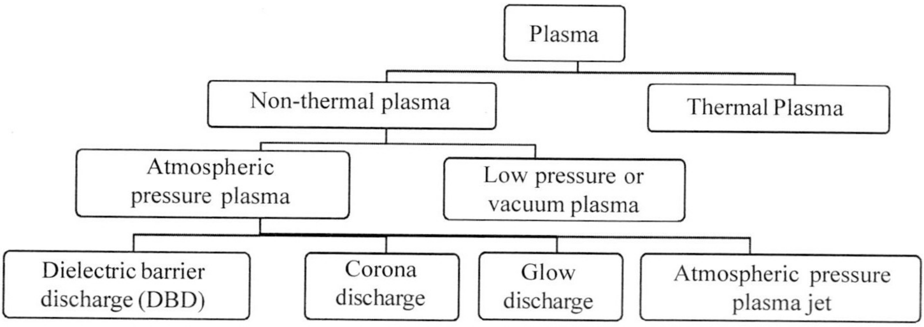

Fig. 1: Classification of plasma

3.3 Atmospheric plasma sources

The plasma courses are basically the different configurations and components of the physical or benchtop set-up having commercial and technological applications. A detailed classification of the different plasma sources is presented in Fig. 2.0. However, the brief description of commonly employed plasma sources for food processing applications has been presented in this chapter.

3.3.1 Direct current (DC) and low frequency discharges

The continuous or pulsed mode under which the direct current and low frequency discharges principally works. While the system warming up is limited, a pulsed working mode permits the injection of high energy values in the discharge. However, the pulsed power supply is simple in DC sources otherwise more complex (Tendero et al., 2006).

3.3.1.1 Continuous working mode (Arc plasma torches)

DC power supply is used to generate the arc plasma torches. The current carrying arc and transferred arc are the major two categories of arc plasma torches and its schematic diagram has been shown in Fig. 3 (Tendero et al., 2006). There are three different parts having distinct functions. A cathode is mainly responsible for emitting the electrons. Other two components are plasma gas injection system and nozzle. Nozzle mainly confines the plasma.

The nozzle works as an anode as it is positively charged and this is in case of current carrying arc torch. On the other hand, nozzle is at a floating potential and the treated material works as anode and, this is the case of a transferred arc torch. The arc ignited in between the anode and cathode ionizes the plasma gas. The temperature at which plasma exists is very high and ranges between

8000K (plasma envelope) to 15000 K (plasma core) and thereby facilitates high temperature application. Pencil-like torches are innovative arc plasma torches with powered flexibility. These torches are a low powered, homogeneous, current carrying arc plasma jet and generate very little heat as compared to classical plasma torches. This property allows treatment of the surface of various materials and polymers including food products which get degraded at high temperature (Tendero et al., 2006).

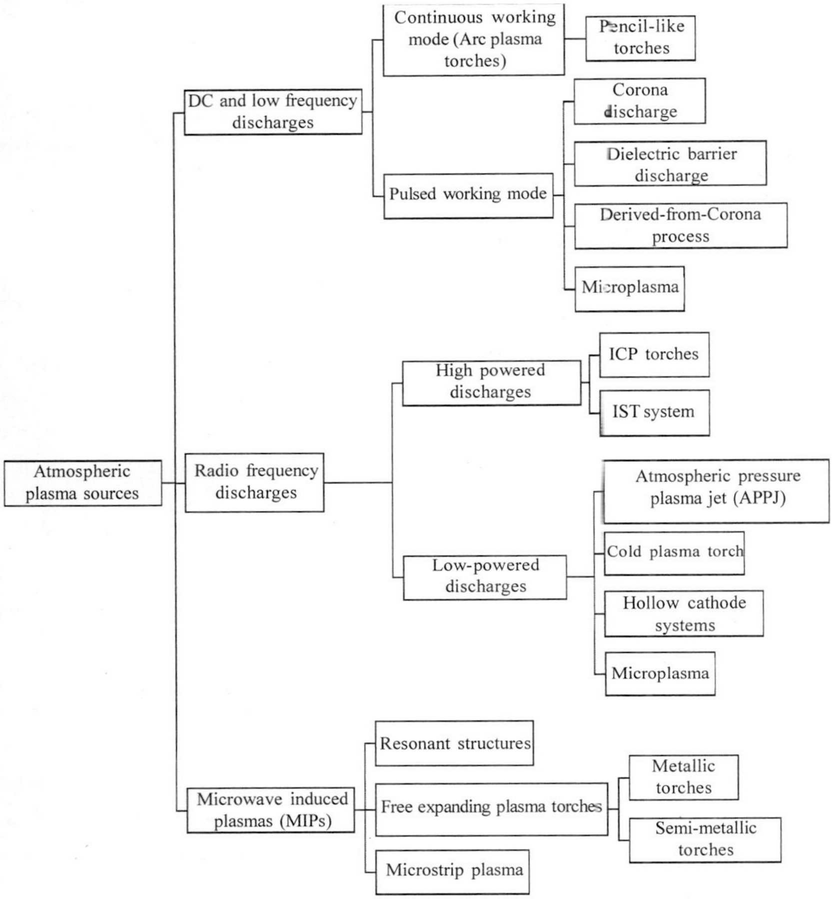

Fig. 2: Classification of the different plasma sources (Tendero et al., 2006)

3.3.1.2 Pulsed working mode

3.3.1.2.1 Corona discharge

Corona discharge is also based on non-LTE discharge having low current density and consists of a cathode wire, the treated material as anode and DC power supply (Fig. 3) (Tendero et al., 2006). It is named as "Corona" since a lighting crown is created around the wire. The corona is generated by a strong electric field associated with small diameter wires, needles, or sharp edges on an electrode. Negative corona discharge also happens through the sharp edges and wires. However, the positive ions orient and move near to wire where secondary electrons are discharged and accelerated into plasma. In this, the moving front of high-energy electrons of 10eV is followed by a tail of electrons having lower energy (1eV) known as streamers (Tendero et al., 2006). The formation of reactive species occurs by the inelastic collision that occurs between these high-energy electrons and heavy particles. The drawback of the corona discharge is the small volume of plasma which leads to only thin size of surface treatment. The material's surface suffers with non-homogeneous treatment and this is because the streamers always initiate at the same place and a dielectric barrier discharge was devised to overcome this problem.

3.3.1.2.2 Dielectric barrier discharge

Dielectric barrier discharge is made up of two plane-parallel metal electrodes as dielectric plates filled with working gas (a dielectric layer). The gap between the electrodes is limited to a few millimetres for plasma operation and plasma gap flows in the gap as represented in Fig.3. The discharge is created by means of sinusoidal or pulsed power source (Tendero et al., 2006). The discharge can either be filamentary or glow which is formulated on the basis of frequency excitations, working gas composition and voltage. Micro-discharges or a steamer favours a filamentary discharge. However, helium is used as plasma gas which forms and favours a glow discharge.

3.3.2 Radio frequency (RF) discharges

The atmospheric pressure plasma jet is analogous to RF discharge as it is a small RF plasma torch which works at low power (Tendero et al., 2006). The system includes two concentric electrodes through which the working gas flows such as a mixture of helium, oxygen and other gases (Fig.3). When a voltage between 100-150V is applied to the inner electrode, the gas discharge is ignited. The ionized gas produced exits through a nozzle and directed towards the substrate. The ionized gas velocity is approximately $12ms^{-1}$ and the injected power having low values permits the torch to generate a stable discharge and prevent the arc transition.

3.3.3 Microwave induced plasma

Microwave induced plasma is gaining popularity in recent years for food application and a schematic diagram of microwave plasma torch is shown in Fig.3. It works on a simple principle that microwaves are conveyed along the system and energy is transmitted to the gas in the chamber converting into plasma gas electrons. During this process, elastic collision occurs amidst electrons and heavy particles resulting in bouncing of collided electrons as it consists of considerable mass of heavy particles, but the heavy particles persist to be static. The electrons start accelerating due to kinetic energy and the heavy particles are heated to some extent. Subsequently, with numerous elastic collisions i.e. following the probabilistic laws, the electrons receive adequate energy to attain inelastic exciting or even ionizing collisions. This results in partial ionization of the gas leasing to plasma formation which assist microwave propagation (Laroussi, 2005).

4. Factors influencing the efficacy of cold plasma

The efficiency of cold plasma is mainly influenced by various processing variables such as type of gas, frequency and input voltage impact the nature and quantity of active species generated in the discharge. Moreover, the mode of exposure (direct/indirect) also has an impact on the efficacy of cold plasma. Another factor impacting the adequacy of cold plasma during inactivation of microbes in food products is the internal characteristics of the microorganisms as suggested by (Chizoba Ekezie et al., 2017). The different strains of microorganisms exhibit different sensitivity to treatments. As reported in study, bacteria in the stationary phase are prone to various inactivation treatments as compared to microbes in exponential phase. Similarly, gram positive bacteria due to the presence of peptidoglycan layer exhibits less sensitivity to plasma treatment than gram negative bacteria. Further, bacterial spores and fungi exhibit stronger resistance to cold plasma due to the variation in internal structure.

In cold plasma treatment, the milieu factors such as relative humidity (RH), pH and nature of the food matrix determines the efficacy of the cold plasma treatment (Chizoba Ekezie et al., 2017). The increase in RH during cold plasma treatment results in higher inactivation of microbes due to the augmentation in the composition of hydroxyl radicals. Moreover, lower pH activity has shown higher log reduction in microorganisms as compared to higher pH. The different food matrices i.e. solid and liquid divulge in different ways with reactive species as liquid food tend to vaporize during treatment. Further, nature of matrix contributes significantly to the process efficiency in microbial inactivation which can be illustrated by a study wherein inactivation in agar plate or filter by cold plasma showed higher reduction in microorganisms but in case of fresh fruits and products microorganisms can migrate from the outer to inner tissues at a definite velocity resulting in lower decimal reduction.

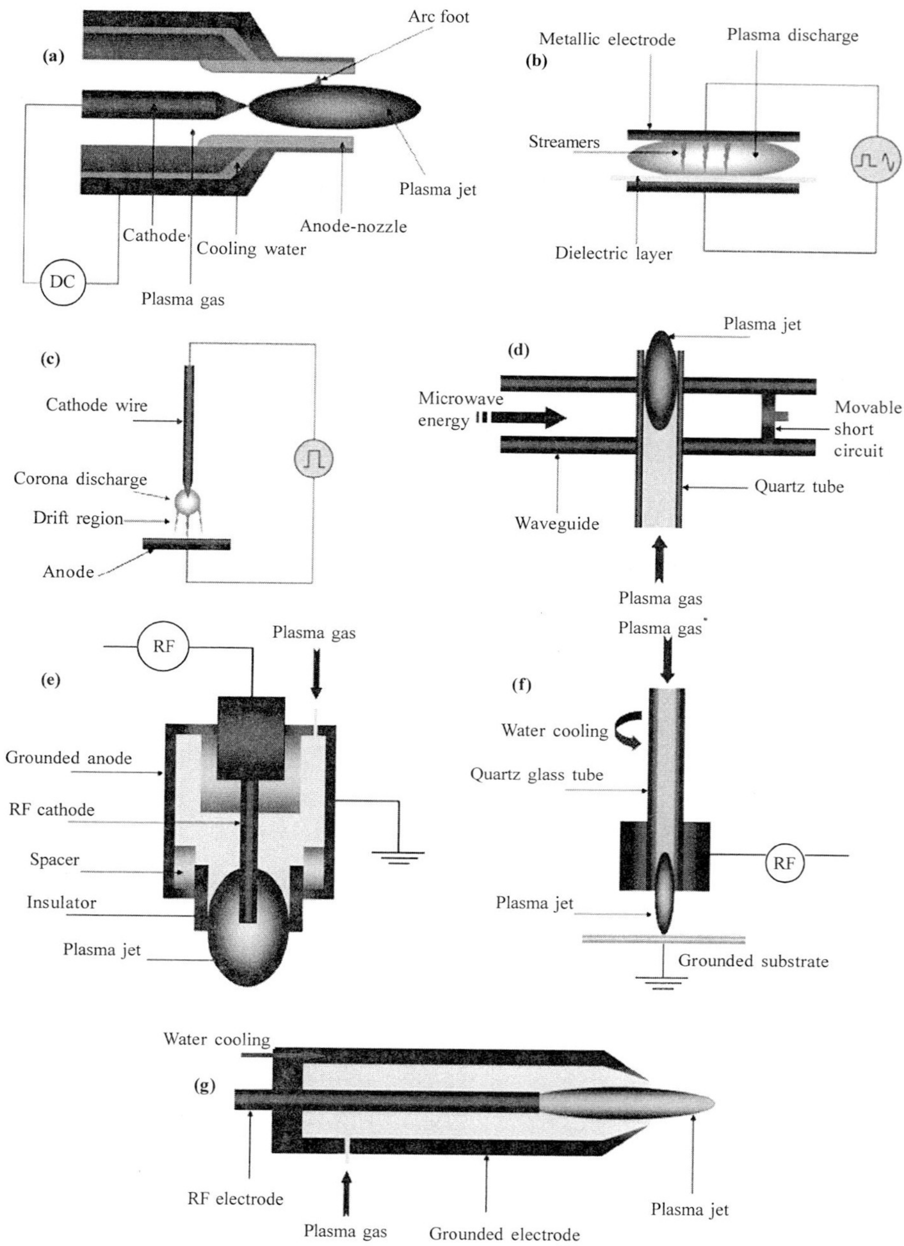

Fig. 3. Some plasma sources (a) Arc plasma torches (b) Dielectric barrier discharge (DBD) (c) Corona discharge (d) Microwave plasma discharge (e) Cold plasma torch discharge (f) Barrier torch design (g) Atmospheric pressure plasma jet (APPJ) (Tendero et al., 2006)

5. Mechanistic understanding of inactivation mechanism

5.1 Inactivation agents and its mechanism on microorganisms

The utilization of sterilization properties was first introduced towards the end of the 60's and since then significant research has been conducted on the inactivation mechanism of microorganisms by plasma agents. The various inactivation agents involve UV radiation, charged particles (electrons and ions) and reactive species and their mechanism is shown in Fig.4.

5.2 Effect of UV radiation

The application of ultraviolet (UV) photons produced from cold plasma sources is still in research and requires comprehensive study. In food processing, the UV emissions are confined to the near UV region (300-400 nm, also called UV-A) (Misra & Jo, 2017). The main mechanism proposed to cause bacterial cell damage involves the formation of thymine dimers in the DNA (Fig.4). This restricts the ability of the bacterial cell to replicate (Laroussi, 2005). According to various reports there are certain drawbacks of UV radiation since no significant UV emission occurs below 285 nm. First drawback is that the most of UV energy is absorbed by molecules of the cell leading to excitation and rapid relaxation. Secondly, the DNA repair mechanism of microbial cells is adequately smart enough and thus, a high quantum of energy is required for cell damage (Misra & Jo, 2017). Moreover, the power density of the emitted UV radiation in the 200-300nm wavelength region is below 50 ìW.cm^{-2}. Therefore, at these conditions, the UV radiation would not be expected to play a significant role in the sterilization process.

5.3 Charged particles

The charged particles (electrons and ions) can play a considerable role in the destruction of microorganisms. The bombardment of charged particles with bacterial cells can cause rupture of the outer membrane. The charge accumulated on the outer surface of the cell membrane overcomes the tensile strength of the membrane and causes its rupture as shown in Fig. 4. According to reports, when a bacterial cell of size in µm range is charged, it experiences an outward electrostatic force due to each charge being subjected to repulsive forces of all the similar charges accumulated on the cell surface (Laroussi, 2005). This condition is more effective against gram negative bacteria due to their irregular surface with small radii of curvatures which causes localized high outward electrostatic forces. Various microscopic observations support this mechanism of microbial inactivation.

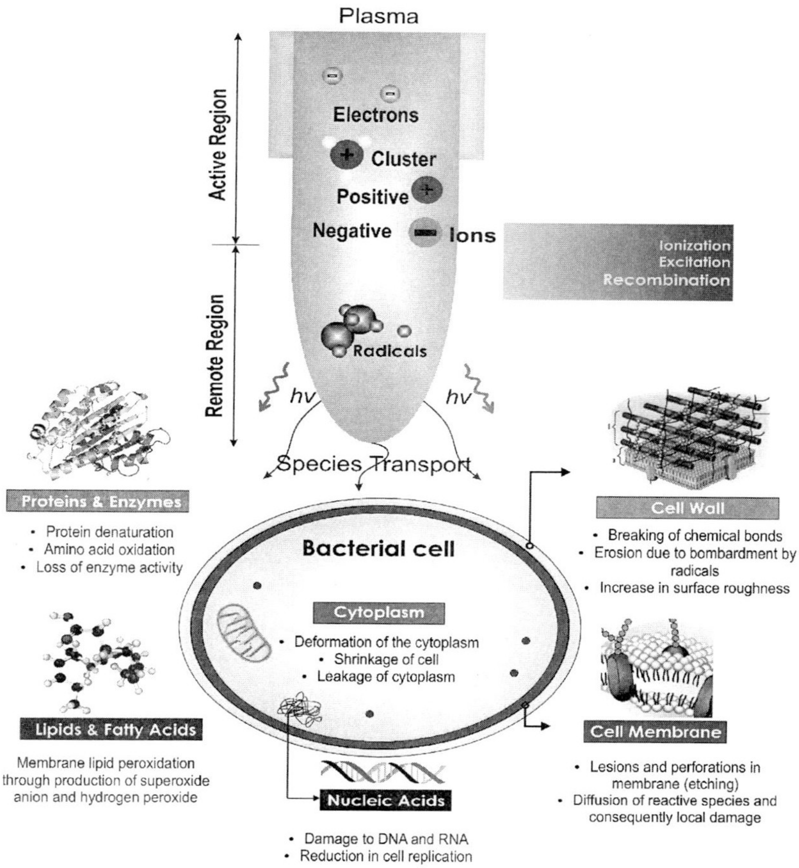

Fig. 4. Action of cold plasma on microbial cell structures (Misra & Jo, 2017)

5.4 Reactive species

Reactive species play a significant role in all plasma surface interactions. Their main mechanism involves oxidation of the outer surface of microbial cells. Air plasma are excellent sources of reactive oxygen-based and nitrogen-based species such as O2•, O3, OH•, NO• etc (Laroussi, 2005). The bacterial cell membrane is made up of lipid bilayers mainly consisting of unsaturated fatty acid. This layer felicitates the movement of biochemical by-products across the membrane. The unsaturated fatty acids are susceptible to attack by the hydroxyl radical (OH•) leading to disruption of its main function i.e. transport of ions and polar compounds in and out of the cell and the process is shown in Fig. 4. Sun et al., (2007) studied the inactivation of *Staphylococcus aureus* and *Escherichia coli* were treated by using dielectric barrier discharge (DBD) and observed that *E.coli* was more prone to reactive oxygen species (ROS) than *S.aureus* since *E.coli* possess thinner peptidoglycan layer. The reactive species also cause breakage of structurally important bonds including C-O, C-N and C-C. The double bonds of unsaturated lipids are observed to be more vulnerable to ozone attack (Misra & Jo, 2017). Moreover, the membrane lipids are located along the surface of the cell which makes it more susceptible to bombardment of oxidizing agents. Protein molecules which also play a role in controlling the passage of various macromolecules in and out of cells are affected by atomic oxygen or metastable oxygen molecules (Laroussi, 2005).

Spores are highly resistive states of cells and are composed of several coats surrounding a genetic core. The inner spore membrane is very important as it possesses receptors for various germinants that assist in germination and growth of spore. However, access to these membranes is limited due to the various outer coats which require penetration in order to reach the inner core. The inner core also involves various protection mechanisms which involve small acid soluble proteins that are tightly associated and protect DNA from external physical or chemical attack. Excited inert gas (metastable gas) molecules and radicals may be a potent source to penetrate into the interior of the spore and therefore attack the DNA at the core. If the inner membrane is damaged by the radicals, dipicolic acid (DPA) leaks out which gives a clear indication of both structural spore damage and initial germination (Shintani et al., 2010)

6. Application of atmospheric pressure plasma

6.1 Treatment of solid wet produce

6.1.1 Fruits and vegetables

Fruits and vegetables are the essential components of the human diet due to the presence of high levels of micronutrients and fibres. Minimally processed fruits and vegetables are gaining popularity among consumers due to their fresh like characteristics. However, these minimally processed fruits and vegetables undergo various unit operations such as selection, cleaning, washing, trimming, peeling, cutting and shredding, sanitizing and packing. During these operations (cutting and shredding specially) the surface gets exposed and contamination may occur. The outer protective epidermal layer gets ruptured which provide larger surface area and increases the nutrition availability that assist in microbial growth and eventually decreases the shelf life. Moreover, the mechanical damage caused to cells during processing may increase the rate of tissue senescence reducing their resistance to microbial spoilage (Ramos et al., 2013). The microflora of vegetables and fruits consists of various microbes including *Pseudomonas* spp, *Erwinia herbicola*, *Flavobacterium*, *Xanthomonas*, and

Enterobacter agglomerans as well as various moulds, *Alternaria*, *Penicillium*, *Fusarium* and *Aspergillus*. Cold plasma treatment was observed to have significant effect on the removal of surface human pathogens such as *E.coli* 0187:H7 and *Salmonella* spp (Ramos et al., 2013). The total mesophiles and yeast/mould population in cold plasma treated strawberries were observed to reduce by 12-85% and 44-95% (Misra, et al., 2014). Similarly, cherry tomatoes containing *Escherichia coli, Salmonella typhimurium*, and *Listeria monocytogenes* by up to 3.5, 3.8, and 4.2 log CFU/tomato, respectively were successfully decontaminated when treated with dielectric barrier discharge (DBD) plasma at 80 kV for 5 min (Misra, et al., 2014). This is attributed to the action of reactive species and charged particles on the microorganisms.

Another major issue for quality degradation in cut fruits and vegetables is enzymatic browning. Cutting and shredding allows interaction between the polyphenol oxidase (PPO) and polyphenolic substance in presence of oxygen leading to melanin accumulation. Application of cold plasma has been reported to inactivate enzymes in cut fruits and vegetables. The generated reactive species are mainly responsible for the inhibitory effect of cold plasma on browning enzymes. The main hypothesis involves action of OH, O2–, HOO and NO radicals that induce chemical modification of chemically reactive side-chains of the amino acids viz. cysteine, aromatic rings of phenylalanine, tyrosine and tryptophan that eventually leads to the loss of enzyme activity (Tappi et al., 2014). This leads to the enhancement of the shelf life of cut fruits and vegetables.

6.1.2 Meat and egg

The common enteric pathogens that contaminate poultry are *Salmonella, Campylobacter* and *Listeria monocytogenes*. Post slaughter contamination of poultry is much higher which may be due to the development of rigor mortis during which the muscle fibres undergo radial shrinkage and pull away the surrounding connective tissues. This results in a fissure which allows bacterial penetration (Noriega et al., 2011). Although various decontamination processes are applied such as steam treatment, chlorination but its uses are limited because of adverse effects on quality. Cold plasma can be employed for sterilization for meat surfaces. (Noriega et al., 2011) studied the efficacy of cold plasma decontamination on chicken skin and muscle inoculated with *Listeria innocua* and observed that 8 min treatment gave a 1-log reduction on skin and 4 min treatment gave > 3-log reduction on muscles. A significant reduction of *L. innocua* was reported after cold plasma treatment on sliced ready-to-eat meat product (Bresaola) packed inside sealed LDPE bags containing 30% oxygen and 70% argon (Rød et al., 2012)

Eggs are mainly contaminated by the *Salmonella* spp and this can be a potential health hazard for the consumers. The conventional methods to sanitization cannot be employed for eggshells and other alternative methods such as pulsed electric field, infrared radiation, UV and ozone treatment possess certain limitations. (Ragni et al., 2010) evaluated the feasibility of resistive barrier discharge (RBD) plasma for decontamination of eggshells. The eggshell after inoculation with *Salmonella enteritidis* and *Salmonella typhimurium* (5.5–6.5 Log CFU/ eggshell) were treated with plasma at different RH for different time intervals. The results indicated reduction of 3.5-log CFU/eggshell in *S. typhimurium* and 3.5-4.5 log/CFU in *S. enteritidis* after 90 min of treatment at higher RH (65%). The microbial inactivation was enhanced at higher RH due to the presence of oxygen species. Moreover, eggs treated with cold plasma maintained the cuticle which is considered as the first line of defence from microbial invasion.

6.2 Treatment of solid dry produce

6.2.1 Spices

Spices are an important ingredient in most of the dishes worldwide. They are mostly naturally contaminated during the harvesting, handling and drying process. The commonly found microorganisms on the surface of spices are Salmonella and bacillus which causes food borne diseases and intoxication. The decontamination of dried products like spices is much more difficult than moisture rich products due to the low water activity (a_w) which causes higher resistance of microorganisms. Therefore, alternative non-thermal techniques such as cold plasma may be employed for decontamination of spices. (Hertwig et al., 2015) studied the efficacy of a non-thermal remote plasma application on different herbs and spices (pepper seeds, crushed oregano and paprika powder) with different surface to volume ratio. It was observed that microflora of the pepper seeds and the paprika powder reduced by > 3 $\log_{10}$ and that of oregano reduced by 1.6-$\log_{10}$ due to the generation of various reactive species as detected by FT-IR in this study. Thus, cold plasma can be applied in particulate materials such as powders for decontamination with maximum retention of colour and flavour.

6.2.2 Nuts

Nuts are produced all over the world and rich in nutrients. It is mostly contaminated by aflatoxins produced by *Aspergillus spp* which is responsible for serious health implications. Basaran et al. (2008) inoculated hazelnuts, peanuts, and pistachio nuts with *A. parasiticus* which were then treated with air gases plasma and SF6 plasma for up to 20 min duration. Results showed that SF6 plasma was more effective against *A. parasiticus* (5-log decrease) as compared to air gas plasma (1-log decrease). A reduction of 50% was observed in total aflatoxins (AFB1, AFB2, AFG1, and AFG2) after treatment with 20 min air gases plasma treatment whereas only a 20% reduction in total aflatoxin was observed after 20 min SF6 plasma treatment. Thus, cold plasma can be potentially applied for disinfection of dry food surfaces without compromising their quality attributes.

6.3 Treatment of liquid food products

Liquid packed food products such as juice and milk are consumed worldwide. Conventionally thermal treatments (pasteurisation or ultra-high temperature) are applied for preservation. With growing demand for natural and nutritious food an alternative non-thermal technology such as cold plasma can be employed to treat liquid products at room temperature and atmospheric pressure. The reactive species and photons generated by plasma have a lethal effect on the microorganisms as they cause rupture of the cytoplasmic membrane. In apple juice a 5 log cycle reduction of *C.freundii* was reported when treated for 480s with cold plasma produced from argon with 0.1% oxygen. It was also observed that direct contact of bacterial cells and plasma is not required as H_2O_2 generated by plasma causes cell permeabilization and RNA damage of microbes present in liquid (Surowsky et al., 2014). (Gurol et al., 2012) investigated the ability of cold plasma to decontaminate *E. coli* in milk with different fat content. It was observed that plasma treatment reduced the number of colonies in different types of milk (whole, semi-skimmed, skimmed) with relatively no negative effect on pH and color measurements.

6.4 Treatment of packaging material

Food packaging material serves the purpose of preserving food and food products and protecting it from damage, outside contamination during transportation, distribution and storage. Contamination of the packaging material may lead to spoilage of food resulting in health hazards. Conventionally sterilization of bottles, cans and lids are obtained by utilization of oxidizing chemical germicidal liquids such as hydrogen peroxide, per acetic acid, ozonated water etc. (with or without mild heat treatment). Cold plasma can be effectively used as an alternative technique to these chemical methods. It allows a safe and fast sterilization of plastic bottles, lids and films without creating any adverse effect on packaging material. This was observed from the study of Muranyi et al. (2010) where Polyethylene terpthalate (PET) was treated with cold plasma and resulted in inactivation of several microorganisms. Plastic trays showed no viable count after 90 and 120 s plasma treatments and only three decimal reductions on paper cups and aluminium foils under identical process parameters. Plasma deposition of heat sensitive materials such as vitamins, antioxidants and antimicrobials into the packaging material may be sought as potential alternatives in the emerging field of antimicrobial and active packaging (Misra et al., 2011).

6.5 Application of plasma activated water (PAW)

In plasma treatment of food, reactive oxygen species are contemplated as the major bactericidal agent in plasma inactivation (Thirumdas et al., 2018). This active composition of cold plasma can be utilized to react with water for initiation of plasma activated water (PAW) which possesses excellent antimicrobial effect and can also be applied for various food purposes. The effect of PAW on microbial inactivation and post-harvest quality of Chinese bayberry was determined and observed that a 0.5 min PAW treatment decreased the microbial load (1.11 log CFU/g) i.e. both bacteria and fungi during the 8 days storage along with retention of quality (high firmness, colour index etc) (Ma et al., 2016) . Similarly, PAW treatment of strawberries inoculated with Staphylococcus aureus showed a reduction of 1.7 to 34 log at 4-day storage. PAW has been also employed for seed germination, plant growth etc (Ma et al., 2015).

7. Conclusions

Cold plasma is a novel non-thermal emerging technology which offers an alternative sterilization technique for surface decontamination of raw or fresh produce, dried and particulate food material and packaging materials. The reactive species and charged particles generated by plasma is mainly responsible for inactivation of microorganisms that brings about change at both cellular and morphological level. Although this technology is accepted by various food processors mainly for surface decontamination, its application in the food industry is still in its infancy. Cold plasma can also be applied for sensitive food materials such as lipids, vitamins, volatile and organic compounds but it requires further investigation to find wider application in the food industries.

Bibliography

Basaran, P., Basaran-Akgul, N., & Oksuz, L. (2008). Elimination of Aspergillus parasiticus from nut surface with low pressure cold plasma (LPCP) treatment. *Food Microbiology*, *25*(4), 626–632. https://doi.org/10.1016/j.fm.2007.12.005

Chizoba Ekezie, F. G., Sun, D. W., & Cheng, J. H. (2017). A review on recent advances in cold plasma technology for the food industry: Current applications and future trends. *Trends in Food Science and Technology*, *69*, 46–58. https://doi.org/10.1016/j.tifs.2017.08.007

Gurol, C., Ekinci, F. Y., Aslan, N., &Korachi, M. (2012). Low Temperature Plasma for decontamination of E. coli in milk. *International Journal of Food Microbiology*, 157(1), 1–5. https://doi.org/10.1016/j.ijfoodmicro.2012.02.016

Hertwig, C., Reineke, K., Ehlbeck, J., Erdo‡du, B., Rauh, C., & Schlüter, O. (2015). Impact of remote plasma treatment on natural microbial load and quality parameters of selected herbs and spices. *Journal of Food Engineering*, *167*, 12–17. https://doi.org/10.1016/j.jfoodeng.2014.12.017

Laroussi, M. (2005). Low temperature plasma-based sterilization: Overview and state-of-the-art. *Plasma Processes and Polymers*, *2*(5), 391–400. https://doi.org/10.1002/ppap.200400078

Ma, R., Wang, G., Tian, Y., Wang, K., Zhang, J., & Fang, J. (2015). Non-thermal plasma-activated water inactivation of food-borne pathogen on fresh produce. *Journal of Hazardous Materials*, *300*, 643–651. https://doi.org/10.1016/j.jhazmat.2015.07.061

Ma, R., Yu, S., Tian, Y., Wang, K., Sun, C., Li, X., Zhang, J., Chen, K., & Fang, J. (2016). Effect of Non-Thermal Plasma-Activated Water on Fruit Decay and Quality in Postharvest Chinese Bayberries. *Food and Bioprocess Technology*, 9(11), 1825–1834. https://doi.org/10.1007/s11947-016-1761-7

Misra, N. N., & Jo, C. (2017). Applications of cold plasma technology for microbiological safety in meat industry. *Trends in Food Science and Technology*, 64,74–86. https://doi.org/10.1016/j.tifs.2017.04.005

Misra, N. N., Keener, K. M., Bourke, P., Mosnier, J. P., & Cullen, P. J. (2014). In-package atmospheric pressure cold plasma treatment of cherry tomatoes. *Journal of Bioscience and Bioengineering*, *118*(2), 177–182. https://doi.org/10.1016/j.jbiosc.2014.02.005

Misra, N. N., Martynenko, A., Chemat, F., Paniwnyk, L., Barba, F. J., & Jambrak, A. R. (2018). Thermodynamics, transport phenomena, and electrochemistry of external field-assisted nonthermal food technologies. *Critical Reviews in Food Science and Nutrition*, *58*(11), 1832–1863. https://doi.org/10.1080/10408398.2017.1287660

Misra, N. N., Patil, S., Moiseev, T., Bourke, P., Mosnier, J. P., Keener, K. M., & Cullen, P. J. (2014). In-package atmospheric pressure cold plasma treatment of strawberries. *Journal of Food Engineering*, *125*(1), 131–138. https://doi.org/10.1016/j.jfoodeng.2013.10.023

Misra, N. N., Tiwari, B. K., Raghavarao, K. S. M. S., & Cullen, P. J. (2011). Nonthermal Plasma Inactivation of Food-Borne Pathogens. *Food Engineering Reviews*, 3(3–4), 159–170. https://doi.org/10.1007/s12393-011-9041-9

Muranyi, P., Wunderlich, J., & Langowski, H. C. (2010). Modification of bacterial structures by a low-temperature gas plasma and influence on packaging material. *Journal of Applied Microbiology*, 109(6), 1875–1885. https://doi.org/10.1111/j.1365-2672.2010.04815.x

Noriega, E., Shama, G., Laca, A., Díaz, M., & Kong, M. G. (2011). Cold atmospheric gas plasma disinfection of chicken meat and chicken skin contaminated with Listeria innocua. *Food Microbiology*, *28*(7), 1293–1300. https://doi.org/10.1016/j.fm.2011.05.007

Ragni, L., Berardinelli, A., Vannini, L., Montanari, C., Sirri, F., Guerzoni, M. E., & Guarnieri, A. (2010). Non-thermal atmospheric gas plasma device for surface decontamination of shell eggs. *Journal of Food Engineering*, 100 (1), 125–132. https://doi.org/10.1016/j.jfoodeng.2010.03.036

Ramos, B., Miller, F. A., Brandão, T. R. S., Teixeira, P., & Silva, C. L. M. (2013). Fresh fruits and vegetables - An overview on applied methodologies to improve its quality and safety. *Innovative Food Science and Emerging Technologies*, 20, 1–15. https://doi.org/10.1016/j.ifset.2013.07.002

Rød, S. K., Hansen, F., Leipold, F., &Knøchel, S. (2012). Cold atmospheric pressure plasma treatment of ready-to-eat meat: Inactivation of Listeria innocua and changes in product quality. *Food Microbiology*, *30*(1), 233–238. https://doi.org/10.1016/j.fm.2011.12.018

Shintani, H., Sakudo, A., Burke, P., & McDonnell, G. (2010). Gas plasma sterilization of microorganisms and mechanisms of action. *Experimental and Therapeutic Medicine*, 1(5), 731–738. https://doi.org/10.3892/etm.2010.136

Sun, Y., Qiu, Y., Nie, A., & Wang, X. (2007). Experimental research on inactivation of bacteria by using dielectric barrier discharge. *IEEE Transactions on Plasma Science*, 35(5 II), 1496–1500. https://doi.org/10.1109/TPS.2007.905947

Surowsky, B., Fröhling, A., Gottschalk, N., Schlüter, O., & Knorr, D. (2014). Impact of cold plasma on Citrobacter freundii in apple juice: Inactivation kinetics and mechanisms. *International Journal of Food Microbiology*, 174, 63–71. https://doi.org/10.1016/j.ijfoodmicro.2013.12.031

Tappi, S., Berardinelli, A., Ragni, L., Dalla Rosa, M., Guarnieri, A., & Rocculi, P. (2014). Atmospheric gas plasma treatment of fresh-cut apples. *Innovative Food Science and Emerging Technologies*, *21*, 114–122. https://doi.org/10.1016/j.ifset.2013.09.012

Tendero, C., Tixier, C., Tristant, P., Desmaison, J., &Leprince, P. (2006). Atmospheric pressure plasmas: A review. *Spectrochimica Acta - Part B Atomic Spectroscopy*, *61*(1), 2–30. https://doi.org/10.1016/j.sab.2005.10.003

Thirumdas, R., Kothakota, A., Annapure, U., Siliveru, K., Blundell, R., Gatt, R., & Valdramidis, V. P. (2018). Plasma activated water (PAW): Chemistry, physico-chemical properties, applications in food and agriculture. *Trends in Food Science and Technology*, *77*(May), 21–31. https://doi.org/10.1016/j.tifs.2018.05.007

Zhang, Z. H., Wang, L. H., Zeng, X. A., Han, Z., & Brennan, C. S. (2019). Non-thermal technologies and its current and future application in the food industry: a review. *International Journal of Food Science and Technology*, 54(1), 1–13. https://doi.org/10.1111/ijfs.13903

24

Packaging

Kshitiz Kumar and Vijay Singh Sharanagat

1. Introduction

Packaging is the integral part and cornerstone of food processing industry. The development of food processing industry goes hand in hand with advancement of packaging technology. Packaging Institute International has defined packaging as "enclosure of products, items or packages in a wrapped pouch, bag, box, cup, tray, can, tube, bottle or other container form to perform one or more of the following functions: containment, protection, preservation, communication, utility and performance". Lockhart has defined packaging as "a socio-scientific discipline that operates in society to ensure the delivery of goods to the ultimate consumer of those goods in the best condition intended for their use". Food packaging technology has came a long way from use of anything in hand for wrapping up a product to becoming a key element in product development and design.

2. Levels of Packaging

Packaging arrangement of food materials are classified as distinct levels depending upon how they are protecting the product during shipping and storage. During designing of product packaging it is important to consider all the levels of packaging to ensure product safety and branding.

2.1 Primary package

It is in direct contact with the contained product. It provides the initial protection and usually all the function of the package. Examples: metal cans, paperboard cartons, glass bottles and plastic pouches.

2.2 Secondary package

It contains a number of primary packages. It provides the physical protection to food product as well as primary package. It also serves to display the product in retail outlets. Example: paperboard, corrugated case or box. Secondary packaging can sometime overlap with primary packaging or tertiary packaging.

2.3 Tertiary package

It is also called distribution package. It is made up of a number of secondary packages. It enables efficient handling and protection of secondary package. Example: stretch-wrapped pallet of corrugated cases.

2.4 Quaternary package

It refers to the unit load used to facilitate the handling and storage of tertiary packages in supply chain. This is generally a metal container.

3. Functions of Packaging

There are four basic functions of packaging: containment, protection, convenience and communication.

3.1 Containment

The food product must be contained by the "package" first and then only the package can perform other function successfully. The size of the package with respect to food quantity becomes important factor for this function.

3.2 Protection

It is the primary function of the package. The protection is to be provided from outside environmental influences such as water, water vapor, gases, odors, microorganisms, dust, shocks, vibrations and compressive forces. Mechanical strength, moisture and gas barrier properties of packaging film become important to perform this function.

3.3 Convenience

Today's lifestyle requires the food package to be convenient in terms of its handling, holding, opening, pouring, using and disposing by the users at all levels of the supply chain especially the end user. The uses of zip locking, resealable pouch, easy open aluminium can are some example of convenient function of packaging.

3.4 Communication

The package inform the user about the product, quantity, ingredients, nutritional value, usage method, use by date and many other necessary information about the product contained in it. Package acts as "silent salesman" and develops the brand value of the product.

4. Types of Packaging Material

4.1 Paper and Paperboard

Approximately 10% of all paper and paperboard consumption is used for packaging. Over 50% of the paper and paperboard used for packaging is used by the food industry. Paper and paperboard vary in appearance, strength and many other properties depending on the type(s) and amount of fibre used and how the fibres are processed in paper and paperboard manufacture. The amount of fibre is expressed by the weight per unit area (grams per square metre, g/m^2), thickness (microns and thou (0.001 inch), also referred to as points). Paperboard is thicker than paper and has a higher weight per unit area. Paper over thickness 300 μm and weight (grammage) 200 g/m^2 is known as paperboard or board. The manufacture of paper and paperboard includes the steps such as pulping, stock preparation, sheet forming, pressing, drying, coating, reel-up and finishing.

Advantage of paper packaging includes its low cost, lightweight printability, biodegradable, recyclable properties. The limitation of paper packaging materials includes: poor moisture and gas barrier properties, viscoelasticity i.e. tendency to distort over time as force is applied, lack of resistance to pests, limited formability, ability to absorb aroma from packed food materials.

Paper packaging is used in packaging of loose and inexpensive material such as flour, sugar. Paperboard constitute the secondary packaging materials of products such as cereals, fruit juice, egg boxes, chocolate etc.

4.2 Glass

Glass is defined as an amorphous, inorganic product of fusion that has been cooled to a rigid condition without crystallizing. Glass is neither a solid nor a liquid but exists in a vitreous or glassy state. At ambient temperature it is a supercooled liquid. The basic structural unit is the silicon-oxygen tetrahedron in which a silicon atom is tetrahedrally coordinated to four surrounding oxygen atoms. The principal ingredient of glass is silica derived from sand, flint or quartz. Silica is combined with other raw materials in various proportions to form glass. The percentage of various component in glass are: Silica (SiO_2)

68%–73%; Calcia (CaO) 10%–13%; Soda (Na_2O) 12%–15%; Alumina (Al_2O_3) 1.5%–2% Iron oxides (FeO) 0.05%–0.25%.

The advantage of glass as packaging materials are transparency, inertness to chemicals, excellent barrier to moisture and gas, smoothness, strength, resistance to high internal pressure, high heat resistance, reusable, recyclable. The limitation includes fragility, brittleness, heavy weight and safety issue related to its fragility.

4.3 Metals

Four metals are commonly used for the packaging of foods: steel, aluminum, tin and chromium. The composite of tin and steel is called tinplate and electrolytically chromium-coated steel (ECCS) is referred to as tin-free steel (TFS).

4.3.1 Tin Plate

Tin plate is a thin steel sheet coated by tin. It has low-carbon mild steel sheet varying in thickness from around 0.15–0.5 mm with a coating of tin between 2.8 and 17 gsm (0.4–2.5 μm thick) on each surface of the material. Tinplate has excellent fabrication qualities such as ductility, drawability, weldability, lacquerability, nontoxicity, lubricity and corrosion-resistant surface.

4.3.2 Three-Piece and Two-piece Cans

Three piece cans are made from ECSS or aluminum sheets rolled into cylinder and two end pieces. The large steel sheets are cut into strip, lacquered one side, dried and slit into small strip one for each can body. Each strip is rolled into cylinder, edges welded and flanged outward at both end. One end is then double seamed onto cylindrical body and then walls are beaded to give added strength. The other end is also double seamed after filling the product to give a hermetic seal. Two piece cans are formed by ECSS or aluminum sheets rolled into cylinder and one end pieces. Two piece cans are formed through draw-and-redraw and draw-and-iron processes.

4.3.3 Electrolytically Chromium-coated Steel (ECCS)/ tin-free steel (TFS).

The production of ECCS is similar to electrotinning but flow melting and chemical passivation are not involved. The process involves cathodic deposition in a dilute chromium plating electrolyte (e.g., 50 gL^{-1} CrO_3 and 0.5 gL^{-1} H_2SO_4) at a temperature in the range of 50°C–70°C. ECCS consists of a duplex coating of metallic chromium : 0.07 and 0.15 gsm and chromium sesquioxide : 0.03 and 0.06 gsm. Finally a oil coating of either DOS or butyl stearate oil (BSO) is applied. ECSS differs from tin plate in following ways:

- Much thinner.
- Surface of ECCS is more acceptable for protective enamel coatings or printing inks and varnishes than tinplate.
- ECCS is less resistant to corrosion than tinplate.
- ECCS containers cannot be soldered.

4.3.4 Metal/Aluminum Foil

It is a thin-rolled sheet of alloyed aluminum varying in thickness from about 4–150 μm. Foil can be produced by two methods. In first method heated aluminum sheet ingot is passed between rollers in a mill under pressure and then rerolled on sheet and plate mills until the desired gauge is obtained. The second method is by continuously casting and cold rolling. Aluminum foil is used in flexible packaging such as retort pouch, aseptic pouch and fruit juice box. Aluminum foil is essentially impermeable to gases and water vapor when it is thicker than 15 μm.

4.3.5 Metallized Films

Metallized films are prepared by physical vapour deposition method where bulk material (metal) is transported to the substrate's surface in vapor form under vacuum where it condenses. Aluminum, nickel and chromium used for metallization. Oriented polypropylene and PET are the most common films used for metallization. These films have good barrier property against moisture, gases, oils and light. They are cheaper than aluminum foil.

4.3.6 Advantage and Limitations of Metal Packaging

The advantage of metal packaging includes: high mechanical strength and resistance to working, low toxicity, superior barrier properties, thermal stability, ability to withstand wide extremes of temperature, ideal surfaces for decoration and lacquering and recyclability. The limitations are its weight, reactivity with food, corrosion and increased cost.

Major application of metal cans are in packaging of thermally processed foods such as fruit, vegetables and meat.

4.4 Plastic

Plastics is a generic term for macromolecular organic compounds obtained from molecules with a lower molecular weight or by chemical alteration of natural macromolecular compounds. The building blocks of plastic are called monomers, the resulting compound being called a polymer.

4.4.1 Processing of Plastic packaging Materials

4.4.1.1 Extrusion

It is defined as continuously forcing a molten material through a shaping device. Monolayer extrusion is done by flat film process (also known as cast film or slit die process) and tubular process (or blown film process).

4.4.1.2 Coextrusion

Coextrusion consists of coupling two or more extruders feeding different resins to a single die head to simultaneously extrude polymers that fuse at the point of film formation into a single web. Coextruded film has combined properties of two or more thermoplastic materials in one composite.

4.4.1.3 Coating and Laminating

Coating is the process of applying one or more layers of a fluid or melt to the surface of a material, while laminating is the bonding of two or more webs. Coating is mainly done through extrusion coating where a molten web of synthetic resin is coated onto a substrate material. Nitrocellulose and PVdC copolymer are the most common surface coatings used.

A laminate is defined as any combination of distinctly different plastic film materials or plastic plus nonplastic materials where each major web is generally thicker than 6 μm. Laminating can be done in following ways: a) Thermal laminating: b) Wet bond laminating: c) Dry bonding: d) Solventless laminating: e) Extrusion laminating:

4.4.1.4 Surface Treatment

The objective of surface treatment is to increase the surface energy of the film surface to improve wet-out and adhesion of coatings, inks and adhesives. The increased surface energy results in the substrate surface being more compatible with a freshly oxidized polyethylene surface, thus promoting adhesion between the two. The different surface treatment method used are: a) corona treatment; b) flame treatment; c) priming; d) chemical treatment.

4.4.1.5 Blending

In blending the ingredients are brought together in right proportion and homogeneously mixed mainly by extrusion. In comparison to multilayer extrusion polymer blending is more economical option.

4.4.1.6 Orientation

Orientation involves stretching the material to line up the molecular chains. Once lined up, the ordered arrangement is frozen in the strained condition. Orientation may be uniaxial or biaxial. Common commercially oriented films are PET, PA, PVdC copolymers, PP and LDPE. The effect of orientation are development of superior tensile and impact strengths, improved flexibility, clarity, stiffness and toughness and increased shrinkability. Orientation is done either by tenter frame or double-bubble method.

4.4.1.7 Cross-Linking

Cross-linking interferes with molecular packing, reduces the level of crystallization and, consequently, results in a polymer with a lower hardness and yield strength. Three main approaches are used for cross-linking polyethylene are: Radiation cross-linking, Peroxide cross-linking, Vinyl silane cross-linking. Radiation cross-linking is normally carried out using an electron beam accelerator that produces high energy β rays. The properties that can be improved by cross-linking include mechanical properties such as tensile strength, scratch resistance, performance at higher temperatures, increase in the melting temperature, resistance to chemicals, gas permeation reduction and shape memory retention.

4.4.1.8 Injection Molding

It is used for producing tubs , jars, caps, spouts and dispensers. This process is also used to manufacture PET performs. It is a three-step process. First a stock of granules are thermally plasticized. Second injecting it under high pressure into a relatively cool mold. Third cooling and ejecting the molded component.

4.4.1.9 Blow Molding

It is process to produce hollow objects. It is done in three stages: a) Melting the resin b) Forming the parison c) Blowing the parison to produce the final shape. Blow molding can be done in following ways: a) Extrusion blow molding b) Injection blow molding c) Stretch Blow Molding

4.5 Types of Plastic and Their Properties

4.5.1 Low Density Polyethylene (LDPE)

It is the largest volume single polymer used in food packaging in both the film and blow molded form. It is tough, slightly translucent material that can be blow extruded into tubular film. It has good tensile strength, burst strength, impact resistance and tear strength, retaining its strength down to “60°C. While it is an excellent barrier to water and water vapor, it is not a good barrier to gases.

4.5.2 Linear Low Density Polyethylene (LLDPE)

LLDPE combines the main features of both LDPE and HDPE. The advantages of LLDPE over LDPE are: improved chemical resistance, improved performance at low and high temperatures, higher surface gloss, higher strength at a given density, better heat sealing properties and greater resistance to environmental stress cracking.

4.5.3 Very Low Density Polyethylene (VLDPE)

The density is less than 915 kg m^{-3}. It exhibits excellent stretchability. It is used for blown film requiring a combination of excellent optical properties, outstanding tear and impact strength and very good sealability.

4.5.4 High Density Polyethylene (HDPE)

HDPE film is stiffer and harder than LDPE and densities range from 940 to 975 kg m^{-3}. Tensile and bursting strengths and resistance to oils and greases are higher but impact and tear strengths are both lower than LDPE. The film offers excellent moisture protection, a much decreased gas permeability compared with LDPE film, but is much more opaque. It has a white and translucent appearance.

4.5.5 Polypropylene (PP)

PP has a lower density (900 kg m^{-3}) and a higher softening point (140°C–150°C) than the PEs. It has low water vapor transmission, medium gas permeability, good resistance to greases and chemicals, good abrasion resistance, high temperature stability, good gloss and high clarity (ideal for reverse printing). PP is used to produce closures for HDPE, PET and glass bottles, as well as thin-walled pots and crates.

4.5.6 Ethylene–Vinyl Acetate (EVA)

EVA have three advantages over LDPE: the heat sealing temperature is lower, the barrier properties are better, they have excellent stretch properties, the first 50% of extension at room temperature being elastic. EVA is used as a stretch film for food packaging (particularly fresh meat) and have replaced PVC for many stretch-wrapping applications.

4.5.7 Ethylene–Vinyl Alcohol (EVOH)

It posses excellent processability and are superior barriers to gases, odors, aromas and solvents. Its use in a packaging structure to enhances flavor and quality retention. It is used in MAP to retains the CO_2 or N_2.

4.5.8 Polystyrene

It is stiff and brittle at room temperature but has excellent optical properties. One of its largest uses is for packaging fresh produce because it is highly permeable to O_2 and water vapour.

4.5.9 High Impact Polystyrene (HIPS)

HIPS has high impact strength, stiffness, flexibility, extensibility, glossiness and ease of processing. The negative effect of copolymerization are reduction in thermal resistance, tensile strength and transparency. It is used in manufacture of injection moulded tub.

4.5.10 Polyvinyl Chloride (PVC)

After polyethylene PVC is second most widely used polymer. In plasticized form it is in thin film with excellent gloss and transparency and used in stretch wrapping and shrink wrapping. In unplasticized form PVC is rigid and used to form boxes and trays for biscuits, chocolate etc.

4.5.11 Polyvinylidene Chloride (PVdC)

PVdC copolymers have high barrier to water vapour, gases, odour and exhibit good environmental stress crack resistance. PVdC copolymers are used in the packaging of processed dairy product such as cheese, meat and bakery. PVdC can withstand hot filling and retort temperatures.

4.5.12 Polyethylene Terephthalate (PET)

In film form PET has good tensile strength, resistance for chemical reactions, wide temperature tolerance from -60 ^{0}C to 220 ^{0}C, light weight and elastic. It is also widely used in coextruded and metallized form. They can be microwaved and through stretch blow molding shaped into bottles used in beverage industry.

4.6 Permeability determination of Polyfilms

Diffusion Flux (J): The amount passing through a plane (surface) of unit area normal to the direction of flow during unit time

$$J = \frac{Q}{At} \qquad (1)$$

Where Q is total amount of permeate (kg) which have passed through area A (m^2) during time t (s). The unit of diffusion flux (J) is kg m^{-2} s^{-1}. The total amount of permeate, Q can also be expressed in mole and volume units.

The steady state diffusion equation for gas through a polymer gives the value of permeate, Q as:

$$Q = \frac{DSAt\Delta p}{X} \tag{2}$$

Where, D is diffusion coefficient ($m^2 s^{-1}$) which reflects the speed at which the permeant diffuses through the polymer. It is function of temperature and may be a function of concentration. The quantity, S is solubility coefficient (kg m^{-3} Pa^{-1}) and reflects the amount of permeant in the polymer. The quantity, X is thickness of polymer (m).

Now, a term 'Permeability coefficient' (P) which represents the ease with which a gas permeates through a polymer when subjected to a pressure gradient is defined as:

$$P = DS \tag{3}$$

Hence, equation (2) can be written as:

$$Q = \frac{PAt\Delta p}{X} \tag{4}$$

Which can be rearranged as:

$$P = \frac{Q}{At}\frac{X}{\Delta p} \tag{5}$$

The unit of permeability coefficient (P) will be:

$$P = \frac{cm^3 (STP) \times cm}{cm^2 \times s \times Pa}$$

In the above equation, cm^3 (STP) represents the molar quantity of gas rather than a true volume and is the quantity of gas that would take up 1 cm^3 at STP as calculated via the ideal gas law. The SI unit of permeability coefficient (P) is kg m m^{-2} s^{-1} Pa^{-1}.

The term Q/At is known as transmission rate (TR). Hence equation (5) becomes:

$$P = TR\frac{X}{\Delta p} \tag{6}$$

The unit of transmission rate, (TR) is same as diffusion flux (J) i.e., kg m^{-2} s^{-1}, if Q is expressed in mass term, kg. Transmission rate (TR) if used for oxygen it is called oxygen transmission rate (OTR), for water vapour it is called water vapour transmission rate (WVTR) and for gas it is termed gas transmission rate (GTR). The term, P/X is known as permeance (R) and has unit, kg m^{-2} s^{-1} Pa^{-1}.

The permeability coefficient for laminates and multilayer films when flux of permeate is at steady state and area across which the permeation takes place are equal, can be written as:

$$P_T = \frac{X}{\left(\frac{X_1}{P_1}\right) + \left(\frac{X_2}{P_2}\right) + \ldots\ldots + \left(\frac{X_n}{P_n}\right)} \tag{7}$$

Where X_1, X_2, ...X_n are thickness of different layers of multilayer film and P_1, P_2, ...P_n are the pressure at different layers of multilayer film.

5. Shelf Life Calculation for Packaged Food

One the most widely used equation for calculation of shelf life of packaged food is:

$$\ln \frac{m_e - m_i}{m_e - m_c} = \frac{P}{X} \frac{A}{W_s} \frac{P_0}{b} \theta_s \tag{8}$$

Where m_e is equilibrium moisture content (dry basis), m_i is initial moisture content (dry basis), m_c is critical moisture content (dry basis), P_0 is vapour pressure of pure water at the storage temperature (Pa), W_s is weight of dry solid enclosed (kg), θ_s is shelf life and b is slope of the moisture sorption isotherm (kg H_2O/kg solid/unit a_w).

Shelf life of packaged food can be calculated by determining the rate of deteriorative reaction of any component (A) in food placed inside package. In food most of the deteriorative reaction follows either zero or first order reaction. The zero order reaction is:

$$A_e = A_0 - k\theta_s \tag{9}$$

Where, A_e and A_0 is the value of component *A* at end of shelf life and at beginning respectively, θ_s is the shelf life, k is the zero order rate constant.

Similarly for first order reaction the equation is:

$$A_e = A_0 e^{-k\theta_s} \tag{10}$$

Here k is the first order rate constant.

6. Testing for Packaging Materials

6.1 Paper

6.1.1 Grammage

The grammage specifies the weight of the paperboard per unit area in g/m2. Measured with digital balance.

6.1.2 Thickness

It is the perpendicular distance between the two surfaces of the board/paper, expressed in mm or μm, measured with a micrometer. Plastic and metal packaging material thickness is also measured in same way.

6.1.3 Tearing Strength

The tear strength (reported in gm) is the force necessary to continue tearing of a sample after a nick has been made. It is measured by tearing strength tester (Elmendorf Type). It is also done for plastic films.

6.1.4 Stiffness

It is measurement of the force which must be applied to deflect a defined piece of material through a defined distance or angle. The resistance to that bending force is measured by the machine. Stiffness tester (Taber Type) is used to measure the stiffness of paper/board.

6.1.5 Folding Endurance

Folding endurance is an empirical test which measures the amount of folding the paper will endure before its tensile strength falls below a standard value of 1 kg. Folding endurance tester (Kohler Molin Type) is used to measure this parameter.

6.1.6 Bursting Strength

Bursting strength of a material is defined as the maximum hydrostatic pressure required to produce rupture of the material when a controlled and constantly increasing pressure is applied through a rubber diaphragm to a circular area, 30.5 mm (1.20 in.) diameter.

6.1.7 Grease Resistance

It is a method for testing the degree of repellency and/or the antiwicking characteristics of paper or paperboard treated with fluorochemical sizing agents. It is also known as kit test.

6.2 Glass

6.2.1 Thermal Shock Resistance

The actual temperature change or thermal shock measured in degree Celsius, which a container can withstand without breaking. The tests used to measure the thermal shock resistance are pass test, progressive test and high-level test.

6.2.2 Thermal Shock Endurance

An interpolated thermal shock resistance value at which 50 % of the containers will probably fail.

6.3 Metal/Tin

6.3.1 Clark's Test

It is used to determine the thickness or weight per unit area of tin coatings on steel through chemical analysis.

6.4 Plastic

6.4.1 Tensile Strength

It measures the maximum load that a material can support without fracture when being stretched, divided by the original cross-sectional area of the material.

6.4.2 Yield Strength

It is the tensile stress at which the first sign of a nonelastic deformation occurs.

6.4.3 Elongation

It is measured at the point where the film breaks. It is expressed as the percentage of change of the original length of the material between the grips of the testing machine.

An instrument called Good Brand Tensile Tester is used for the measurement of above mentioned three tests.

6.4.4 Water Vapor Transmission Rate (WVTR)

WVTR (water vapor transmission rate) is the steady state rate at which water vapor permeates through a film at specified conditions of temperature and relative humidity. Values are expressed $g/m^2/24$ hr in SI units.

6.4.5 Oxygen Transmission Rate

OTR (oxygen transmission rate) is the steady state rate at which oxygen gas permeates through a film at specified conditions of temperature and relative humidity. Value are expressed cc/m^2/24 hr in SI units. Standard test conditions are 73°F (23°C) and 0% RH.

7. Additional Packaging Types

7.1 Vacuum Packaging

Vacuum packing is a method of packaging that removes air from the package prior to sealing. It involves placing items in a plastic film package, removing air from inside and sealing the package. Vacuum packing reduces atmospheric oxygen.It is used to store dry foods over a long period of time, such as cereals, nuts, cured meats, cheese, smoked fish, coffee, and potato chips (crisps).

7.2 Retort Packaging

Retort pouch is a flexible package made from multilayer heat resistant plastic films suitable for processing in retorts at temperature of around 121°C normally encountered in thermal sterilization of food. The aim of retort packaging is thermal processing of food with the package itself and to cook food ready for consumption. These are constructed with four laminate consisting PET as outside layer, nylon and aluminium foil as middle layer and polypropylene as inside food contact layer.

7.3 Shrink Packaging

It is a process that uses relatively soft polyolefin film in which a loose sheath of film is created around the product, often sealed on two sides. The loosely sheathed product passes through a heat source, which shrinks the film to the dimensions of the product. The result is a very tight wrap, conforming to the product shape. Shrink films are composed of three basic categories: polyolefins, PVC and PVdC copolymer.

7.4 Stretch Wrapping

In stretch wrapping, the film is stretched around the article and the loose end "wiped" against the underlying film, where the film-to-film adhesion or cling is sufficient to hold it in place. The main polymers used in stretch/cling films are LDPE, LLDPE, VLDPE, PVC, EVA copolymer and PP.

7.5 Aseptic Packaging

Aseptic packaging (AP) is the filling of sterile containers with a commercially sterile product under aseptic condition, and then sealing the containers so that reinfection is prevented, i.e, so that they are hermetically sealed. It enables the containers to be used that are unsuitable for in-package sterilization. Application of aseptic packaging includes packaging of presterilized and sterile product such as milk and dairy products, puddings, desserts, fruit and vegetable juices, soups, sauces etc. A typical structure of a paperboard laminate carton for aseptic filling consists of outside to inside as polyethylene-bleached paperboard-laminated polyethylene-aluminium foil-two inner polyethylene layers.

7.6 Modified Atmosphere Packaging (MAP)

MAP is defined as 'the packaging of a perishable product in an atmosphere which has been modified so that its composition is other than that of air'. The three main gases used in MAP are O_2, CO_2 and N_2. For example MAP for apple consists of 1-3% O_2, 0-6% CO_2 and rest N_2 at temperature 1-4 ^{0}C and relative humidity of 90-95%.

7.7 Controlled Atmosphere Storage (CAS)

CAS involves maintaining a fixed concentration of gases surrounding the product by careful monitoring and addition of gases.

7.8 Active Packaging

Active packaging is the inclusion of subsidiary constituents into the packaging material or the package headspace with the intent of enhancing the performance of the package system. Packaging may be termed active when it performs some desired role in food safety or quality other than providing an inert barrier to external conditions. Example of active packaging are oxygen (O_2) scavengers or absorbers, carbon dioxide (CO_2) absorbers or generators, ethylene absorbers or emitters and moisture absorbers.

7.9 Intelligent Packaging

Intelligent Packaging systems monitor the condition of the packaged foods to give information about the quality of the packaged food during transport and storage. Example are time-temperature indicators, microbial growth or freshness indicator, oxygen Indicators and carbon dioxide Indicators.

Examples

Example 1. Calculate the rate of oxygen permeating through PVdC copolymer having cross sectional area of 400 cm^2 and thickness 0.2 cm. Given permeability coefficient of PVdC copolymer to O_2 is 6.42 x 10^{-20} kg m m^{-2} s^{-1} Pa^{-1} and the partial pressure of O_2 across the film is 20 kPa.

Solution

Using the equation (4): $\frac{Q}{t} = \frac{PA\Delta p}{X}$

$$\frac{Q}{t} = \frac{6.42\ \text{x}10^{-20}\left(\text{kg m m}^{-2}\ \text{s}^{-1}\ \text{Pa}^{-1}\right)0.040\left(\text{m}^2\right)20\times10^3\left(\text{ Pa}\right)}{0.02(m)}$$

= 2.56 x 10^{-15} kg s^{-1}

Example 2. A multilayer film of cross sectional area 400 cm^2 is made up of LDPE, PP and nylon 6,6. The thickness of LDPE, PP is 0.05 cm each while nylon layer thickness is 0.1 cm. Calculate how much water vapour would permeate per day if the pressure difference across the film is 0.2 atm. Given $P_{\text{water vapour (LDPE)}}$ = 800 x 10^{-11} cm^3.cm.cm^{-2}.s^{-1}.cmHg^{-1}; $P_{\text{water vapour (PP)}}$ = 680 x 10^{-11} cm^3.cm.cm^{-2}.s^{-1}.cmHg^{-1}; $P_{\text{water vapour (Nylon 6,6)}}$ = 7000 x 10^{-11} cm^3.cm.cm^{-2}.s^{-1}.cmHg^{-1}.

Solution

Approach: First we calculate the overall permeability coefficient of the film using equation (7) and then calculate the permeation rate using equation (4)

Pressure difference across the film is 0.2 atm = 15.2 cmHg

$$P_T = \frac{0.2(cm)}{\left(\frac{X_1}{P_1}\right)+\left(\frac{X_2}{P_2}\right)+\ldots..+\left(\frac{X_n}{P_n}\right)}$$

$$P_T = \frac{0.2}{\left(\frac{0.05}{800\times10^{-11}}\right)+\left(\frac{0.05}{680\times10^{-11}}\right)+\left(\frac{0.1}{7000\times10^{-11}}\right)}$$

P_T = 1.33 x 10^{-8}

Now using equation, $\frac{Q}{t} = \frac{P_T A\Delta p}{X}$

$$\frac{Q}{t} = \frac{1.33\text{x}10^{-8}\left(\text{cm}^3\text{cm cm}^{-2}\ \text{s}^{-1}\ \text{cmHg}^{-1}\right)400\left(\text{cm}^2\right)15.2\left(\text{ cmHg}\right)}{0.2(cm)}$$

$= 40.43 \times 10^{-3}$ cm^3 s^{-1}

$= 3.49 \times 10^3$ cm^3 day^{-1}

Problem 3. Check the suitability of 0.1 cm thick LDPE and PET films for six month storage of fried product at 25 ^{0}C and 90% RH if the cut off point for oxygen permeability is 2.2 cm^3. The surface area of the film is 10 cm x 10 cm and the partial pressure difference across the film is 15 cmHg. Given the permeability coefficient of different films at 25 ^{0}C and 90% RH is: $P_{oxygen\,(LDPE)} = 55 \times 10^{-11}$ cm^3.cm.cm^{-2}.s^{-1}.cmHg^{-1}; $P_{oxygen\,(LDPE)} = 0.22 \times 10^{-11}$ cm^3.cm.cm^{-2}.s^{-1}.cmHg^{-1}

Solution:

Approach: We use equation (4) to estimate the value of 't' when Q = 2.2 cm^3 for each film.

Using the equation $t = \dfrac{QX}{PA\Delta p}$

For LDPE film,

$$t_{LDPE} = \frac{2.2\left(cm^3\right)0.1\left(cm\right)}{55\text{x}10^{-11}\left(\text{cm}^3\text{cm cm}^{-2}\text{ s}^{-1}\text{ cmHg}^{-1}\right)100\left(\text{cm}^2\right)15.0\left(\text{ cmHg}\right)}$$

$t_{LDPE} = 2.66 \times 10^5$ s

$t_{LDPE} = 3.08$ days

For PET film,

$$t_{PET} = \frac{2.2\left(cm^3\right)0.1\left(cm\right)}{0.22\text{x}10^{-11}\left(\text{cm}^3\text{cm cm}^{-2}\text{ s}^{-1}\text{ cmHg}^{-1}\right)100\left(\text{cm}^2\right)15.0\left(\text{ cmHg}\right)}$$

$t_{PET} = 6.66 \times 10^7$ s

$t_{PET} = 771.6$ days

Thus, it can be concluded that LDPE is unsuitable and PET is suitable for packaging the fried product in above example.

Example 4. A extruded snack having initial moisture content of 3 % (db) is to be packed in a 60 μm OPP. The snack loses its acceptance at critical moisture content of 7 % (db) due to loss of crispness. The equilibrium moisture content at 25 ^{0}C is 12 % (db); the slope of the moisture sorption isotherm line is 0.150 g H_2O/g solid/unit a_w. Calculate the shelf life of the snack. The weight of the dry snack in the package is 150 g and the dimension of the bag are 15 cm x 10 cm. The packed product is to be stored at 25 ^{0}C and 75% RH. Given WVTR at 25 ^{0}C and 75% RH for 50 μm OPP = 1.3 g m^{-2} day^{-1} and saturated vapour pressure at 25 ^{0}C = 2.376 cm Hg.

Solution:

Using the equation $\ln\frac{m_e - m_i}{m_e - m_c} = \frac{P}{X}\frac{A}{W_s}\frac{P_0}{b}\theta_s$ (i)

From given data, X= thickness of film = 60 μm

A= area = 15 cm x 10 cm = 0.015 m^2

m_e = equilibrium moisture content = 12%

m_i = initial moisture content = 3%

m_c = critical moisture content = 7%

P_0 = Saturated vapour pressure of pure water at 25 ^{0}C = 2.376 cm Hg

and relative humidity is 75 %, hence Vapour pressure of water at 25 ^{0}C/75 % RH i.e.

Driving force, Δp = 2.376 x 0.75 = 1.782 cmHg

W_s is weight of dry solid enclosed = 150 g

θ_s = shelf life (To be calculated)

b = slope of the moisture sorption isotherm = 0.150 g H_2O/g solid/unit a_w

Since permeability coefficient (P) value is not given and water vapour transmission rate is given we will use the equation $P = TR\frac{X}{\Delta p}$ to evaluate P/X used in equation (i)

$$\frac{P}{X} = \frac{TR}{\Delta p} = \frac{1.3\left(g\,m^{-2}day^{-1}\right)}{1.782\left(cmHg\right)} = 0.729\,g\,H_2O\,m^{-2}day^{-1}cmHg^{-1}$$

Put all these in equation (i):

$$\ln\frac{12-3}{12-7} = 0.729\frac{0.015}{150}\frac{2.376}{0.150}\theta_s$$

Solving for shelf life $\theta_s = \frac{\ln 1.8}{0.0011} = 534$ days

References

Brown, W.E. (1992). *Plastics in Food Packaging: Properties, Design and Fabrication*. New York: Marcel Dekker.

Fellows, P. (2000). Freezing. In: *Food Processing Technology Principles and Practice Second Edition* (pp. 418-439). CRC Press, Boca Raton, Florida, USA.

Hanlon, J.F., Kelsey, R.J., Forcinio, H.E. (1998). *Handbook of Package Engineering, 3rd edn.* CRC Press, Boca Raton, Florida, USA.

Lange, J, Wyser, Y. (2003). Recent innovations in barrier technologies for plastic packaging: A review. *Packaging Technology and Science* 16: 149–153.

Lund, P.R., McCaul, J.P. (2009). Nitrile polymers. In: Yam, K.L. (Ed.), *The Wiley Encyclopedia of Packaging Technology, 3rd edn.*,New York: John Wiley & Sons, pp. 819–822.

Mauer, L. J., Ozen, B.F. (2004). Food Packaging. In: Smith, J.S. and Hui, Y.H. (Ed.), *Food Processing Principles and Applications* (pp. 101-132). Blackwell Publishing, 2121 State Avenue, Ames, Iowa 50014, USA

Morris, S.A. (2011). *Food and Package Engineering*. Chichester, England: Wiley-Blackwell.

Robertson, G.L. (Ed.) (2010). *Food Packaging and Shelf Life: A Practical Guide*. CRC Press, Boca Raton, Florida, USA.

Robertson, G.L. (Ed.) (2013). *Food Packaging Principles and Practice Third Edition*. CRC Press, Boca Raton, Florida, USA.

Singh, R.P. & Heldman, D.R. (2009). Food Freezing. In: *Introduction to Food Engineering Fourth Edition* (pp. 501-539). Academic Press, USA

Guillet, M., Rodrigue, N. (2010). Shelf life testing methodology and data analysis. In: Robertson G.L. (Ed.), *Food Packaging and Shelf Life*, (pp. 31–53). CRC Press, Boca Raton, Florida, USA.